Computer-aided Pattern Design and Product Development

Other titles of interest

Metric Pattern Cutting
Third Edition
Winifred Aldrich
0 632 03612 5

Pattern Cutting for Women's Tailored Jackets
Winifred Aldrich
0 632 05467 0

Metric Pattern Cutting for Menswear
Third Edition
Winifred Aldrich
0 632 04113 7

Metric Pattern Cutting for Children's Wear and Babywear
Third Edition
Winifred Aldrich
0 632 05265 1

Fabric, Form and Pattern Cutting
Winifred Aldrich
0 632 03917 5

Fashion Source Book
Kathryn McKelvey
0 632 03993 0

Fashion Design: Process, Innovation and Practice
Kathryn McKelvey and Janine Munslow
0 632 05599 5

Illustrating Fashion
Kathryn McKelvey and Janine Munslow
0 632 04024 6

Fashion Marketing
Edited by ME Easey
0 632 03459 9

Fashion Buying
Helen Goworek
0 632 05584 7

Master Patterns and Grading for Women's Outsizes
G Cooklin
0 632 03915 9

Introduction to Clothing Manufacture
G Cooklin
0 632 02661 8

Carr and Latham's Technology of Clothing Manufacture
Third Edition
David J Tyler
0 632 02896 3

Computer-aided Pattern Design and Product Development

Alison Beazley and Terry Bond

Editorial Offices:
9600 Garsington Road, Oxford OX4 2DQ, UK
Tel: +44 (0)1865 776868
108 Cowley Road, Oxford OX4 1JF, UK
Tel: +44 (0)1865 791100
Blackwell Publishing Inc., 350 Main Street, Malden, MA 02148-5018, USA
Tel: +1 781 388 8250
Iowa State Press, a Blackwell Publishing Company, 2121 State Avenue, Ames, Iowa 50014-8300, USA
Tel: +1 515 292 0140
Blackwell Munksgaard, 1 Rosenørns Allé, P.O. Box 227, DK-1502 Copenhagen V, Denmark
Tel: +45 77 33 33 33
Blackwell Publishing Asia Pty Ltd, 550 Swanston Street, Carlton South, Victoria 3053, Australia
Tel: +61 (0)3 9347 0300
Blackwell Verlag, Kurfürstendamm 57, 10707 Berlin, Germany
Tel: +49 (0)30 32 79 060
Blackwell Publishing, 10 rue Casimir Delavigne, 75006 Paris, France
Tel: +33 1 53 10 33 10

First published 2003

A catalogue record for this title is available from the British Library

ISBN 1-4051-0283-7

Library of Congress
Cataloging-in-Publication Data
Beazley, Alison.
Computer-aided pattern design and product development/Alison Beazley and Terry Bond.
p. cm.
Includes bibliographical references and index.
ISBN 1-4051-0283-7 (softcover: alk. paper)
1. Dressmaking–Pattern design–Data processing.
2. Clothing and dress measurements–Data processing.
3. Computer-aided design. I. Bond, Terry, 1954–
II. Title.

TT520.B384 2003
646.4′072′0285–dc21

2002043912

Set in 9.25/11pt Times
by DP Photosetting, Aylesbury, Bucks

For further information on
Blackwell Publishing, visit our website:
www.blackwellpublishing.com

Contents

Preface

The computer is a very useful tool that when used correctly can increase accuracy and productivity, and manage information. This removes the time consuming tasks of cutting card patterns and planning and drawing markers by hand, and the duplication of hand-written instructions. This computer technology has enabled the clothing design, pattern construction and product development to be integrated into a more continuous process.

Computer design systems enable designers to illustrate and visualise their designs both two and three dimensionally. The pattern technologist can construct and grade the patterns simultaneously. The sample garment can be viewed three dimensionally in simulated fabric. The cost of the garment can be calculated from a computer lay plan of the pattern pieces on the fabric for fabric utilisation. Having an easy access to a database assists the clothing technologist to calculate the cost of the garment for their specifications.

However, it is essential that the preparatory work is based on knowledge of the principles and techniques of pattern construction, grading and pattern lay planning and marker making. Initially the preparatory work may appear somewhat time consuming, but once the correct data has been put into the computer it can be operated with confidence.

The definition of design in the context of this book is not the creation of fashionable styles but the procedure of developing a style suitable for production that is influenced by the body dimensions, the fabric and the production methods. Originally computer programs were developed to grade garment patterns into a range of sizes for lay planning and marker making. Today there are major systems that have the further facility for designing patterns and altering patterns to the size and fit for an individual customer.

The pattern construction and design in this book is based on well-tried and proven methods used in the past by well-respected pattern makers. These methods have been adapted for today's computer systems but can also be drafted manually and then digitised into the computer. All the patterns constructed for this book were developed using a computer pattern design system and are for women's garments. They have all been tested by producing sample garments.

The intention of this book is:

- To introduce and explain the wide range of computer programs available to the clothing industry for pattern design and product development.
- To give guidance to those operating or planning to use computer programs for pattern designing, grading and lay planning by combining a theoretical understanding with the practical application.
- To give a reference source to students following courses in Pattern Design, Clothing Technology and Clothing Management.
- To give knowledge and understanding of the principles for developing garments to those conversant with operating computers but lacking experience in clothing product development.
- To assist those experienced in clothing technology with the transition from manual methods to operating computer systems.

The book is divided into six parts for ease of reference:

INTRODUCTION: Developments within computer-aided apparel systems.
The clothing industry has changed profoundly in recent years. Globalisation, speed of information and communication has stimulated competition. While manufacturers offer unlimited designs, the problem is how to bring products to the market quickly and achieve up-to-date information that is easily obtainable. CAD systems are now the essential tools required to integrate and achieve success taking the role of the 'configurator' between manufacture and retail. Utilising a full range of electronic tools, from the ubiquitous internet to the new powerful CAD systems can deliver clothing at relatively short cycles. Integration and communication utilising the internet become the new systems designed to achieve this.

PART 1: Pattern construction
The obtaining of body measurements and how they are formulated into size charts is explained. These size charts are requisite to all the pattern designing, grading and customisation that follow in Part 2, Part 3, Part 4 and Part 5. The various techniques of constructing and manipulating patterns manually and by computer are compared. The drafting of basic block patterns is described and their adaptation into

secondary blocks. This knowledge assists in the calculation of grading increments discussed in Part 2. These block patterns are the foundation for designing pattern for styles described in Part 3.

PART 2: Computer pattern grading
The principles of pattern grading to produce a range of sizes are explained. How they are applied to computerised grading is covered in detail. This is illustrated by the grading of the block patterns constructed in Part 1. The benefits of grading block patterns prior to computer pattern design is that the size increments are transferred on to the new style. This eliminates grading as a separate process. Alternatively, the digitising and grading of manually produced patterns is also explained.

PART 3: Pattern designing and grading
This covers the pattern construction or adaptation of a variety of designs for skirts, bodices, collars and sleeves. Suitable methods of grading are suggested. Details for completing the pattern with seams, hems and facings for production are also given.

PART 4: Pattern modification for garment size and fit
This section gives an introduction to the assessing of the figure shape and garment fit related to the stature, posture, body size and contour. The identification of fitting faults is explained and the appropriate pattern corrections. This information is in preparation for pattern alteration systems and made-to-measure.

PART 5: Computerised marker making systems
It has long been recognised that improvements at the front end of cutting can show substantial fabric savings. Fabric and trim account for about half the total costs of goods manufactured, and in a competitive situation, the first place that cost reduction can be achieved is in fabric utilisation. It is generally understood that 40% of the finished garment cost is fabric; it is also recognised that 90% of cutting room costs are fabric. Parameters relevant to lay planning and marker making will be identified.

PART 6: Product data management systems
Product data management (PDM) systems have been developed to improve the product and the process of the product development cycle. These systems provide an automated means to control and facilitate the flow of up-to-date information to authorised participants throughout the organisation.

PDM acts as a communication tool between design, retail and manufacture, containing details about patterns, garment construction, fabric and trims, packaging costs, quality and measurement specifications. It is the direct interface between CAD/CAM systems and management information systems.

APPENDICES
References and further reading are listed on p. 205. Appendix I gives a comprehensive index of technical terms and abbreviations related to the text. Appendix II shows reduced size basic block patterns for use as exercises in digitising, grading and pattern design. These are at 33.3% of the original and can be plotted full-scale.

Great improvements have been made to computer systems since the early 1990s. They are much more 'user friendly' today and are being continually updated. The content of this book is not specific to one specific system; the authors have used various systems for testing the illustrations. The reader should become conversant with the system they will be using, preferably by training from the supplier.

Acknowledgements

We wish to express our appreciation and thanks to the following people: Poppy Thomason, Penny Preddle and Anita McAdams for the time that they freely gave to reading the draft and constructive criticism of the text and diagrams. We wish to thank Aileen Jefferson and Julie Vernon for their practical contribution. We appreciate the following companies allowing the use of published photographs of their equipment and quote details of their products: Gerber Garment Technology, Wicks and Wilson, Telmat Informatique and the Textile/Clothing Technology Corporation. The preparation of this book could not have been undertaken if we had not had the support, use of computer facilities and permission to use some illustrations from Manchester Metropolitan University.

Abbreviations and symbols

bk	back
BP	bust point
CB	centre back
CF	centre front
fr	front
FS	face side of the fabric
GRL	grade reference line
NP	neck point
R	grade rule
SS	side seam
UP	underarm point
WS	wrong side of the fabric

Construction line	- - - - - - - - - - - - - - - -
Construction line (secondary) and alteration line	— — — — — — — —
Drill hole	✈
Grade direction	↱
Grain line	←——————→
Notch	——————┴——————
Pattern parameter	————————————
Square corner at 90°	∟
Stitch line	....
Style line	━━━━━━━━━━━━
Zero point for grading	◯

Introduction

Developments within computer-aided apparel systems

Developments within computer-aided design for fashion, clothing and visualisation have been realised using 3D software. Offering the designer a virtual prototyping system has been an active research area for many years. Despite being applied in other commercial industries, the development of 3D imaging for use within the clothing industry has met many research challenges. However, by presenting recent developments within this virtual environment the 3D picture becomes much clearer.

In relation to pattern design the ability to move from 2D to 3D is perhaps the area of most interest. The creation of 2D patern shapes that can be wrapped around a virtual mannequin fits nicely within the 2D CAD pattern development application used within the industry, and development from this is the most likely way forward for the designer and pattern technologist.

3D software developed by Pad systems was one of the first commercial packages available to the clothing industry offering further integration between the pattern technologist and designer. This modular-based software allows 2D patterns to be modified and, following a sequence of assigning sew points, a 3D simulation on a virtual mannequin can be created. Fabric models within the module allow simulations of garment drape which can be linked to objective measurement data.

Gerber Technology now offers commercially their APDS-3D virtual garment draping system, which enables pattern technologists to view 2D garments assembled and draped on a virtual mannequin. Both the viewing and lighting angles of images are user defined; the horizontal and vertical cross sections of the mannequin can be viewed, offering the pattern technologist the ability to verify fit and ease allowance. Modifications in either 2D or 3D mode take immediate effect with results displayed. The system also allows integration of fabric images from the Artworks studio module, another of the family of software solutions offered by Gerber Technology. Current 3D software provides the pattern technologist and designer with a toolset to review the design and construction of their garments.

Lectra Systemes has expanded its organisation to embrace all areas of CAD/CAM. The developments of internet, intranet and virtual reality technologies are given high priority, the aim being to improve the products through brand building and to increase sales with leading technologies. These developments will allow Lectra to incorporate their pattern module solutions and offer commercially four key components: E-Design, E-Manufacturing, E-Sales and Lectra on-line. By maximising the potential of these technologies it may become possible to view an entire garment collection on a virtual reality catwalk.

CAD vendors with this developing technology bring a more structured and systematic approach to the pattern cutting and garment construction processes. At this stage of development, the 3D tools require improvement if they are to fulfil their promise and acceptability.

Developments within 3D body scanning systems capable of producing anthropometrics data offer a direct link to 3D design and pattern making. There are a small number of companies involved within body scanning: TC2, Tecmath, Telmat, Hammatsu and Wicks & Wilson. Telmat, the French company, have developed a 3D body scanning system, the Symcad Flash 3D system, which offers instant 3D automatic body measurements and open connectivity to CAD systems. In formation it can be directly linked to a made-to-measure CAD module either from a Symcad system or via the internet or an ISDN line. Developments within this area allow integration into the manufacture and retailing interface, offering individual service.

Among the software solutions offered by CAD vendors, resurgence in made-to-measure (MTM) allows manufacturers and retailers to develop into the rapidly growing area of mass customisation. With new technologies developed to simplify the customisation of a garment it is now possible to automate the garment development through to the point of manufacture. This gives the ability to manufacture single garments at mass production speeds and avoids the high cost usually associated with single garment production. MTM software is designed to integrate with existing CAD modules, allowing quick and easy entry of customer details, body measurements and customer orders. Information is linked to pattern-making software, marker planning, plotters and single-ply cutters.

Product data management software is a CAD tool which aims to reduce development time, increase quality and improve communications between manufacturer and clothing retailers. Its function is to organise information in the product development phase, to ensure technical specifications are followed to the last detail into the production phase of the garments. More specifically, product data management (PDM) systems contain information about patterns, garment construction, costs, quality and measurement specifications. New PDM systems are now Web enabled, allowing the major CAD vendors products to be internet, intranet and extranet enabled. The ability to transfer/share reliable information, ease of communication, is of the utmost importance.

3D visual merchandising is the new media promoted by major CAD vendors offering the ability to quickly simulate apparel collections in any virtual 3D retail environment. The ability to create and control the retail environment defining store layout, selected garments, style, colour, assortment and retail space offers the ultimate assortment planning 3D visual merchandising system for apparel brands and retailers. This is made possible by a powerful database encompassing a catalogue of 3D fixtures, dressed mannequins and custom objects importable from a 2D CAD media.

The success achieved by other industries in internet e-commerce has not so far extended into apparel, although the continued developments in visualisation technologies along with the ability to use 3D scanning data constantly improve the representation of garments on-line. In the near future from the comfort of home it will be possible to select a garment, use the data from the 3D body scan to try it on your own digital model, view around 360°, select the size that fits best or have the garment altered to your own specific measurements for a customised fit, then sit back and await the delivery of the selected garment.

Part 1
Pattern construction

The emphasis in Part 1 is on the preparatory work to be undertaken before the final patterns are constructed. This part covers:

- Taking body measurements manually
- Computerised measuring systems
- Size chart formulation
- Pattern construction techniques
- Block pattern construction
- Primary block construction
- Secondary block construction

The first requirement is obtaining body measurements that have to be formulated into the size charts for garments. These size chart measurements are then used for constructing block patterns that are adapted for the final garment patterns. The size charts and block patterns are also required for pattern grading explained in Part 2, pattern design and grading in Part 3 and computer 'made-to-measure' systems in Part 4.

Patterns represent the two-dimensional component parts of a garment. They are used as a guide for cutting the fabric, which when sewn together forms a three-dimensional garment. The creation of these patterns is the technique of pattern construction. In the past this was often termed pattern cutting, but with the advent of computers the cutting of individual patterns by hand is less essential. This is why in this book the term 'pattern cutting' is replaced by 'pattern construction'.

Pattern construction is part of the garment design process and product development. The pattern can also be considered as a foundation for garment production. The complexities of developing and grading a pattern are often underestimated. The designer or pattern technologist is creating a three-dimensional garment which is made in unstable two-dimensional fabric to be worn by a flexible body. The wearer has not only to feel physically comfortable in her garment but also psychologically confident and socially accepted.

To construct patterns by computer requires two skills: knowledge of pattern construction and how to operate the computer program. A skilled pattern technologist has also to be both mathematical and creative. Only guidance can be given for constructing patterns by computer as each computer system has different sets of commands and methods of operating.

BODY AND GARMENT MEASUREMENTS

Body measurements are a prerequisite to pattern construction. The size and fit of a garment depends upon their accuracy. At present we are in the transition between traditional manual measuring by tape measure and computerised scanning or photographic systems. Manual measuring requires a high degree of skill and is time-consuming. The techniques of computerised measuring have recently improved considerably and will supersede manual methods in the future. Whichever method is used the first consideration is to decide which measurements are required. Are they needed to develop size charts for garment production or for an individual. The age range of the group or individual being measured is important as this has great bearing upon the body proportion and size charts. Different types of garment also require a different set of measurements. Detailed below is the procedure for manual measuring. This is followed by an account of the development of three-dimensional body scanning and computerised measuring systems. Either methods can be used for measurement surveys of women to develop size charts or for altering standard patterns used in the computer 'made-to-measure' systems (see Part 4). A more detailed account of how to undertake a survey of body measurement will be found in an article by Beazley (1997).

Taking body measurements manually

PREPARATION

- The preparation is very important for accuracy. Good rapport must be developed between the measurer and the person being measured. The person being measured should feel at ease and be relaxed.
- They should remove thick outer garments and only wear the underclothes to be worn beneath the garments to be made.

- The person being measured should stand normally and evenly on both feet, with relaxed shoulders, arms hanging at either side and head erect.
- Locate nape of the neck (at top of seventh cervical) with masking tape or soft removable marker.

EQUIPMENT

Tape measure

A fibreglass tape measure is recommended. It should be clearly marked and have brass tips at both ends. Alternatively, a retractable metal tape measure can be used, as it is firmer for taking girth measurements.

Metre rule

A metal rule is recommended to measure the hip level parallel to the ground.

Tapes

Adjustable elastic tapes are useful to attach around major girth measurements to locate the measuring position (these must not indent the body). Alternatively, for the neck base measurement, a fine chain can be positioned.

Mirror

A full-length mirror positioned behind the person being measured is useful for checking girth measurements that need to be parallel to the ground.

Record sheet

A record sheet lists all the measurements required in the measuring sequence. Posture details are also useful (e.g. shoulders square or sloping; posture erect or stooping; height short, medium or tall).

MEASURING TECHNIQUE

- The measurer should stand slightly to one side when facing the person being measured.
- Hold the tape measure close to the body and taut, but not pulled tightly to indent. Do not add any extra to the measurements for ease allowance. These extra allowances can be added when constructing garment patterns.
- Be discreet about the measurements. Do not let the person being measured move to look down at the tape measure.

MEASURING METHOD FOR BODICE AND SLEEVE

(a) Bust girth: The person being measured stands facing the measurer. The measurement is taken horizontally around the fullest part of the bust and approximately parallel to the ground to incorporate the shoulder blades.

(b) Waist girth: The waist elastic should sit comfortably in the natural position of the waist (not parallel to the ground). The tape measure is held firmly, but not indenting, over the waist level elastic. This can be checked in the mirror for the correct position.

(c) Neck girth: The base of the neck should be measured in a suitable position for a close fitting collar. Starting from the nape position place a narrow cord or chain around the base of the neck. When this is straightened the distance is measured against a tape measure.

(d) Upper arm girth: Measure the thickest girth of the right upper arm, either at the armpit or biceps level.

(e) Elbow girth: Position the tape in the bend of the right elbow, and then have the arm bent across the front waist. The measurements have to be taken over the bone of the elbow.

(f) Wrist girth: While the arm is still bent measure the wrist around the widest part.

(g) Nape to waist: The top of the tape measure is positioned at the nape (seventh cervical) and placed vertically down the centre back to the lower edge of the waist level elastic tape.

(h) Front length to bust: The tape measure is positioned from the nape over the right shoulder at neckline, then diagonally to the prominence of the right breast. (Half the back neck measurement of the garment is then subtracted from this measurement.)

(i) Front waist level: Measure as 'h' and continue the tape measure from the bust prominence vertically down to the lower edge of the waist level tape. (Half the back neck measurement of the garment is then subtracted from this measurement.)

(j) Elbow level: The person being measured stands with her right side to the measurer and right arm bent across her front waist. The tape measure is positioned from the nape over the end of the shoulder, diagonally to the point of the elbow. (The garment centre back neck to end of shoulder measurement is subtracted to give the final elbow level.)

MEASURING POSITIONS FOR BODICE AND SLEEVE

(a) Bust girth
(b) Waist girth
(c) Neck girth
(d) Upper arm girth
(e) Elbow girth
(f) Wrist girth
(g) Nape to waist
(h) Front length to bust
(i) Front waist level
(j) Elbow level
(k) Sleeve length
(l) Across back
(m) Across front
(n) Shoulder length
(o) Bust prominence width

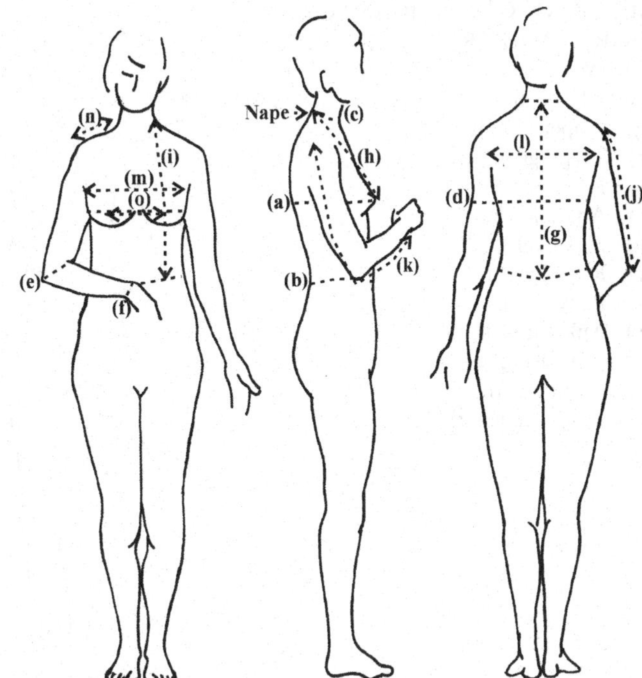

(k) Sleeve length: Measure as 'j' and continue the tape measure to the end of the wrist bone at the 'little finger' side of the hand. (The garment centre back neck to end of shoulder measurement is subtracted to give the final sleeve length.)

(l) Across back: This width measurement is taken horizontally and gauged just above the skin folds where the arms connect to the torso.

(m) Across front: This width measurement is taken horizontally between the centre front neck and bust level. The width is gauged at the skin folds where the arms connect to the torso.

(n) Shoulder length: The highest part of the shoulder is located and measured from the base of the neck to the bone at the end of the shoulder.

(o) Bust prominence width: Measurement horizontally between the most prominent part of the left and right breasts.

MEASURING POSITIONS FOR SKIRTS AND TROUSERS

(b) Waist girth
(p) Hip girth
(q) Upper hip girth
(r) Thigh girth
(s) Knee or calf girth
(t) Ankle girth
(u) Hip level
(v) Knee length
(w) Back length
(x) Ankle length
(y) Outside leg length
(z) Front length
(zz) Inside leg length

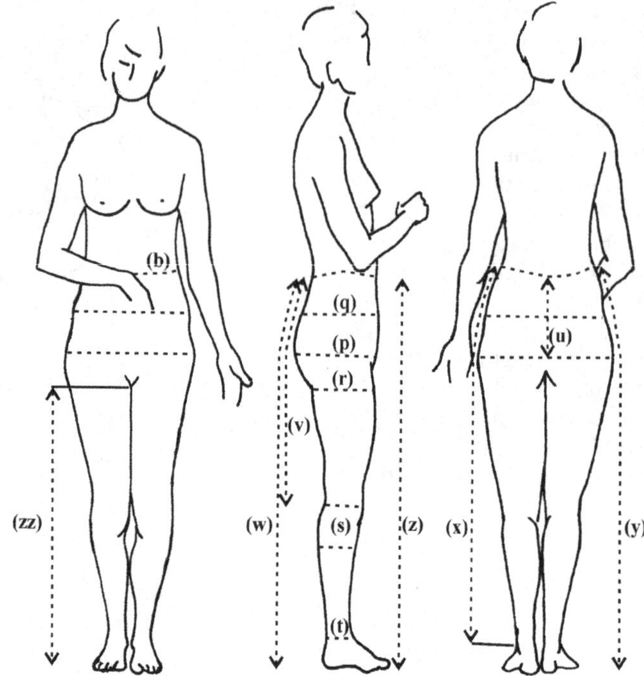

MEASURING METHOD FOR SKIRT AND TROUSERS

(b) Waist girth: See measuring method for bodice and sleeve.

(p) Hip girth: The tape measure is positioned horizontally around the fullest part of hips and buttocks and parallel to the ground (optionally an elastic tape can be positioned and levelled by using a metre rule). A note can be made of the measurement between the centre back waist and hip level (see (u)).

(q) Upper hip girth: Measurement midway between the waist and hip levels and parallel to the ground. The correct position can be checked in the mirror.

(r) Thigh girth: The person being measured stands with her legs slightly apart. The measurement is taken horizontally around the thickest part of the right upper thigh just below the crutch level.

(s) Knee or calf girth: Measurement horizontally around the thickest part of the right knee or calf, whichever is the largest.

(t) Ankle girth: Measurement around the thickest part of the right ankle.

(u) Hip level: Measure vertically from the centre back waist and the hip level.

(v) Knee length: Measure as (u) and continue down vertically to the crease at the back of the knee. (This measurement can be used as a guide for skirt lengths.)

(w) Back length: Measure as (u) and continue vertically down the back to the ground.

(x) Ankle length: Measure vertically from the side waist, over the side hipbone down to the lower edge of the anklebone.

(y) Outside leg length: Measure as (x) and continue the tape measure vertically down to the ground.

(z) Front length: Measure vertically from the centre front waist down to the ground.

(zz) Inside leg length: The person being measured can position the end of the tape measure between the legs at the crutch level. The measurer then places the tape down the inside of the leg to the ground.

Computerised body measuring systems

Various computerised body measuring systems have been developed since the early 1980s. These automatic systems operate by using scanning or photographic equipment linked to a computer. Developments since the early 2000s have realised great improvements and there are several systems now in commercial use. These systems record the body shape and posture two or three dimensionally, then calculate the body measurements. These systems can also be linked to a computer pattern alteration system or made-to-measure system (see Part 4). This enables the computer to find the nearest size pattern to the individual and alter the pattern according to the new measurements. These altered patterns can be transferred to an automatic marker making system and a single ply computer controlled cutter (see Part 5).

In Britain, Loughborough University pioneered the development of an anthropometrics shadow scanner (LASS). This produced a three-dimensional (3D) model of the human body. This method required a person, minus their outer wear, to stand stationary on a turntable as strips of light were projected vertically on to them while they were rotated 360°. A column of cameras recorded this within three minutes. This curve-fitting process treated the body as a series of 32 horizontal slices that corresponded to specific anatomical landmarks. The computer then processed this data that enabled a 3D image to be projected on the screen and the body measurements to be calculated (Fitting Research 1994 Apparel International).

An improved method of this system is now available commercially. This is the Wicks and Wilson's Triform scanner. The system is operated by the person to be scanned entering a booth wearing their underwear (Figure 1.1). They place their feet in a marked area, their hands holding a support rail which moves their arms away from their sides so that their torso is clearly revealed. They remain stationary while narrow and wide strips of white light are projected on to them and the camera captures one or more views of the illuminated object (Figure 1.2). 'By analysing the way in which the pattern of light is distorted by the shape of the object, the x, y and z co-ordinates can be calculated' (Wicks and Wilson 2000). This system only takes 12 seconds scanning time, as the person being scanned is not rotated as in the original LASS system.

The Triform body scanner processes the information within 2 minutes to produce a body shape image from which measurements can be extracted using the TriBody measuring system. This system requires the end points of the measurements to be placed manually on the scanned image so that the computer can read the distance between them. The system will also calculate the girth measurements around the body, e.g. waist, hips.

In France, Telmat Informatique developed the Systems of Measuring and Creating Anthropometric Data (SYMCAD). This was firstly using a 2D scanner, where a person enters a booth and removes their outerwear. Then a vision system records them standing facing forward and in profile. Two outline images are produced on a computer screen, which indicate their body shape and posture. This system was developed initially for use by the military to improve the allocation of correctly sized uniforms. It was used extensively in France and the UK.

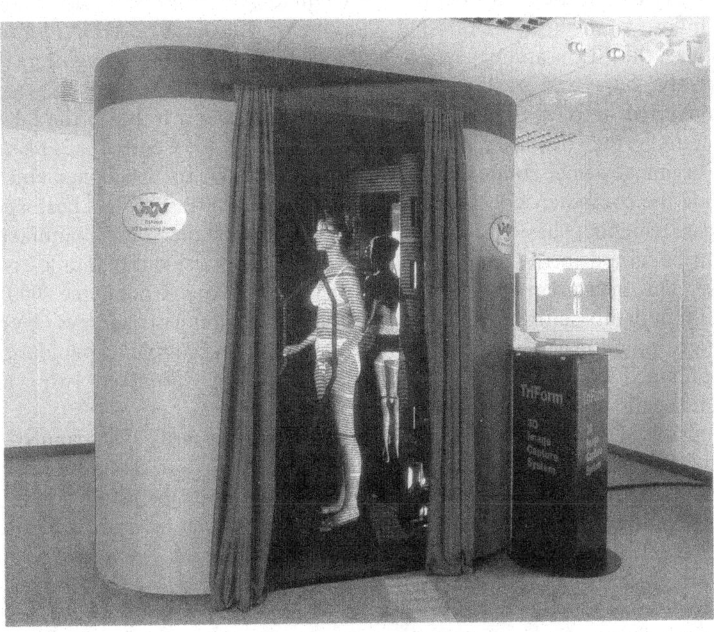

Figure 1.1 3D Scanning booth (by permission of Wicks and Wilson Ltd)

The SYMCAD Turbo Flash/3D is similar in principle to the Wicks and Wilson's Triform scanner and has now superseded the 2D scanner. The 'SYMCAD Turbo Flash/3D takes more than 70 measurements and body figuration (or shape). Every measurement is automatically calculated at pre-defined points according to the ISO 8559 and 3635 standard' (Telmat Informatique 2000).

Another 3D scanning system developed in the USA on a different principle is the $[TC]^2$ (Textile/Clothing Technology Corporation 2000). The scanning system is designed using four stationary surface sensors. 'Each sensor consists of a projector and an area sensing camera, thus forming a vertical triangulation with the object or body' (Textile/Clothing Technology Corporation 2000). These capture an area segment of the surface. These segments are combined to form an integrated surface of the body. The actual scan raw data is further processed to show an image that has the positions of the extracted measurements superimposed within a matter of 53 seconds (Figure 1.3).

One of the major advantages of 3D body scanning is the speed, which has now been reduced to a matter of seconds, when compared with manual measuring with a tape measure. Some companies state that their system gives consistent results even if the person being scanned moves or breathes. Scanning also helps to waylay the apprehension of the person being measured, as it is not so intrusive compared with being measured manually. It is difficult to identify the prominences and hollows of the body when using a tape measure, whereas the 3D scanning system represents the exact body shape and posture. The scanned image is converted into a digital form required by a computer and is then displayed as a 3D image on the screen. This image can be rotated on the screen. From this image the computer will calculate the required body measurements. These 3D images and measurements can be stored in the computer's memory for future use. This measurement information can be transferred to a database from which the distribution of a specific population can be calculated and new size charts developed.

There are still certain difficulties in defining the true body shape when using 3D scanning. There is a problem with defining the person's actual height because of the amount of hair that rises above the scalp. At present the scanners cannot differentiate between the hair and the head. Another problem arises when scanning larger people as some of their flesh can form folds, for example under the chin or the underside of a woman's bust. Other areas that are difficult to scan are the armpits and the crutch. To overcome this the stance required while being scanned is with the legs slightly apart and the arms away from the sides. This is not really a normal posture. At present those being scanned are bare footed. Most people wear shoes when fully dressed and the height of the heel has an influence on their posture and length measurements. Consequently this can affect the balance or hang of their clothes.

It can be difficult to calculate from the scanned image some body measurements required for constructing or altering garment patterns. The scanner cannot accurately locate landmark positions on the skeleton that are used when measuring manually. For example, the seventh cervical at the nape of the neck or the bone at the end of the shoulder, which a measurer finds by feel.

Humans do not remain static in the present scanning positions. Allowing for the change in body size when moving is critical for the customer's comfort. There is considerable scope for further research into the changing body shape and measurements when, for instance, the arms and legs are lifted; also, the amount that knees and elbows increase in size when bent, and the hip size when seated.

The main current use of 3D body scanning systems is for surveys to give comprehensive information on the body shape of a specific population. This will aid the defining of the body size, shape and posture for developing size charts, and will improve the fit of clothing. There are two surveys being undertaken at the time of writing this book. In the USA there is the Civilian American and European Surface Anthropometry Resource (CAESAR) Project for which the Society of Automotive Engineers is gathering data of approximately 8400 men and women in the USA, Netherlands and Italy. In the UK a national survey organised by the Centre for 3D Electronic Commerce measured 8000 to 10 000 men and women during the autumn of 2001. This has been sponsored by a consortium of retailers and manufacturers and a grant from the Department of Trade and Industry (Centre for 3D Electronic Commerce 2000). One problem for body measurement surveys is obtaining a truly representative sample of population. This is because they rely on volunteers who are willing to be measured.

Another use of 3D scanned images is in the manufacture of workroom stands or mannequins. At present many of the stands are of an idealistic body shape. However, a more realistic contour can be produced from an average of scanned images or for one individual.

The generated body measurements for an individual's scanned image can be electronically compared with garment specifications. This can be useful for the computer pattern alteration or made-to-measure systems (see Part 4). In the future these customers could have a 'smart card' containing their measurements. This could assist them in obtaining the correct size from a retailer, a catalogue or the internet. A further possibility is for the customer to view the garment's appearance and fit on a simulation of a specific standard size or their own silhouette. Such facilities may in future be in general use in retail stores.

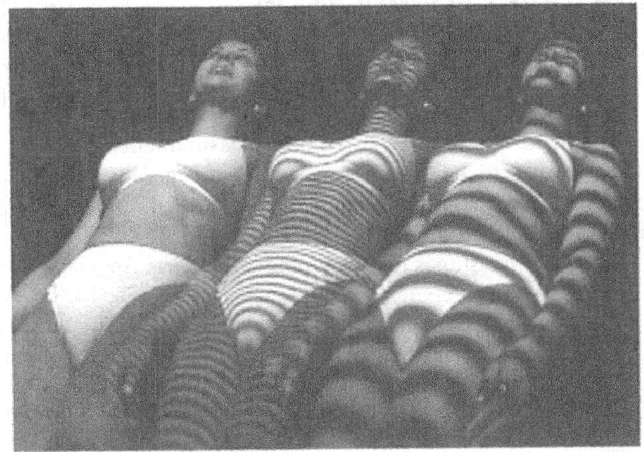

Figure 1.2 Camera images of narrow and wide strips of white light projection on to the body (by permission of Wicks and Wilson Ltd)

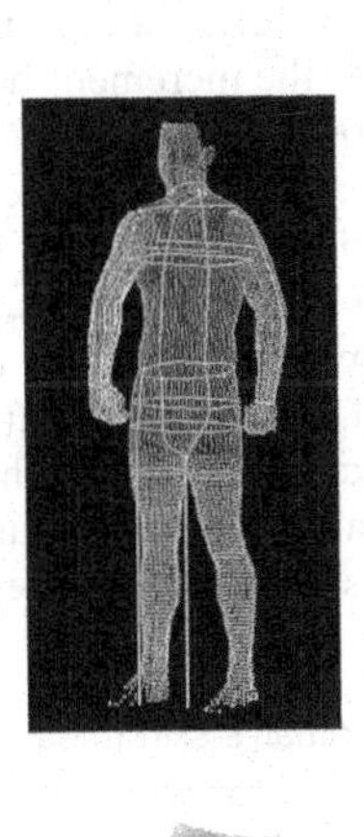

Figure 1.3 Scanning process and measurement extraction time (by permission of Textile/Clothing Technology Corporation)

SIZE CHART FORMULATION

A size chart is the dividing of average body or garment measurements artificially into categories to form a range of sizes. These average measurements are obtained from surveys of body measurements. Each size has to be given a code that is generally recognised by the public, such as 10, 12, 14, or labelled small, medium, large.

There are five stages in developing size charts for garments:

- Obtaining body measurements
- Statistically analysing the measurements
- Adding ease allowances
- Formulating the size charts
- Fitting trials to test the size charts

Firstly the body measurements have to be obtained generally through surveys taken manually or with the use of computerised equipment. (The details are described in the earlier section 'Body and garment measurements'). This is followed by the second stage of statistical analysis. Details of methods are described in an article by Beazley (1998). Often the statistical analysis of the body measurements runs to several decimal points of a centimetre, which are inconsistent and inconvenient to use for clothing manufacture. This may require the raw data to be rounded up or down to a whole centimetre or to one decimal point. Therefore the second stage is to round the measurement data to produce tables of body measurement (see Chart 1.1).

In the third stage a tolerance is added to the body measurements that is generally known as **ease allowance**. This is because garments have to be larger than the wearer to allow for movement and expansion. Three other factors which influence the ease allowance are:

- The function of the garment and whether it is worn over other garments, e.g. a coat requires extra width
- The style of the garment and whether it is close or loose fitting which depends upon the current fashion
- The type of fabric, whether it is stable or extensible, e.g. woven or knitted

Chart 1.1 illustrates the adding of ease allowances to the rounded body measurements to produce garment measurements for a straight skirt in woven fabric. Figure 1.4 illustrates the positions for the ease allowance to women's body measurements for a fitted bodice, semi-fitted sleeve and straight skirt for woven fabric. Initially the appropriate amount of ease allowance to be added in the correct position has to be estimated. The correct amount can only be confirmed after fitting trials of sample garments. More details concerning ease allowances can be found in an article by Beazley (1999).

The fourth stage is the formulation of size charts. These can be for either body measurements or garment measurements. It is difficult to manufacture a garment to an exact measurement due to dimensionally unstable fabric and sewing production. This requires a **production tolerance** to be calculated which is a measurement added to, or subtracted from, a garment measurement but still giving an acceptable size. When formulating size charts care has to be taken that the increment between the sizes is not the same as or less than the production tolerance.

The final, fifth, stage is testing the new size chart by constructing and grading patterns to the measurements, from which sample garments are cut and made. The sample garments are tested by fitting trials on groups of women of similar size. These trials confirm the correct sizes and also the amount and position of the ease allowances. If adjustments have to be made the charts and patterns are revised and re-tested.

Chart 1.1 Example of the three stages of formulating a size chart for a woman's skirt (measurements in centimetres)

SIZE		8	10	12	14	16
To fit	Waist	62.0	66.0	70.0	74.0	79.0
	Hip	88.0	92.0	96.0	100.0	105.0
Waist	Raw data	62.3	66.4	69.7	73.5	78.6
	Rounded	62.0	66.0	70.0	74.0	79.0
	Plus ease	66.0	70	74.0	78.0	84.0
Hip	Raw data	88.0	92.5	96.0	99.8	104.6
	Rounded	88.0	92	96.0	100.0	105.0
	Pluse ease	92.0	96.0	100.0	104.0	110.0

Women's size charts

The British Standards Specification for Size Designation of Women's Wear BS 3666 was last updated in 1982. These size designations are out of date when compared with retail sizing of today. Each successive generation has grown taller and wider in waist and hip girth. Very little corsetry is worn today compared with earlier generations (Beazley 1999). The most recent National Sizing Survey has not yet been published. For this survey it is planned to measure 10 000 women by a body scanning system. The following size charts are based on small-scale research undertaken by the Department of Clothing Design and Technology at Hollings Faculty, Manchester Metropolitan University between 1992 and 1998. The block patterns developed from these sizes have undergone numerous fitting trials. Therefore the sizes and patterns in this book represent the contemporary women's figure. However, any other satisfactory size chart can be used.

SIZE CHARTS FOR THE RANGE 8 TO 16

Charts 1.2 and 1.3 illustrate the development of the body measurement tables into garment size charts by adding the ease allowances for the base size 12. The amount added is the minimum suitable for woven fabric. It is advisable to have the base size as the central size from which the larger and smaller sizes are graded. This maintains accuracy in the pattern shape for each size.

The last column gives the grading increments for two size ranges. These size ranges are designated by the grade of key measurements to which all the other grades are proportional. The key measurement in Chart 1.2 is the bust girth with a grade of 4 cm or 5 cm (in parenthesis); in chart 1.3 it is the hip girth with a grade of 4 cm or 5 cm (in parenthesis). The key body measurement is often quoted at the top of size charts or on garment labels:

e.g. Size 12 to fit bust 88 cm
to fit hip 96 cm

Size charts with equal size intervals are restricted to four or five sizes as the body proportion changes shape beyond this number. Only five sizes are quoted in the charts from size 8 to 16 with the central size 12 as the base or sample size. Some extra measurements are included in Chart 1.2 for bodice and sleeve measurements only. These are not included in most size charts, but are useful for constructing patterns. Also included within Chart 1.2 (column headed 'Dress') are the suggested amounts for dart suppression relative to back/front shoulder and waist darts.

In the garment size charts, Chart 1.4 for dresses

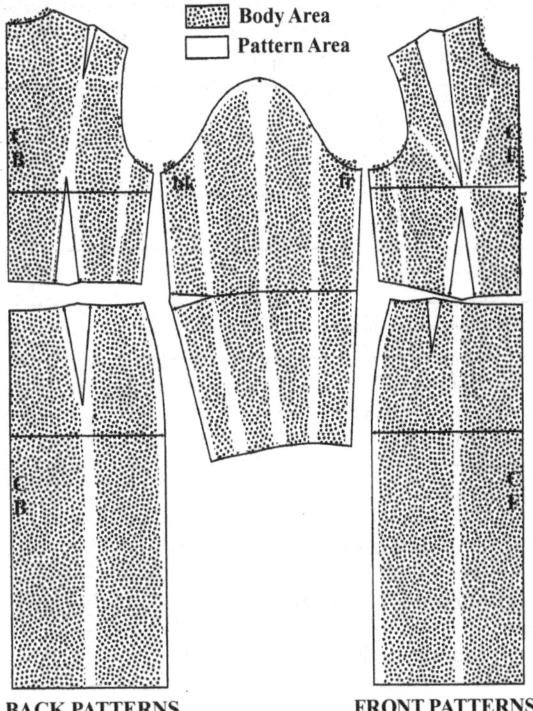

Figure 1.4 The positions for adding ease allowances to women's body measurements for a fitted bodice, semi-fit sleeve and straight skirt

and Chart 1.5 for skirts and trousers have a 4 cm grade for the key measurements bust, waist and hips, whereas Charts 1.6 and 1.7 have a 5 cm grade. The base size 12 is the same size for both ranges but the other sizes vary slightly. For example, the garment bust grade:

	4 cm grade	*5 cm grade*
size 8	86 cm	84 cm
size 16	102 cm	104 cm

This gives a 2 cm difference on the largest and smallest sizes. All the other grading increments are proportional to whether the major girths of bust, waist and hips have a 4 cm or 5 cm grade.

SIZE CHARTS FOR THE RANGE 8 TO 20

The size charts which have been presented so far are for a limited range of four to five sizes with equal increments between the sizes. However, some styles need to cover a much larger range of sizes, as many as seven to ten sizes. These larger size ranges can be approached in two ways. Firstly, a very large size range can be split into two or three short ranges of differing proportions with separate central base size patterns. For example, a women's size range from

Chart 1.2 Development of body measurements into garment measurement by adding ease allowances for size 12 women's bodice and sleeve (measurements in centimetres). Key measurement: bust 4 cm grade (5 cm grade in parenthesis). For average height 164 cm (5 ft 4½ in)

	Body	Ease	Dress	Grade
(a) Bust girth	88.0	6.0	94.0	4.0 (5.0)
(b) Waist girth	70.0	4.0	74.0	4.0 (5.0)
(c) Neck girth	38.0	2.0	40.0	1.0 (1.0)
(d) Upper arm girth	28.0	6.0	34.0	1.0 (1.6)
(e) Elbow girth	26.0	5.0	31.0	0.75 (1.0)
(f) Wrist girth	16.0	2.0	18.0	0.5 (0.5)
(g) Nape to waist	41.0	—	41.0	0.5 (0.5)
(h) Front neck point to bust	27.0	—	27.0	0.5 (0.5)
(i) Front neck point to waist	44.0	—	44.0	0.5 (0.5)
(j) Shoulder to elbow	35.0	—	35.0	0.5 (0.5)
(k) Shoulder to wrist	59.0	—	59.0	0.5 (0.5)
(l) Across back (at mid armhole)	35.0	2.0	37.0	1.0 (1.2)
(m) Across front (at mid armhole)	32.0	1.0	33.0	1.0 (1.2)
(n) Shoulder length	13.0	—	13.0	0.3 (0.4)
(o) Bust prominence width	19.0	—	19.0	0.4 (0.4)
EXTRA MEASUREMENTS				
Width of armhole	10.0	1.5	11.5	1.0 (1.3)
Back shoulder dart width at shoulder	—	—	1.5	— —
Front shoulder dart width at shoulder	—	—	4.5	0.5* (0.5*)
Waist darts width	4.0	—	4.0	— —
Depth of armhole	21.0	3.0	24	0.5 (0.5)
Armhole circumference	40.0	5.0	45.0	1.5 (1.8)
Sleeve head depth (approx. ⅓ armhole circumference	—	—	15.0	0.5 (0.5)

* Grade optional

NB For short women 156 cm (5 ft 1½ in), reduce bodice 3 cm between underarm and waist. Reduce the sleeve length 3 cm between the underarm and wrist.

For tall women 172 cm (5 ft 7½ in) increase bodice length 3 cm between underarm and waist. Increase the sleeve 3 cm between the underarm and wrist.

Chart 1.3 Development of body measurements into garment measurements by adding ease allowances for size 12 skirts and trousers. Key measurement: hip 4 cm grade (5 cm grade in parentheses). For average women, medium height 164 cm (5 ft 4½ in) (measurements in centimetres)

	Body	Ease	Garment	Grade
(b) Waist girth	70.0	4.0	74.0	4.0 (5.0)
(b) Waist band girth	70.0	2.0	72.0	4.0 (5.0)
(p) Hip girth	96.0	4.0	100.0	4.0 (5.0)
(q) Upper hip girth	90.0	4.0	94.0	4.0 (5.0)
(r) Thigh girth – straight leg – slim leg	57.0 57.0	10.0 8.0	67.0 65.0	2.6 (3.2) 2.6 (3.2)
(s) Knee or calf girth – straight leg – slim leg	37.0 37.0	15.0 9.0	52.0 46.0	2.0 (2.0) 2.0 (2.0)
(t) Ankle girth	25.0	9.0	34.0	1.0 (1.0)
(u) Centre back waist to hip	20.0	—	20.0	0.5*
(v) Centre back waist to knee	60.0	—	60.0	0.5*
(w) Centre back waist to ground	105.0	—	105.0	0.5*
(x) Side waist to ankle	100.0	—	100.0	0.5*
(y) Side waist to ground	106.0	—	106.0	0.5*
(z) Centre front waist to ground	105.0	—	105.0	0.5*
(zz) Inside leg (crutch to ankle)	72.0	–1.0	71.0	—
Crutch level (x – zz)	28.0	1.0	29.0	0.5 (0.5)

* Skirt and trouser length grade optional.
NB for short women 156 cm (5 ft 1½ in) reduce the skirt 3 cm and the trousers 6 cm.
For tall women 172 cm (5 ft 7½ in) increase the skirt 3 cm and the trousers 6 cm.

size 8 to 26 could be split into two of different proportions, sizes 8 to 16 with a base size of 12, and sizes 18 to 26 with a base size of 22. Alternatively, for a shorter size range of seven sizes the central base size can be graded with the increments varying between the sizes to change the patterns to the correct proportions.

The following size charts illustrate the change in body proportion from size 8 to 20 with a base size 12. The variations have been based on surveys of body measurements. The smaller sizes of 8 to 14 have fewer size differences than the sizes 16 to 20. Also, the bust, waist and hips do not have the same grading increments as the previous size charts. The pattern grading for larger sizes is more complex as the dart suppression can be changed in width. Chart 1.8 is an example of body measurements for sizes 8 to 20 for dresses, Chart 1.9 for skirts and trousers. The amount of ease allowance also varies between some sizes. This can be seen by comparing with the garment measurements of Chart 1.10 for dresses and Chart 1.11 for skirts and trousers. The ease allowance has been increased for the larger sizes. That is why there is a greater increase from size 14 to size 16. It has been advocated (Cooklin 1997) that the ease allowance should be calculated as a percentage of the major girth measurement. Although this is correct in theory, when the amounts were calculated they became complex requiring several decimal points. This has been simplified by rounding the amount of

Chart 1.4 Garment measurement size chart for women's dresses. Key measurements: bust and waist 4 cm grade (measurements in centimetres). Size range: 8–16, for average height 164 cm (5 ft 4½ in)

SIZE		8	10	12	14	16
To fit bust	cm in approx.	80.0 32	84.0 33½	88.0 35	92.0 36½	96.0 38
To fit waist	cm in approx.	62.0 24½	66.0 26	70.0 27½	74.0 29	78.0 30½
To fit hips	cm in approx.	88.0 35	92.0 36½	96.0 38	100.0 39½	104.0 41
MEASUREMENTS						
(a) Bust girth		86.0	90.0	94.0	98.0	102.0
(b) Waist girth		66.0	70.0	74.0	78.0	82.0
(c) Neck girth		38.0	39.0	40.0	41.0	42.0
(d) Upper arm girth		32.0	33.0	34.0	35.0	36.0
(e) Elbow girth (fitted)		29.5	30.25	31.0	31.75	32.5
(f) Wrist girth (fitted)		17.0	17.5	18.0	18.5	19.0
(g) Nape to waist		40.0	40.5	41.0	41.5	42.0
(h) Front neck point to bust		26.0	26.5	27.0	27.5	28.0
(i) Front neck point to waist		43.0	43.5	44.0	44.5	45.0
(j) Shoulder to elbow		34.0	34.5	35.0	35.5	36.0
(k) Shoulder to wrist		58.0	58.5	59.0	59.5	60.0
(l) Across back (at mid armhole)		35.0	36.0	37.0	38.0	39.0
(m) Across front (at mid armhole)		31.0	32.0	33.0	34.0	35.0
(n) Shoulder length		12.4	12.7	13.0	13.3	13.6
(o) Bust prominence width		18.2	18.6	19.0	19.4	19.8
(p) Hip girth		92.0	96.0	100.0	104.0	108.0
(q) Upper hip girth		86.0	90.0	94.0	98.0	102.0
(w) Centre back waist to ground*		104.0	104.5	105.0	105.5	106.0
(u) Centre back waist to hip*		19.0	19.5	20.0	20.5	21.0
(v) Centre back waist to knee*		59.0	59.5	60.0	60.5	61.0
Depth of armhole (derived)		23.0	23.5	24.0	24.5	25.0

* Grading increment of 0.5 cm optimal.

NB For short women of 156 cm (5 ft 1½ in) reduce the bodice length 3 cm between underarm and waist. Reduce the sleeve length 3 cm between the underarm and wrist. For tall women of 172 cm (5 ft 7½ in) increase the bodice length 3 cm between the underarm and waist. Increase the sleeve length 3 cm between the underarm and wrist.

Chart 1.5 Garment measurement size chart for women's skirts and trousers. Key measurements: hip and waist 4 cm grade (measurements in centimetres). Size range: 8–16, for average height 164 cm (5 ft 4½ in)

SIZE		8	10	12	14	16
To fit waist	cm in approx.	62 24½	66 26	70 27½	74 29	78 30½
To fit hips	cm in approx.	88 35	92 36½	96 38	100 39½	104 41
MEASUREMENTS						
(b) Waist girth		66.0	70.0	74.0	78.0	82.0
(b) Waist band girth		64.0	68.0	72.0	76.0	80.0
(p) Hip girth		92.0	96.0	100.0	104.0	108.0
(q) Upper hip girth		88.0	90.0	94.0	98.0	102.0
(r) Thigh girth	straight leg slim leg	61.8 59.8	64.4 62.4	67.0 65.0	69.6 67.6	72.2 70.2
(s) Knee or calf girth	straight leg slim leg	48.0 42.0	50.0 44.0	52.0 46.0	54.0 48.0	56.0 50.0
(t) Ankle girth	fitted	32.0	33.0	34.0	35.0	36.0
(u) Centre back waist to hip*		19.0	19.5	20.0	20.5	21.0
(v) Centre back waist to knee*		59.0	59.5	60.0	60.5	61.0
(w) Centre back waist to ground*		104.0	104.5	105.0	105.5	106.0
(x) Side waist to ankle*		99.0	99.5	100.0	100.5	101.0
(y) Side waist to ground*		105.0	105.5	106.0	106.5	107.0
(z) Centre front waist to ground*		104.0	104.5	105.0	105.5	106.0
(zz) Inside leg (crutch to ankle)*		71.0	71.0	71.0	71.0	71.0
Waist to crutch (x–zz)		28.0	28.5	29.0	29.5	30.0

* Skirt and trouser length grade optional.
NB For short women of 156 cm (5 ft 1½ in) reduce the skirt 3 cm and the trousers 6 cm.
For tall women of 172 cm (5 ft 7½ in) increase the skirt 3 cm and the trousers 6 cm.

ease allowance increase to the nearest whole centimetre.

It must be emphasised that proven size charts have to be selected before any pattern construction can be undertaken. The following size charts illustrate the development of tables of body measurements into garment size charts with the inclusion of a minimum of ease allowances. These charts will be used in the following sections concerning the development of block patterns, pattern grading into other sizes and pattern alterations for computer made-to-measure systems. The reader is not restricted to using these suggested size charts as the techniques that are explained can be applied to any reliable size chart. Most clothing manufacturers have their own size specifications suitable for their retail customers.

Chart 1.6 Garment measurement size chart for women's dresses. Key measurements: bust and waist 5 cm grade (measurements in centimetres). Size range: 8–16, for average height 164 cm (5 ft $4\frac{1}{2}$ in)

SIZE		8	10	12	14	16
To fit bust	cm in approx.	78.0 31	83.0 33	88.0 35	93.0 37	98.0 39
To fit waist	cm in approx.	60.0 $23\frac{1}{2}$	65.0 $25\frac{1}{2}$	70.0 $27\frac{1}{2}$	75.0 $29\frac{1}{2}$	80.0 $31\frac{1}{2}$
To fit hips	cm in approx.	86.0 34	91.0 36	96.0 38	101.0 40	106.0 42
DRESS MEASUREMENTS						
(a) Bust girth		84.0	89.0	94.0	99.0	104.0
(b) Waist girth		64.0	69.0	74.0	79.0	84.0
(c) Neck girth		38.0	39.0	40.0	41.0	42.0
(d) Upper arm girth		30.8	32.4	34.0	35.6	37.2
(e) Elbow girth (fitted)		29.0	30.0	31.0	32.0	33.0
(f) Wrist girth (fitted)		17.0	17.5	18.0	18.5	19.0
(g) Nape to waist		40.0	40.5	41.0	41.5	42.0
(h) Front neck point to bust		26.0	26.5	27.0	27.5	28.0
(i) Front neck point to waist		43.0	43.5	44.0	44.5	45.0
(j) Shoulder to elbow		34.0	34.5	35.0	35.5	36.0
(k) Shoulder to wrist		58.0	58.5	59.0	59.5	60.0
(l) Across back (at mid armhole)		34.6	35.8	37.0	38.2	39.4
(m) Across front (at mid armhole)		30.6	31.8	33.0	34.2	35.4
(n) Shoulder length		12.2	12.6	13.0	13.4	13.8
(o) Bust prominence width		17.8	18.4	19.0	19.6	20.2
(p) Hip girth		90.0	95.0	100.0	105.0	110.0
(q) Upper hip girth		84.0	89.0	94.0	99.0	104.0
(w) Centre back waist to ground*		104.0	104.5	105.0	105.5	106.0
(u) Centre back waist to hip*		19.0	19.5	20.0	20.5	21.0
(v) Centre back waist to knee*		59.0	59.5	60.0	60.5	61.0
Depth of armhole		23.0	23.5	24.0	24.5	25.0

* Grading increment of 0.5 cm optimal.

NB For short women of 156 cm (5 ft $1\frac{1}{2}$ in) reduce the bodice length 3 cm between underarm and waist. Reduce the sleeve length 3 cm between the underarm and wrist.

For tall women of 172 cm (5 ft $7\frac{1}{2}$ in) increase the bodice length 3 cm between the underarm and waist. Increase the sleeve length 3 cm between the underarm and wrist.

Chart 1.7 Garment measurement size chart for women's skirts and trousers. Key measurements: hip and waist 5 cm grade (measurements in centimetres). Size range: 8–16, for average height 164 cm (5 ft 4½ in)

SIZE		**8**	**10**	**12**	**14**	**16**
To fit waist	cm in approx.	60 23½	65 25½	70 27½	75 29½	80 31½
To fit hips 20 cm below waist	cm in approx.	86 34	91 36	96 38	101 40	106 42
MEASUREMENTS						
(b) Waist girth		64.0	69.0	74.0	79.0	84.0
(b) Waist band girth		62.0	67.0	72.0	77.0	82.0
(p) Hip girth		90.0	95.0	100.0	105.0	110.0
(q) Upper hip girth		84.0	89.0	94.0	99.0	104.0
(r) Thigh girth	straight leg slim leg	60.6 58.6	63.8 61.8	67.0 65.0	70.2 68.2	73.4 71.4
(s) Knee or calf girth	straight leg slim leg	48.0 42.0	50.0 44.0	52.0 46.0	54.0 48.0	56.0 50.0
(t) Ankle girth	fitted	32.0	33.0	34.0	35.0	36.0
(u) Centre back waist to hip*		19.0	19.5	20.0	20.5	21.0
(v) Centre back waist to knee*		59.0	59.5	60.0	60.5	61.0
(w) Centre back waist to ground*		104.0	104.5	105.0	105.5	106.0
(x) Side waist to ankle*		99.0	99.5	100.0	100.5	101.0
(y) Side waist to ground*		105.0	105.5	106.0	106.5	107.0
(z) Centre front waist to ground*		104.0	104.5	105.0	105.5	106.0
(zz) Inside leg (crutch to ankle)*		71.0	71.0	71.0	71.0	71.0
Waist to crutch (x–zz)		28.0	28.5	29.0	29.5	30.0

* Skirt and trouser length grade optional.
NB For short women of 156 cm (5 ft 1½ in) reduce the skirt 3 cm and the trousers 6 cm.
For tall women of 172 cm (5 ft 7½ in) increase the skirt 3 cm and the trousers 6 cm.

Chart 1.8 Body measurement tables for women's dresses. Key measurements: hip and waist proportional grade (measurements in centimetres). Size range: 8–20, for average height 164 cm (5 ft 4½ in)

SIZE		8	10	12	14	16	18	20
To fit bust	cm in	78.0 31	83.0 33	88.0 35	93.0 37	98.0 39	104.0 41	110.0 43
To fit waist	cm in	62.0 24½	66.0 26	70.0 27½	74.0 29	80.0 31½	86.0 34	92.0 36½
To fit hips	cm in	88.0 35	92.0 36½	96.0 38	100.0 39½	105.0 41½	110.0 43½	115.0 45½
DRESS BODY MEASUREMENTS								
(a) Bust girth		78.0	83.0	88.0	93.0	98.0	104.0	110.0
(b) Waist girth		62.0	66.0	70.0	74.0	80.0	86.0	92.0
(c) Neck girth		36.0	37.0	38.0	39.0	40.0	41.0	42.0
(d) Upper arm girth		24.0	26.0	28.0	30.0	32.0	34.0	36.0
(e) Elbow girth		23.0	24.5	26.0	27.5	29.0	31.0	33.0
(f) Wrist girth		15.0	15.5	16.0	16.5	17.0	17.5	18.0
(g) Nape to waist		40.0	40.5	41.0	41.5	41.5	42.0	42.0
(h) Fr N point to bust		26.0	26.5	27.0	28.0	29.0	30.5	32.0
(i) Fr N point to waist		43.0	43.5	44.0	45.0	46.0	47.0	48.0
(j) Shoulder to elbow		34.0	34.5	35.0	35.0	35.0	35.0	35.0
(k) Shoulder to wrist		58.0	58.5	59.0	59.0	59.0	59.0	59.0
(l) Across back		33.0	34.0	35.0	36.0	37.0	38.0	39.0
(m) Across front		30.0	31.0	32.0	33.0	34.0	35.5	37.0
(n) Shoulder length		12.4	12.7	13.0	13.3	13.6	13.6	13.6
(o) Bust prom. width		17.0	18.0	19.0	20.0	21.0	22.0	23.0
(p) Hip girth		88.0	92.0	96.0	100.0	105.0	110.0	115.0
(q) Upper hip girth		82.0	86.0	90.0	94.0	100.0	106.0	112.0
(w) CB waist to ground*		104.0	104.5	105.0	105.5	106.0	106.5	107.0
(u) CB waist to hip*		19.0	19.5	20.0	20.5	21.0	21.5	22.0
(v) CB waist to knee*		59.0	59.5	60.0	60.5	61.0	61.5	62.0
Nape to armhole		20.0	20.5	21.0	22.0	23.0	24.0	25.0

* Grading increment of 0.5 cm optimal.

NB For short women of 156 cm (5 ft 1½ in) reduce the bodice length 3 cm between underarm and waist. Reduce the sleeve length 3 cm between the underarm and wrist.

For tall women of 172 cm (5 ft 7½ in) increase the bodice length 3 cm between the underarm and waist. Increase the sleeve length 3 cm between the underarm and wrist.

Chart 1.9 Body measurement table for women's skirts and trousers. Key measurements: hip and waist proportional grade (measurements in centimetres). Size range: 8–20 for average height 164 cm (5 ft 4½ in)

SIZE		**8**	**10**	**12**	**14**	**16**	**18**	**20**
To fit waist	cm in	62.0 24½	66.0 25	70.0 27½	74.0 29	80.0 31½	86.0 34	92.0 36½
To fit hips	cm in	88.0 35	92.0 36½	96.0 38	100.0 39½	105.0 41½	110.0 43½	115.0 45½
MEASUREMENTS								
(b) Waist girth		62.0	66.0	70.0	74.0	80.0	86.0	92.0
(p) Hip girth		88.0	92.0	96.0	100.0	105.0	110.0	115.0
(q) Upper hip girth		82.0	86.0	90.0	95.0	101.0	107.0	113.0
(r) Thigh girth		51.0	54.0	57.0	59.0	62.0	65.0	69.0
(s) Knee girth		32.0	34.0	36.0	38.0	40.0	42.0	44.0
(t) Ankle girth		22.0	23.0	24.0	25.0	25.0	26.0	26.0
(u) CB waist to hip*		19.0	19.5	20.0	20.5	21.0	21.5	22.0
(v) CB waist to knee*		59.0	59.5	60.0	60.5	61.0	61.5	62.0
(w) CB waist to ground*		104.0	104.5	105.0	106.0	107.0	108.0	109.0
(x) Side waist to ankle*		99.0	99.5	100.0	101.0	102.0	103.0	104.0
(y) Side waist to ground*		105.0	105.5	106.0	107.0	108.0	109.0	110.0
(z) CF waist to ground*		104.0	104.5	105.0	106.0	107.0	108.0	109.0
(zz) Inside leg (crutch to ankle)*		72.0	72.0	72.0	72.0	72.0	72.0	72.0
Waist to crutch (x–zz)		27.0	27.5	28.0	29.0	30.0	31.0	32.0

* Skirt and trouser length grade optional.
NB For short women of 156 cm (5 ft 1½ in) reduce the skirt 3.0 cm and the trousers 6 cm.
For tall women of 172 cm (5 ft 7½ in) increase the skirt 3.0 cm and the trousers 6 cm.

Chart 1.10 Garment measurement size chart for women's dresses. Key measurements: bust, waist and hip proportional grade (measurements in centimetres). Size range: 8–20, for average height 164 cm (5 ft 4½ in)

SIZE		8	10	12	14	16	18	20
To fit bust	cm	78.0	83.0	88.0	93.0	98.0	104.0	110.0
	in	31	33	35	37	39	41	43
To fit waist	cm	62.0	66.0	70.0	74.0	80.0	86.0	92.0
	in	24½	26	27½	29	31½	34	36½
To fit hips	cm	88.0	92.0	96.0	100.0	105.0	110.0	115.0
	in	35	36½	38	39½	41½	43½	45½
MEASUREMENTS								
(a) Bust girth		84.0	89.0	94.0	99.0	106.0	112.0	118.0
(b) Waist girth		66.0	70.0	74.0	79.0	85.0	91.0	98.0
(c) Neck girth		38.0	39.0	40.0	41.0	43.0	44.0	45.0
(d) Upper arm girth		30.8	32.4	34.0	35.6	37.2	40.0	42.8
(e) Elbow girth		29.0	30.0	31.0	32.0	34.0	36.0	38.0
(f) Wrist girth		17.0	17.5	18.0	18.5	19.0	20.0	21.0
(g) Nape to waist		40.0	40.5	41.0	41.5	41.5	42.0	42.0
(h) Fr N point to bust		26.0	26.5	27.0	28.0	29.0	30.5	32.0
(i) Fr N point to waist		43.0	43.5	44.0	45.0	46.0	47.0	48.0
(j) Shoulder to elbow		34.0	34.5	35.0	35.0	35.0	35.0	35.0
(k) Shoulder to wrist		58.0	58.5	59.0	59.0	59.0	59.0	59.0
(l) Across back		35.0	36.0	37.0	38.0	40.0	41.0	42.0
(m) Across front		31.0	32.0	33.0	35.0	37.0	38.5	40.0
(n) Shoulder length		12.4	12.7	13.0	13.3	13.6	13.6	13.6
(o) Bust prom. width		17.0	18.0	19.0	20.0	21.0	22.0	23.0
(p) Hip girth		92.0	96.0	100.0	105.0	111.0	116.0	122.0
(q) Upper hip girth		86.0	90.0	94.0	99.0	106.0	112.0	119.0
(w) CB waist to ground*		104.0	104.5	105.0	105.5	106.0	106.5	107.0
(u) CB waist to hip*		19.0	19.5	20.0	20.5	21.0	21.5	22.0
(v) CB waist to knee*		59.0	59.5	60.0	60.5	61.0	61.5	62.0
Nape to armhole		23.0	23.5	24.0	25.0	26.0	27.0	28.0

*Grading increment of 0.5 cm optimal.

NB For short women of 156 cm (5 ft 1½ in) reduce the bodice length 3 cm between underarm and waist. Reduce the sleeve length 3 cm between the underarm and wrist.

For tall women of 172 cm (5 ft 7½ in) increase the bodice length 3 cm between the underarm and waist. Increase the sleeve length 3 cm between the underarm and wrist.

Chart 1.11 Garment measurement size chart for women's skirts and trousers. Key measurements: hip and waist proportional grade (measurements in centimetres). Size range: 8–20, for average height 164 cm (5 ft 4½ in)

SIZE		8	10	12	14	16	18	20
To fit waist	cm	62.0	66.0	70.0	74.0	80.0	86.0	92.0
	in	24½	26	27½	29	31½	34	36½
To fit hips	cm	88.0	92.0	96.0	100.0	105.0	110.0	115.0
	in	35	36½	38	39½	41½	43½	45½
MEASUREMENTS								
(b) Waist girth		66.0	70.0	74.0	79.0	85.0	91.0	98.0
(b) Waist band girth		64.0	68.0	72.0	77.0	83.0	89.0	96.0
(p) Hip girth		92.0	96.0	100.0	105.0	111.0	116.0	122.0
(q) Upper hip girth		86.0	90.0	94.0	99.0	106.0	113.0	119.0
(r) Thigh girth, straight leg		61.8	64.4	67.0	70.3	74.2	77.4	80.7
(s) Knee girth, straight leg		48.0	50.0	52.0	54.0	56.0	58.0	60.0
(t) Ankle girth, fitted		32.0	33.0	34.0	35.0	36.0	37.0	38.0
(u) CB waist to hip*		19.0	19.5	20.0	20.5	21.0	21.5	22.0
(v) CB waist to knee*		59.0	59.5	60.0	60.5	61.0	61.5	62.0
(w) CB waist to ground*		104.0	104.5	105.0	106.0	107.0	108.0	109.0
(x) Side waist to ankle*		99.0	99.5	100.0	101.0	102.0	103.0	104.0
(y) Side waist to ground*		105.0	105.5	106.0	107.0	108.0	109.0	110.0
(z) CF waist to ground*		104.0	104.5	105.0	106.0	107.0	108.0	109.0
(zz) Inside leg (crutch to ankle)*		71.0	71.0	71.0	71.0	71.0	71.0	71.0
Waist to crutch (x–zz)		28.0	28.5	29.0	30.0	31.0	32.2	33.0

* Skirt and trouser length grade optional.
NB For short women of 156 cm (5 ft 1½ in) reduce the skirt 3 cm and the trousers 6 cm.
For tall women of 172 cm (5 ft 7½ in) increase the skirt 3 cm and the trousers 6 cm.

PATTERN CONSTRUCTION TECHNIQUES

The object of this section is to explain different techniques of constructing patterns. These were originally undertaken manually but are easily adapted to construction by computer. Where appropriate a comparison between manual and computer techniques will be explained.

The positioning of computer patterns

The pattern is identified by the computer on the horizontal **X-axis** and vertical **Y-axis**. The majority of programs consider that the warp grain line of the pattern piece runs horizontally on the X-axis, both on the digitising table and monitor's visual display. Computer grading also requires a horizontal reference line which in most cases is the warp grain line running the length of the garment (pattern grading will be explained in Part 2). Most computer lay planning programs also require the warp threads of the fabric to be displayed horizontally. This positioning of computer patterns may take a while to get used to for those who have worked in the more traditional way with warp lengthways grain vertical.

The back bodice block A with a vertical grain line illustrates the traditional pattern position given in books on pattern construction. The two horizontal patterns with horizontal grain lines show the position for computer use, the difference being that the top of the garment is to the left in pattern B and to the right in pattern C. Either positioning, B or C, can be used. However, for the pattern illustrations in this book position B is generally used.

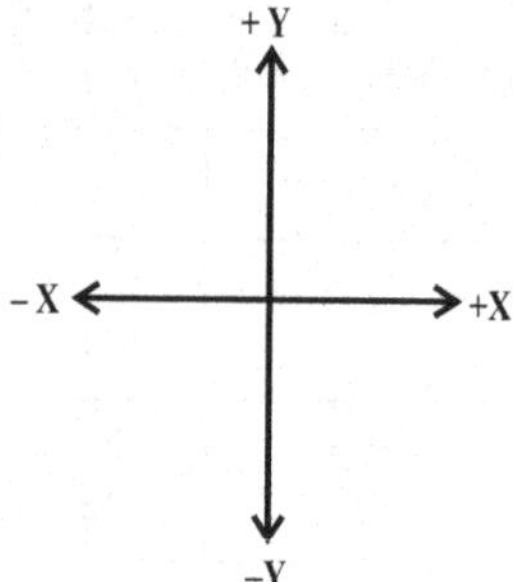

X and Y axis

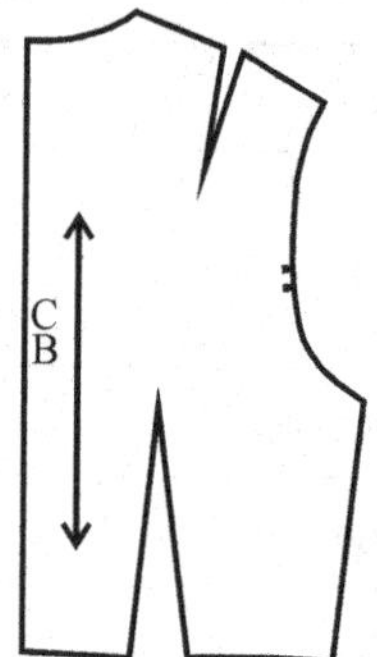

A: Traditional vertical grain line

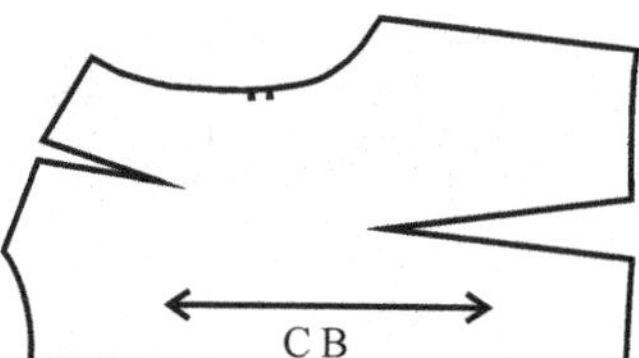

B: Computer horizontal grain line

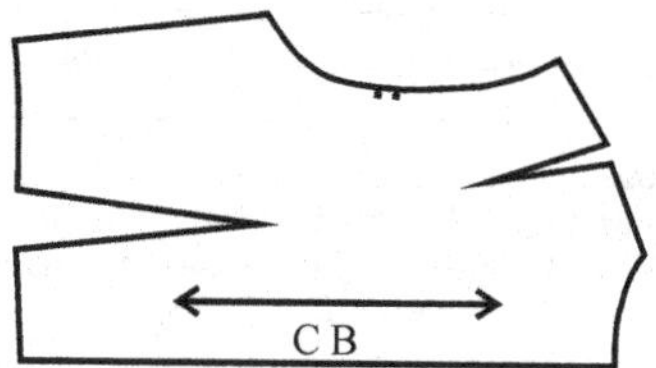

C: Computer horizontal grain line

Creating pattern shapes by computer

Creating a square, rectangle or circle by most computer programs is generally straightforward, requiring only the dimensions or radius. Difficulties can arise where curved lines are not obtained geometrically but require interactive drawing using a mouse or stylus. The shape can be produced by a series of marked points that the computer automatically joins into a curve. If the curve line is not to the required shape or length it can be modified by either moving the points or using a smoothing or similar function.

Some computer programs have a series of tools representing curves similar to a French curve. Useful manual tools can be digitised into the computer. Alternatively the pattern technicians can create their own tools. This is achieved by selecting a proven shaped pattern and copying useful curved lines such as the neck, armhole, sleeve head and skirt side seam. The selected curves are placed within a rectangle. When creating a new pattern these lines can be copied and placed in the new pattern. If the lines are not quite the correct shape they can be modified. They can also be mirrored or flipped over into another direction. The angle of the line can be altered by pivoting or rotating.

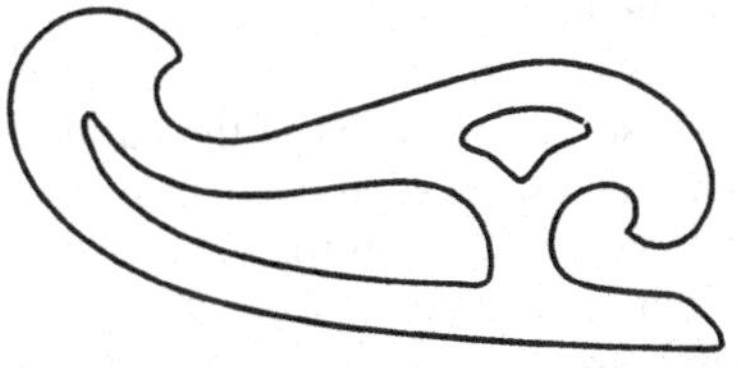

Gerbes tool from Accumark program

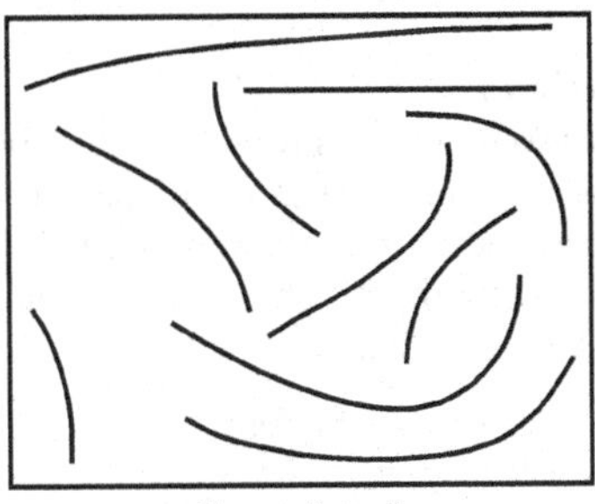

Created tool

Techniques for constructing pattern shapes

There are four basic techniques of constructing a pattern shape: *modelling*, *drafting*, *suppressing* and *flaring*. To explain these techniques four different patterns for a cover of a cone will be constructed. Where shapes are symmetrical it is more accurate and less time consuming to construct a pattern for a section of it rather than the total. For these examples the cover is in two pieces. Therefore only a quarter pattern needs to be constructed and then mirrored for half the cover, which is then duplicated for the second piece. The cone to be covered has a top circumference of 24 cm, lower circumference of 40 cm and height of 10 cm. The following examples can be used as exercises for those wishing to develop computer skills.

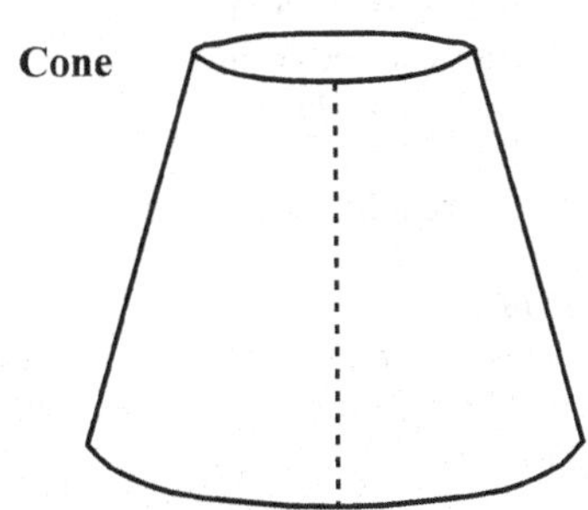

Cone

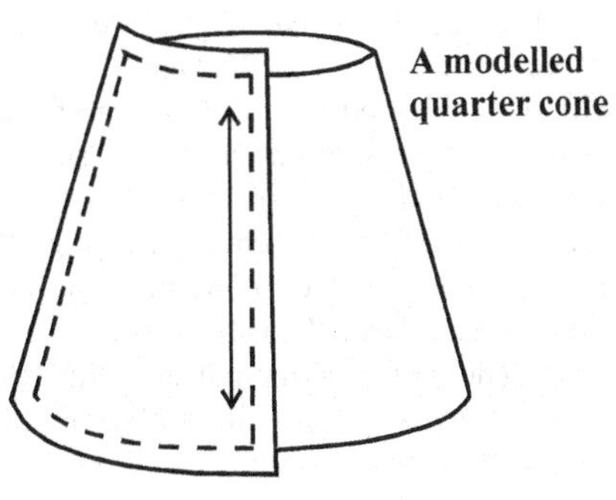

A modelled quarter cone

MODELLING

A quarter of the cone is modelled or draped in fabric by hand. The stitching line and grain line are clearly marked before the modelled fabric is removed from the cone. The fabric is then smoothed out. Alternatively the shape can be traced flat on to pattern paper. Either the draped fabric or traced pattern can be digitised into the computer.

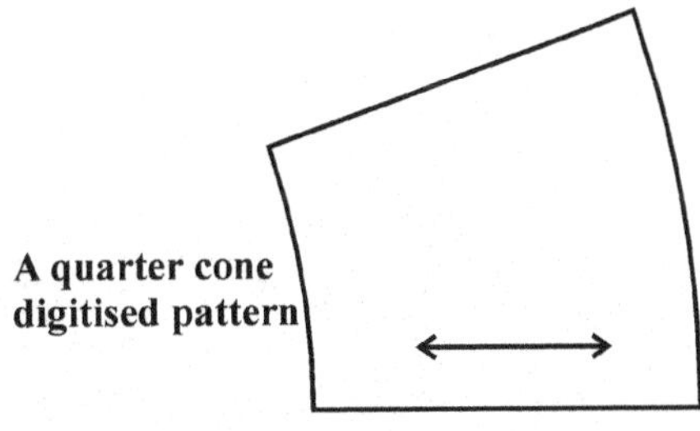

A quarter cone digitised pattern

PATTERN DRAFTING

Once a proven gore shape is obtained instructions can be written so that a replica of the pattern can be drafted. This is useful if the cone is no longer obtainable to measure or drape. The draft is constructed within a rectangle, from which a set of measurements are quoted to produce the required shape. Two alternative methods of drafting, manually or by computer, are illustrated below.

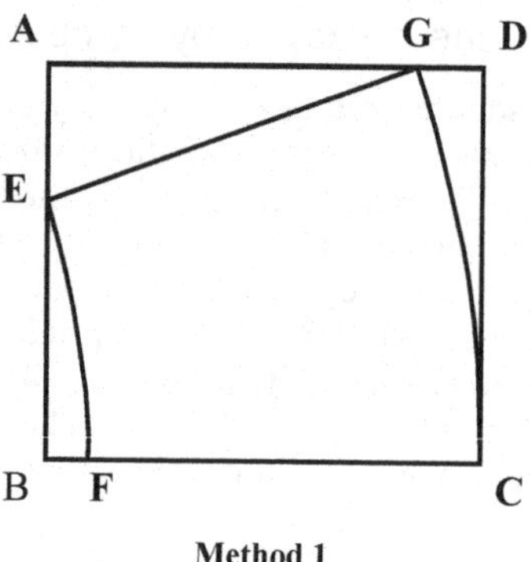

Method 1

Drafting method 1

This is a traditional method found in many books on pattern construction. Often no explanation is given as to why these measurements were used. This is not helpful to the discerning. The shape of the curved lines has to be estimated.

Rectangle	A to B = 9.75 cm
	B to C = 11.25 cm
Construction	A to E = 5.75 cm
	B to F = 1.25 cm
	Curve a line from E to F
	D to G = 2.0 cm
	Join E to G
	Curve a line from G to C

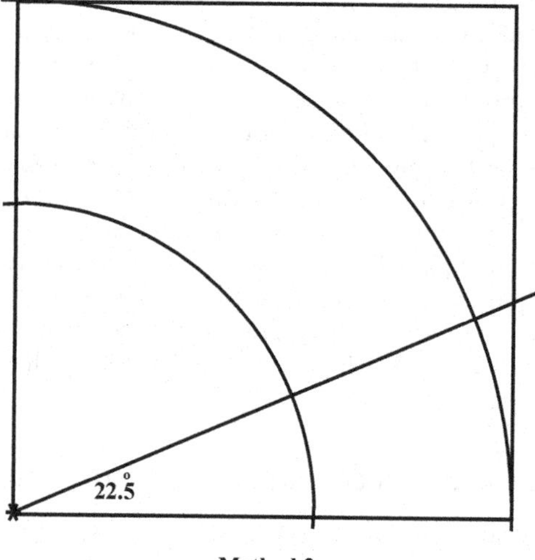

Method 2

Drafting method 2

This quarter cone patterns a $\frac{1}{16}$th segment of a full circle. Therefore the top circumference equals 6 cm × 16 = 96 cm.

$$\text{The radius} = \frac{\text{circumference}}{2 \times \pi} = \frac{96}{2 \times 3.14} = 15.2$$

The height of the cone is 10 cm. Therefore the lower circumference = 15.2 + 10 = 25.2 cm. Construct a square with the sides equal to the largest radius 25.2 cm. From a central point at the corner of a square construct the largest circle with a radius of 25.2 cm and the smallest 15.2 cm. Only $\frac{1}{16}$th of the circle is required for a quarter cone pattern. Therefore another line is rotated 22.5° from the lower side of the square. The quarter cone pattern can then be traced off and mirrored. This method gives better curve lines and is more accurate.

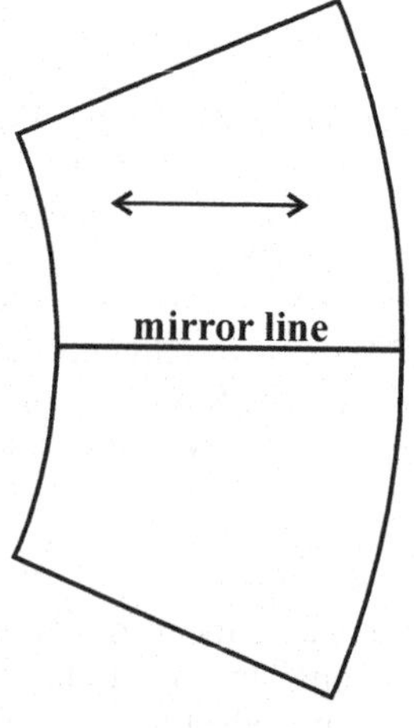

SHAPING A PATTERN BY SUPPRESSION

The technique of **suppression** shapes the pattern by reducing surplus fabric to give a closer fit to the garment. Suppression can be in the form of darts, gathers, pleats or seams. These two examples explain how the quarter cone pattern can be created by reducing the top circumference. Both methods require the construction of a basic rectangle and the calculation of the amount of suppression. The first example is a manual method of closing or pivoting darts. The second method is by computer using a command to reduce the fullness evenly.

Basic rectangle equals:
cone height 10 cm
quarter lower circumference 10 cm

Suppression calculation for a quarter pattern is calculated by dividing the full cone girth measurements by four:

top circumference	24 cm ÷ 4 = 6 cm
lower circumference	40 cm ÷ 4 = 10 cm
suppression of top edge	10 cm − 6 cm = 4 cm

Manual method of suppression by closing darts

The rectangle is divided into four equal sections lengthwise. Three central darts are constructed of 1 cm at the top edge tapering to nothing at the lower edge. Two smaller darts of 0.5 cm are constructed at either side. The top quarter circumference is then suppressed by closing the darts to reduce the measurement to 6 cm. This shape is then traced on to pattern paper and cut to a fold for a half cone pattern.

Computer method of suppression by reducing fullness

The top edge of the constructed rectangle is reduced by 4 cm using the minus fullness or similar command to give the 6 cm measurement. This method gives a smooth curve to both the top and lower edge. The pattern is then mirrored for a half cone pattern.

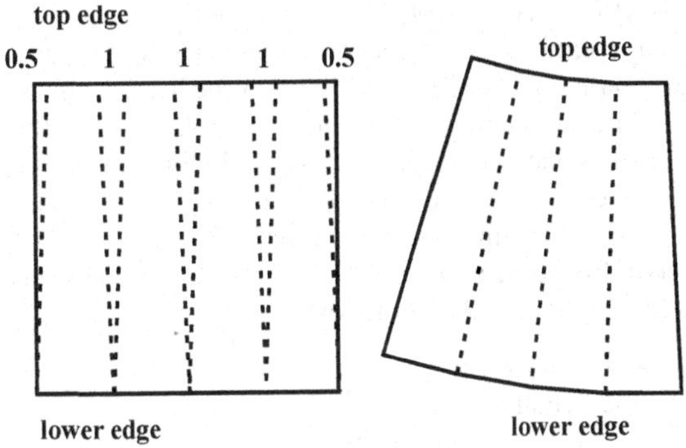

Manual method

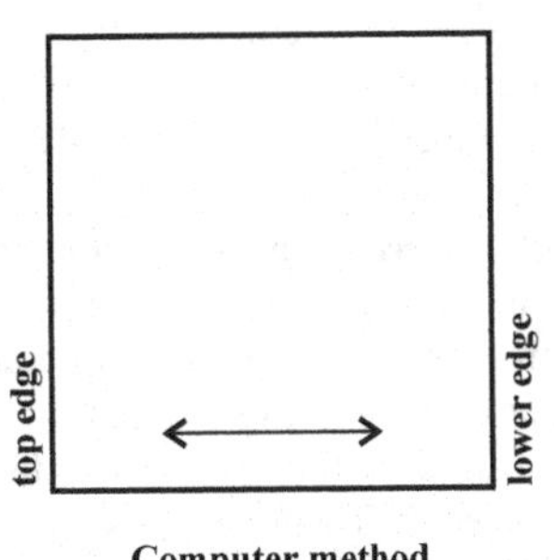

Computer method

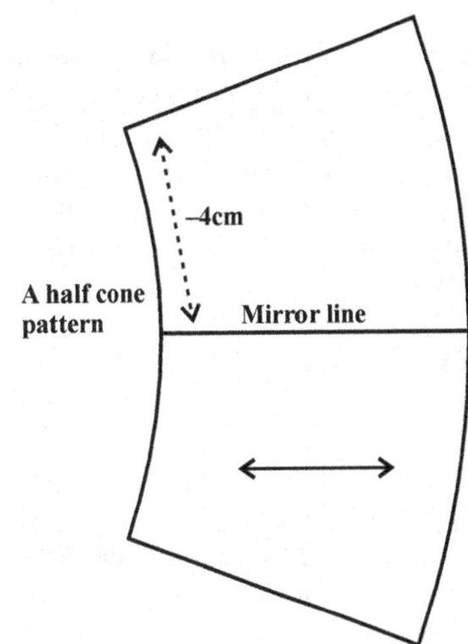

SHAPING A PATTERN BY FLARING

The technique of **flaring** shapes the pattern by gradually increasing the fullness to the lower edge of a pattern. These two examples explain how the quarter cone pattern can be created by increasing the lower circumference. The first example is a manual method of cutting and spreading the lower edge. The second method is by computer using a command to add fullness evenly to the lower edge.

Basic rectangle equals:
cone height 10 cm
quarter top circumference 6 cm

Fullness calculation for a quarter pattern is calculated by dividing the full cone girth measurements by four:

top circumference	24 cm ÷ 4 = 6 cm
lower circumference	40 cm ÷ 4 = 10 cm
increase of lower edge	10 cm − 6 = 4 cm

Manual method of flaring by cutting and spreading
The rectangle is divided into four equal sections lengthwise. Three central lines are cut and spread 1 cm at the lower edge tapering to nothing at the top edge. 0.5 cm is added to the lower sides tapering to nothing at the top edge. The lower edge measurement is now 10 cm. This shape is then traced on to pattern paper and cut to a fold for a half cone pattern.

Computer method of flaring by increasing fullness
The lower edge of the constructed rectangle is increased by 4 cm using the add fullness or similar command to give the 10 cm measurement. This method gives a smooth curve to both the top and lower edge. The pattern is then mirrored for a half cone pattern.

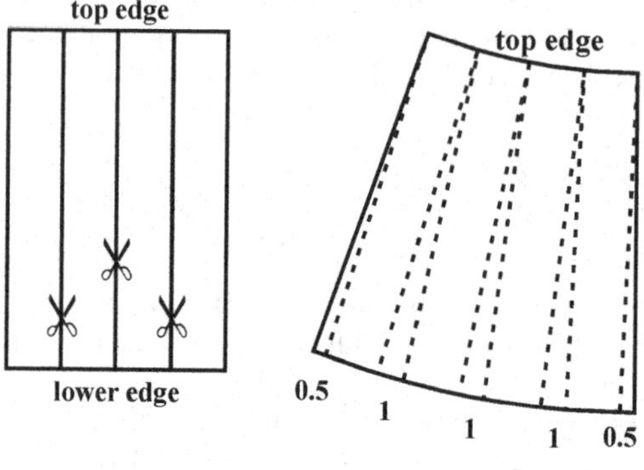

Manual method

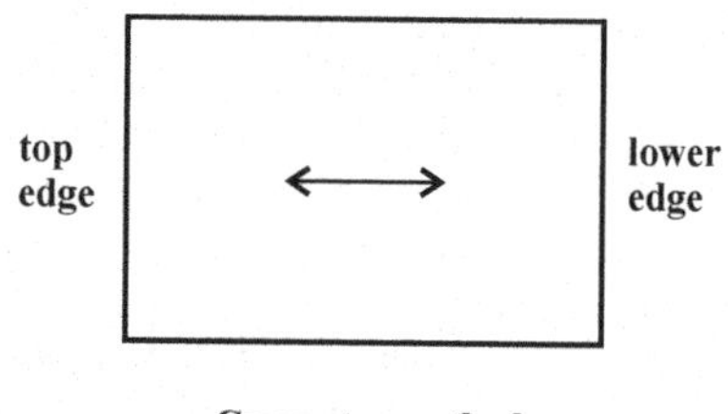

Computer method

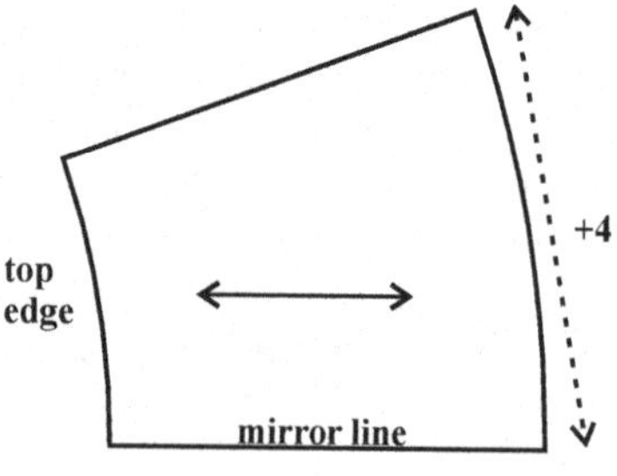

The above illustrates that all the four techniques can produce the same shaped pattern. The method used depends on the skill and experience of the pattern technologist to select the most appropriate method. The suppression or flaring techniques are used mainly when adapting the block patterns into styles. However, experienced pattern technologists may combine several of these techniques within the same pattern.

The final pattern has to be completed by making a tracing from the drafted sections. This new pattern has to have seams added, which is explained in Part 3. For a two-gored cone cover the manual pattern would be cut by hand while folded, whereas the computer pattern is just mirrored. The grain line would be placed on the fold or mirror line. It is advisable to retain all original drafts intact for future reference.

When pattern construction is related to the human form it becomes more complex. The back is a different shape from the front and even the left and right side can vary slightly. However, in wholesale production for symmetrical garments the left side is a mirror image of the right side. It is more accurate and time saving to initially construct patterns for half the garment and mirror the other side to complete the pattern. Generally the total pattern of all the pieces is required for lay planning in preparation for cutting the cloth.

BLOCK PATTERN CONSTRUCTION

Block patterns

A **block** is a foundation pattern that reflects the size, shape and posture of the human figure without the inclusion of style features. It is constructed according to the measurements for the central base size of a size range or an individual person. A copy of the block is adapted to create a garment style. It is advisable to always retain a copy of the block patterns for future use. A newly constructed block has to be tested by making it into a garment. This garment is either fitted on the individual customer, or for wholesale production, on a sample of women who represent the potential customers. If necessary the pattern has to be revised as measurements alone do not reflect the total body shape. The advantage of using block patterns is that they are a permanent record of the correct fit.

The advantage of designing patterns within a computer program is that the blocks can include all the gradings for other sizes. This means that when a new style is created the other sizes are automatically graded. Consequently pattern construction and pattern grading are completed in one operation. In this section the principles and techniques of constructing block patterns are explained. There are several methods that can be applied to computer use. One method is drafting a block pattern using the measurements from a size chart. These drafts can be either constructed directly using a computer system, or manually and then the pattern shapes digitised. For those skilled in modelling or draping on a workroom stand a block can be created by this method and then digitised. Thirdly, a pattern can be taken from an actual garment manually or by the use of a computer.

Finally, if the reader does not have the facilities or skills for creating block patterns there are miniature basic block patterns in Appendix II. These are on a reduced scale (33.33% of the original size) that can be digitised into a computer and then plotted out at 300%.

A detailed description of how to construct the following blocks is given for the average woman size 12.

Primary blocks:

- Straight skirt
- Fitted bodice
- Straight sleeve
- Trouser

Secondary blocks adapted from primary blocks:

- Semi-fitted sleeve
- Fitted sleeve
- Fitted one-piece dress
- Straight one-piece dress
- Semi-fitted one-piece dress
- Fitted dartless top for stretch fabric
- Dartless blouse for woven fabric.

Before constructing block patterns there are several aspects of garment fit that should be understood. These are *garment balance*, *suppression*, *ease allowance* and the *influence of the fabric*. The observing and understanding of the posture, shape and movement of the body is very important. The problem for the pattern technologist is covering a flexible 3-dimensional body with unstable 2-dimensional fabric. For garment production the assumption is made that the body is symmetrical, the left side is a mirror image of the right side, although this is not really the case. Only when making for an individual can figure defects be taken into account.

Garment balance

The hang of the garment around the body is known as **garment balance**. A well balanced garment hangs in the correct relationship with the wearer's size, contour and posture. The posture of the body is determined by the natural stance. Some wearers stand erect, others tend to stoop. This is influenced by both size and age. Those with a fuller figure in the front tend to stand more upright to retain their balance.

Balance marks are positioned at strategic points on the seams to maintain the correct relationship between the back, side and front when the cut garment parts are joined. These balance marks can be in the form of notches on a seam or the matching at the end of a seam. Not all notches are balance marks (e.g. notches for the stitching line of darts). Too many notches can cause confusion when cutting and assembling garments so they should be kept to the minimum. The position of the balance marks will be quoted at the end of each pattern draft. Illustrated below is an example of balance mark positions on a one-piece dress and the related patterns.

A = centre front (CF)
B = centre back (CB)
C = front neck point (fr. NP)
D = back neck point (bk NP)
E = front armhole
F = back armhole
G = underarm point (UP)
H = waist
I = hip

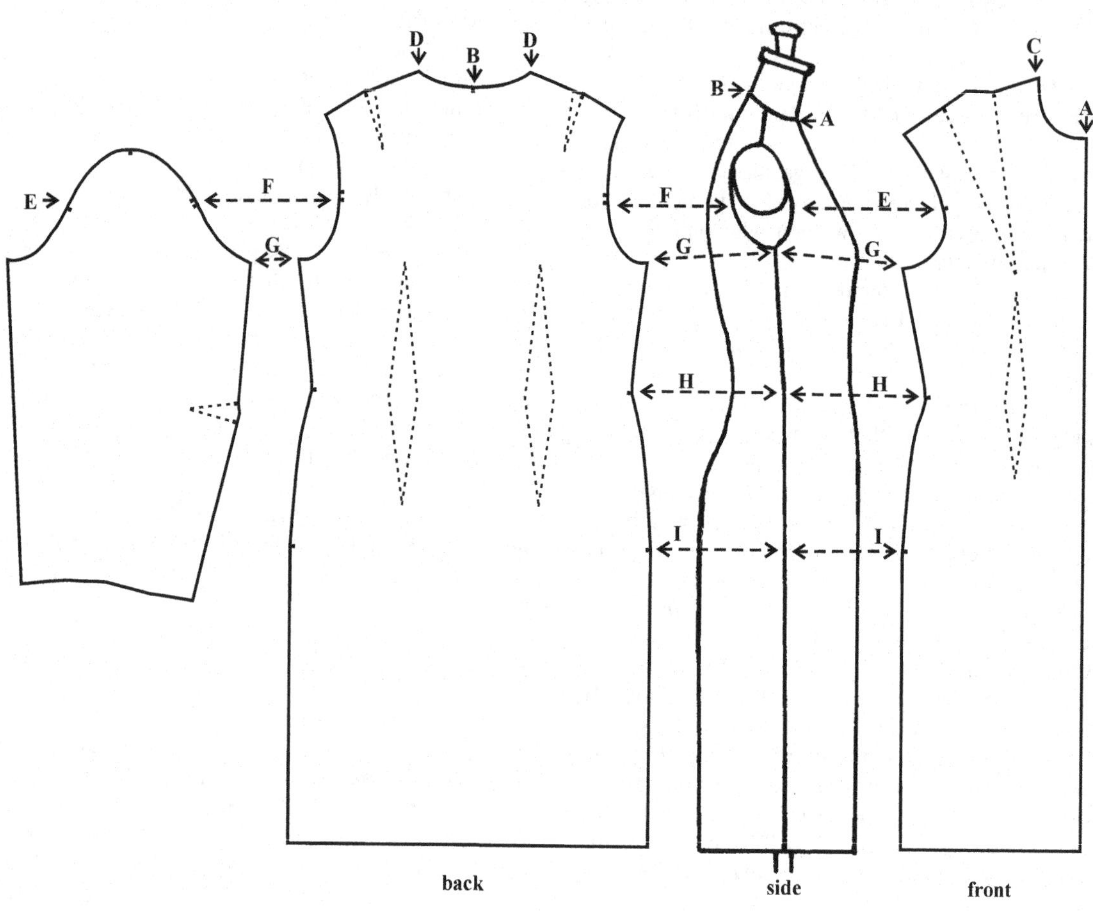

Garment shaping by suppression

Suppression is the reduction of surplus fabric to obtain a closer fit using darts, seams, pleats or gathers. This controls the shape of a garment according to the contour of the body. Suppression is used to reduce a girth measurement that is adjacent to a larger girth measurement; for example, the waist compared with the bust and hips. Suppression always tapers to the prominences of the body.

Suppression for block patterns is not concerned with styling although it can be incorporated later into style features. Illustrated is the position of darts for a fitted bodice and sleeve and the related patterns. The amount suppressed in each dart is not easily calculated. The quantities have been arrived at by modelling on a workroom stand and have been tested by fitting a sample of women.

For block patterns the aim is to have the fabric distributed smoothly around the body. Therefore the dart suppression has to be positioned correctly. The darts A and B radiate from the same point at the bust prominence (BP), but are shortened for a smooth contour. These darts can be pivoted to other positions on the pattern perimeter. For garments with a waist seam the front shoulder dart (A) and waist dart (B) can be combined (see Part 3, Bodice Styling). The shoulder blade prominence is elongated, therefore the shoulder dart (C) and waist dart (D) cannot be combined. The side seam suppression (E) extends from the underarm, increasing at the waist and then tapers to the prominence of the side hip. The prominence of the stomach is higher than the seat, therefore the front dart (B) is shorter than the back dart (D). The elbow dart (F) and wrist dart (G) radiate from the elbow and are shortened for a smooth contour. The wrist dart (G) is positioned on the back arm line.

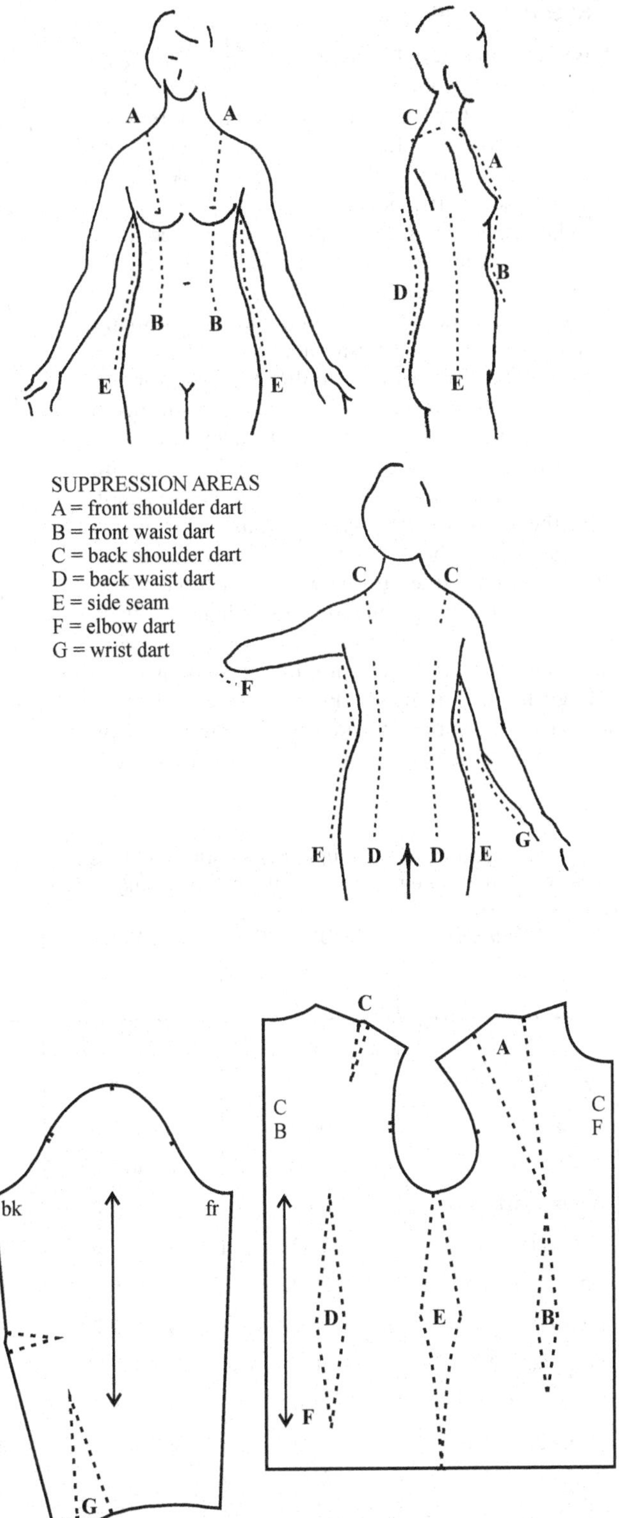

Ease allowance

Ease allowance was briefly mentioned in the section on size chart formulation because ease allowance has to be included in the final garment measurements. The ease allowances that are included in the garment size charts, Charts 1.1 to 1.11, are suitable for primary block construction.

When constructing block patterns the direction of the body movement has to be considered because this influences the amount of ease allowance which has to be added. The activity or occupation of the wearer has also to be taken into account. The greatest movement occurs with the limbs. To allow for this, extra width has to be added to specific positions on the patterns. For example, if both arms are raised forward at shoulder level the across back measurement at the mid armhole has to be increased more than the across front measurement as the arms cannot swing back the same distance. Extra ease allowance has to be added to the waist of a fitted bodice because when both arms are raised high the waist seam rises up towards the ribs to a position wider than the waist. The expansion of the body increases the girth measurements. For example, when the wearer sits down the hip and seat girth measurements increase. For the average size 12 this is between 4 cm and 5 cm. Likewise elbows and knees also increase when bent.

The size and fit of a garment are influenced by the dimensional clearance between the body and garment. It is also important to consider the bulk of the garments worn underneath. An extra measurement has to be added to cover the layer beneath. Thus a coat has to be larger than a dress. This is explained in more detail by Bray (2002b) and Cooklin (1994). Chart 1.12 suggests the minimum amount of ease allowance to be added to body measurements when constructing block patterns for woven fabric.

Influence of the fabric

The dimensional stability of the fabric influences the garment balance, ease allowance and suppression. To maintain the correct garment balance the positioning of the fabric grain is important. With straight cut garments the stronger **warp** threads generally hang perpendicular. Therefore, the **grain line**, which represents the warp threads, is placed running down the length of the garment. The **weft** threads tend to be more elastic and are better positioned around the width of the body for movement.

The type of fabric has a great influence on the amount of ease allowances for body movement,

Chart 1.12 Suggested minimum ease allowance to be added to the body measurements for women's block patterns in woven fabric (measurements in centimetres). Size range 8 to 16 and 16 to 24

MEASUREMENT	BODY SIZE 12	DRESS SIZE 8–16	DRESS SIZE 16–24	SKIRT AND TROUSERS SIZE 8–24	BLOUSE 8–24
Bust girth	88.0	5.0 to 8.0	6.0 to 10.0	—	8.0 to 10.0
Waist girth	70.0	4.0 to 5.0	5.0 to 6.0	2.0 to 4.0	5.0 to 10.0
Hip girth	96.0	4.0 to 5.0	5.0 to 6.0	4.0 to 6.0	5.0 to 6.0
Neck girth	38.0	2.0 to 3.0	3.0 to 4.0	—	2.0 to 4.0
Across back width	35.0	2.0 to 4.0	3.0 to 5.0	—	2.0 to 4.0
Across front width	32.0	2.0 to 3.0	3.0 to 4.0	—	2.0 to 4.0
Upper arm girth	28.0	5.0 to 6.0	6.0 to 8.0	—	4.0 to 8.0
Elbow girth	26.0	4.0 to 6.0	6.0 to 8.0	—	3.0 to 8.0
Wrist semi-fit	16.0	6.0 to 7.0	6.0 to 8.0	—	6.0 to 8.0
Wrist fitted	16.0	3.0 to 4.0	3.0 to 4.0	—	3.0 to 4.0
Depth of armhole	21.0	2.0 to 3.0	3.0 to 4.0	—	2.0 to 5.0
Thigh girth	57.0	—	—	8.0 to 12	—
Knee girth	37.0	—	—	9.0 to 15.0	—
Ankle girth	25.0	—	—	8.0 to 12.0	—

expansion and comfort. The dimensional stability of the fabric is an important factor. The more stable the fabric, the greater the ease allowance. This depends on whether it is woven, knitted or non-woven.

A large proportion of women's wear is made in knitted or stretch fabric. This means that the fabric will stretch and mould around the body. Consequently, the amount of ease allowance can be reduced. The amount of suppression in the form of darts can also be reduced or eliminated. How much reduction depends on the elasticity of the fabric. This is a problem for the pattern technologists as there is so much variation in both the amount and direction of the stretch. Block patterns for knitted fabric can only be an estimate. The construction of the close fitting top secondary block is specifically for knitted fabric. The final pattern for production may have to be modified for a specific fabric. This often depends on the structure of the fabric. Weft knitting is more extensible across, in the weft direction warp knitting is firmer. Woven fabric with a blend of Lycra can have an ability to stretch in either the warp or weft direction or both.

Drafting block patterns

Drafting a pattern is more mathematical as it relies on the correct calculation of measurements from either a size chart or the measurements of an individual. The measurements are generally calculated for half or a quarter of the garment. The longest lengths and widest girth measurements form a rectangle or the section of a circle. Other relevant measurements are positioned geometrically within this rectangle or circle and are known as construction lines. This is the framework from which the final pattern can be shaped.

The method of pattern drafting described in this book is based on a manual system that has been adapted for computer use. Any reliable drafting system familiar to the reader can be likewise adapted. The main variation is the positioning of the draft with the centre back and front lines constructed horizontally. This is because the warp grain line on most computer systems runs horizontally for pattern grading and lay planning. The other difference is that the drafts for the half fronts are for the left side, if viewed as face side up, whereas the back is for the right side. The reason for this is to minimise on the number of grade rules that are applied to these patterns. This will be explained in detail in Part 2.

Testing block patterns

All new block patterns have to be proven by a pattern copied from the blocks. This test pattern has to be completed, with the darts folded to give an underlay, and seams and notches added. (This is explained in Part 3 under 'Production patterns'). A sample garment can be cut on the correct grain, stitched together and fitted on a representative group of women. Once the pattern has been approved the grading rules can be added. (This is explained in Part 2.) Where the block patterns are used for designing styles within the computer program it is advisable to have them without seams or dart underlay. The final fit of the garment is influenced by contemporary fashion and reflects the current silhouette; this necessitates revision periodically to the block patterns.

CONSTRUCTION OF PRIMARY BLOCK PATTERNS

Straight skirt block size 12

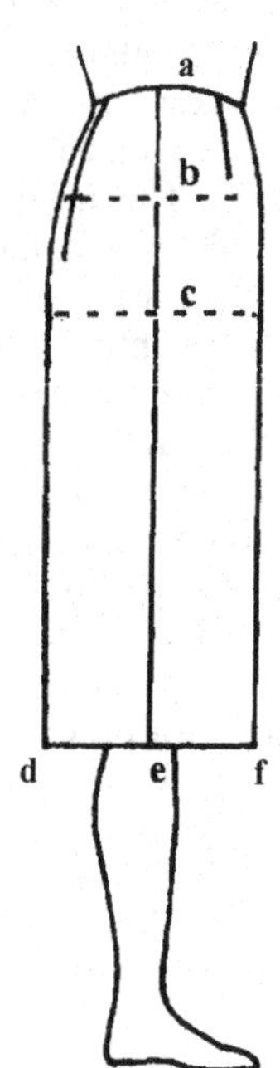

MEASUREMENTS REQUIRED

Measurements	*Body*	*Ease*	*Half pattern*
(a) Waist	70.0	+4.0	37.0
(separate skirt waist band)	70.0	+2.0	36.0
(b) Upper hip (10.0 below waist level)	90.0	+4.0	47.0
(c) Hip (20.0 below waist level)	96.0	+4.0	50.0
(d) Centre back (from waist to knee)	60.0	—	60.0
(e) Side seam (from waist to knee)	61.0	—	61.0
(f) Centre front (from waist to knee)	60.0	—	60.0

The measurements for this draft are for size 12 quoted in Chart 1.3. Other size charts or measurements for an individual can be used.

CONSTRUCTION LINES
Construct a **basic rectangle, horizontal length 61 cm** (side seam), **vertical length 25 cm** (quarter hip girth of skirt pattern).

For the following construction lines the basic rectangle lines can be copied and off-set.

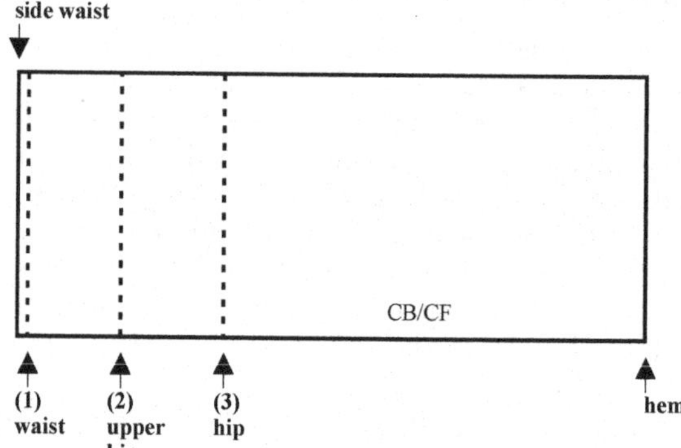

(1) Construct the **waist level** vertically **1 cm** from the side waist level.

(2) Construct the **upper hip level** vertically **10 cm** from the waist level.

(3) Construct the **hip level** vertically **20 cm** from the waist level.

WAIST LINE SUPPRESSION
Calculate the waist suppression by subtracting half the waist plus ease from the half the hip plus ease **(50 – 37 = 13 cm)**. The size of the darts can vary according to the figure shape. The suppression suggested here for half the skirt is for an average proportioned size 12.

Back dart = 4.5 cm Front dart = 2.5 cm
Side seam = 6.0 cm

This is for one dart in the front and back of the half skirt pattern. If the darts appear too large for some figure types or fabrics they can be divided into two darts. The side seam suppression is divided equally at 3.0 cm on both the back and front skirt. Again this can be varied according to the figure shape.

SIDE SEAM SHAPING

(4) From the side waist level mark a point for the **side waist at 3 cm** towards the CB or CF. Mark another point from the side seam at the **upper hip level at 1 cm**. Construct a line connecting these two points to nothing at the hip level. To guide the final side seam shape find a point midway between the side seam level and upper hip level and square out 0.5 cm. Curve the side seam from the side waist level to the upper hip then straight to the hip level and straight from the hip to the hem.

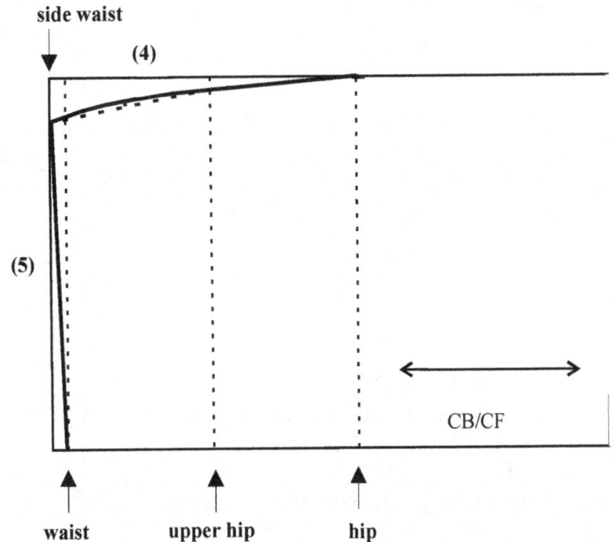

WAISTLINE

(5) Draw a straight line from the top of the side waist level seam to the centre waist level. This will give the same slant to the waistline for both back and front skirts. If required this can be varied after fitting trials for wholesale production, or to an individual's measurements.

BASIC STRAIGHT SKIRT SHAPE

(6) Trace from this draft two copies of the basic straight skirt shape. One is for the back skirt, the other for the front. Include the upper hip and hip level lines.

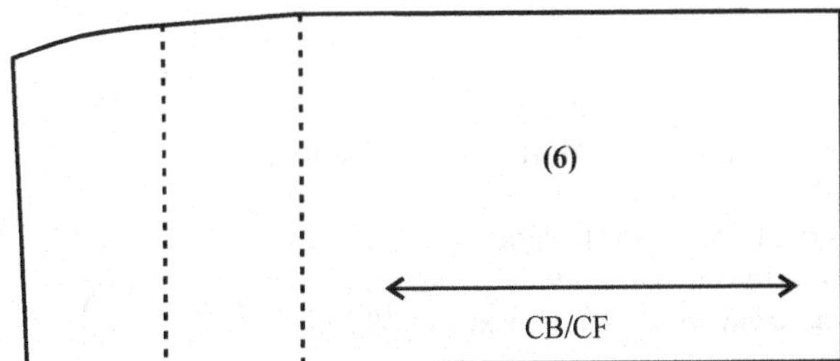

BACK WAIST DART

(7) Insert a dart at a mid point on the back waist line **(width 4.5 cm, length 14 cm)**. The centre of this dart should be a right angle to the waistline.

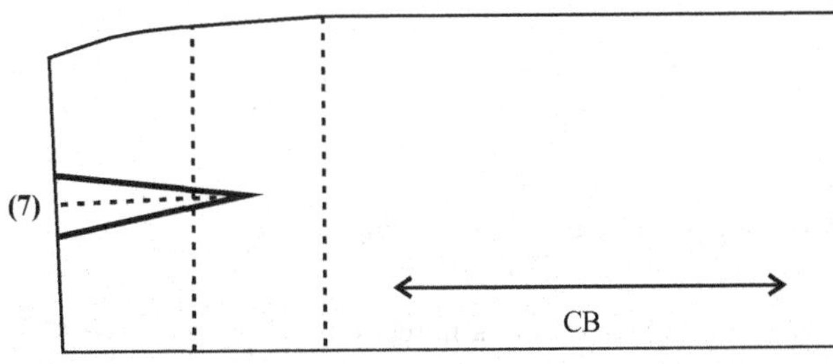

(8) Transfer this dart temporarily to the side seam so that the waistline can be curved smoothly by adding points to the straight line and moving them if necessary. **The back waist should measure 17.5 cm.**

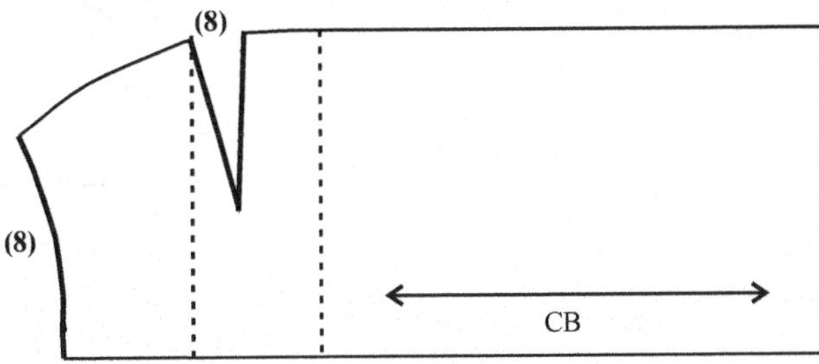

(9) Return the dart to its original position.

(10) Position balance mark notches on the side seams at the hip level.

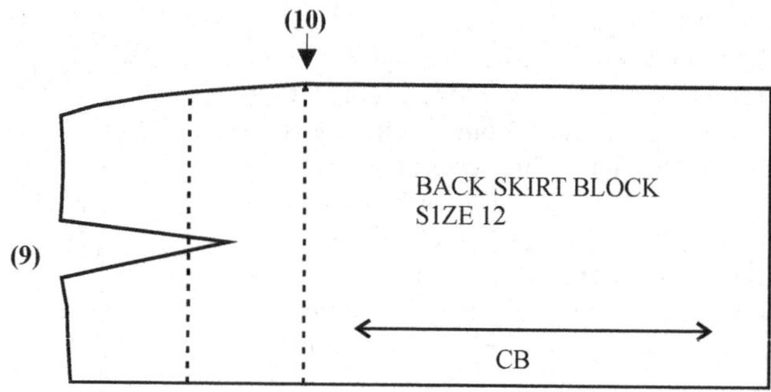

FRONT WAIST DART

(11) Insert a dart **(width 2.5 cm, length 8 cm)** on the waistline **at 14.5 cm** from the CF (two-thirds of the front waist). The centre of this dart should be at right angles to the waistline.

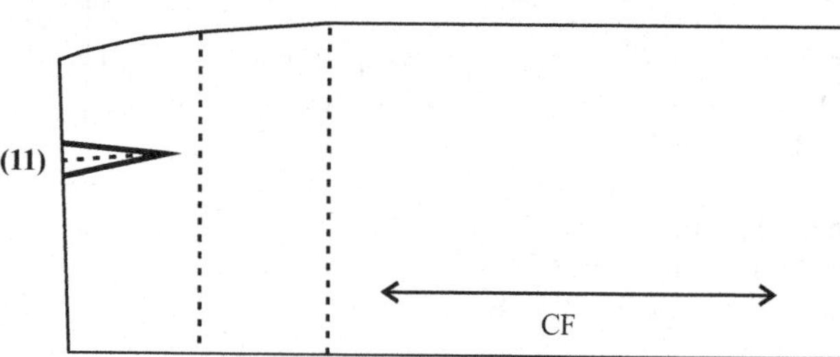

(12) Transfer this dart temporarily to the side seam so that the waist line can be curved smoothly by adding points to the straight line and moving them if necessary. **The front waist should measure 19.5 cm.**

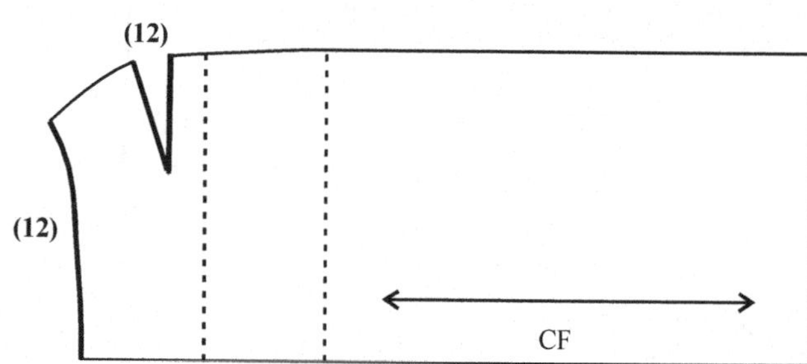

(13) Return the dart to its original position.

(14) Position balance mark notches on the side seams at the hip level to match the back side seam.

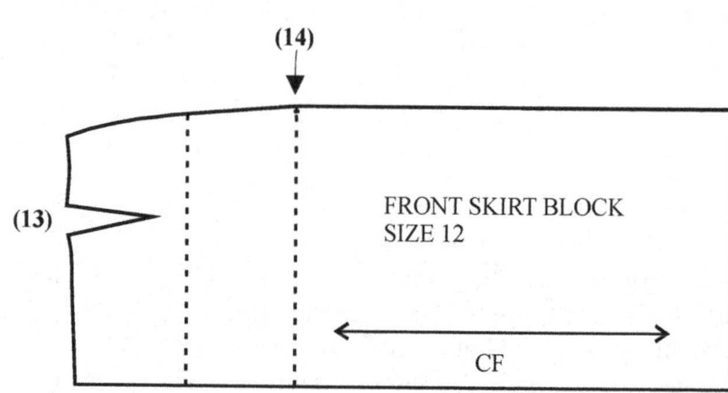

Fitted bodice block size 12

MEASUREMENTS REQUIRED

Girth measurements	*Body*	*Ease*	*Half bodice*
(a) Bust	88.0	+6.0	47.0
(b) Waist	70.0	+4.0	37.0
(c) Neck	38.0	+2.0	20.0
Vertical measurements			
(d) Back neck rise	2.0	—	2.0
(e) Nape to waist	41.0	—	41.0
(f) Armhole depth	21.0	+3.0	24.0
(g) Fr neck point to bust point	27.0	—	27.0
(h) Fr neck point to waist	44.0	—	44.0
Width measurements			
(i) Across back	35.0	+2.0	18.5
(j) Across front	32.0	+1.0	16.5
(k) Shoulder	13.0	—	13.0
(l) Width of bust prominence	19.0	—	19.0
(m) Width of armhole	10.0	+1.5	11.5

The measurements for this draft are for size 12 quoted in Chart 1.2. Other size charts or measurements of an individual can also be used.

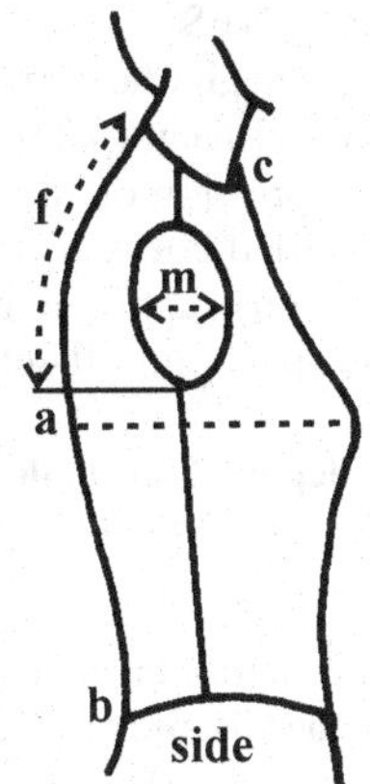

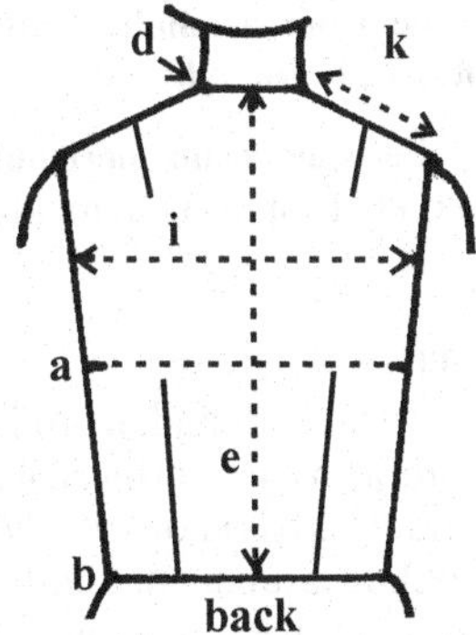

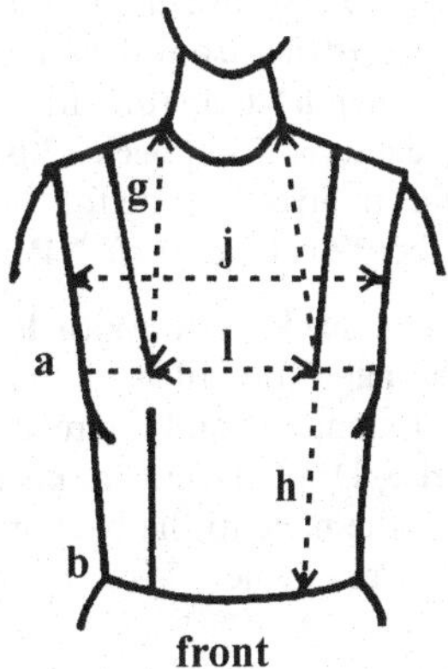

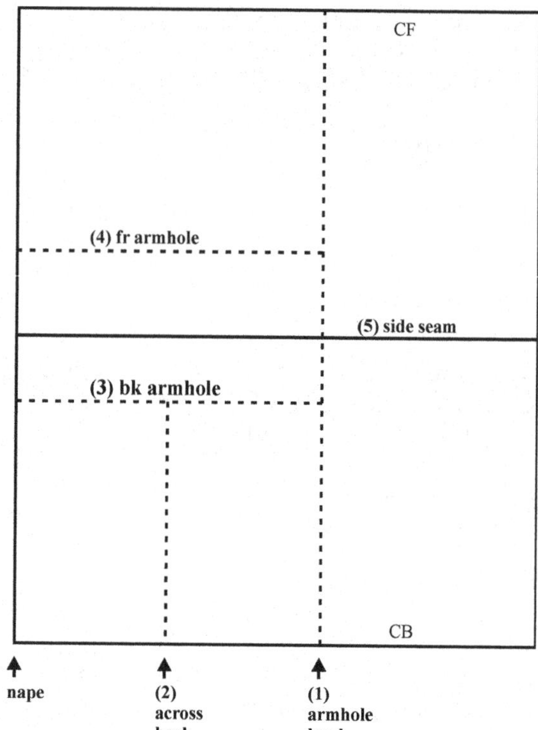

CONSTRUCTION LINES

Construct a **basic rectangle, horizontal length 41 cm** (nape to waist), **vertical length 48.5 cm** (half bodice bust + 1.5 cm for suppression of back waist dart and side seam). The lower left corner is the **nape**.

For the following construction lines the basic rectangle lines can be copied and off-set.

(1) Construct the **depth of armhole level vertically 24 cm from the nape**. This extends from the CB to the CF.

(2) Position the **half across back** measurement vertically **18.5 cm from the CB** midway between the nape and depth of armhole (12 cm).

(3) Construct the **back armhole position horizontally at 18.5 cm from the CB.**

(4) Construct the **front armhole width horizontally 11.5 cm from the back armhole.**

(5) Construct the **side seam horizontally 23.5 cm** (quarter of the body bust measurement + 1 cm) **from the CB.**

NECK CONSTRUCTION

To measure the neck width and depth is difficult. To overcome this problem a **proportional drafting system** is used. This formula is based on calculating one-fifth of the bodice neck measurement and then adding or subtracting a constant measurement.

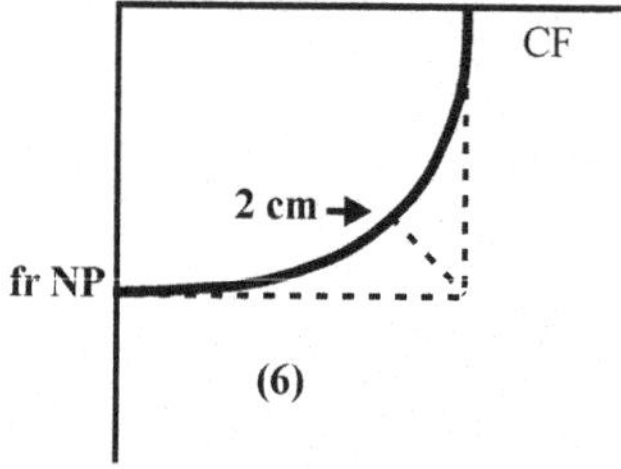

(6) Construct **half the front neck width horizontally 6.5 cm from the CF** (one-fifth of the bodice neck minus 1.5 cm). Construct from the CF a vertical line for the **front neck depth 8.0 cm** (one-fifth of the bodice neck). Draw the curved **front neck 12.0 cm** (the right angle bisect of 2 cm gives a guide for the curve) from the CF to the front neck point (**fr NP**).

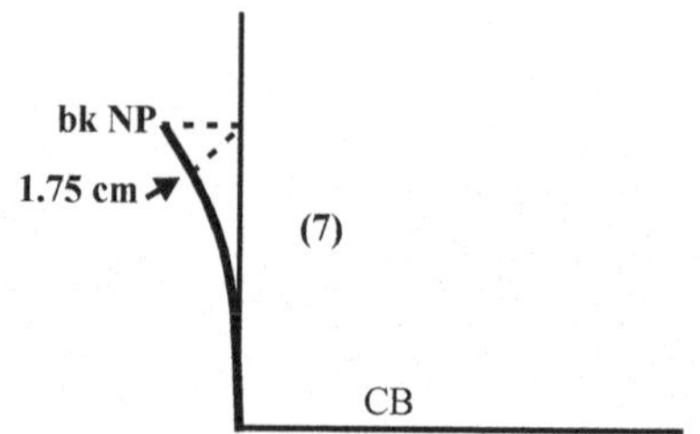

(7) Construct the **back neck width horizontally at 7.5 cm from the nape** (one-fifth of the bodice neck minus 0.5 cm). Extend beyond the rectangle **2 cm for the back neck rise**. Draw the curved **back neck line to finish 8 cm** from the nape to the back neck point (**bk NP**). (Bisect the right angle 1.75 cm as a guide for the curve.)

SHOULDER LEVEL
The shoulder level is for the average slant of approximately 22°. (For variation in shoulder levels see Part 4, Shoulder slant.)

FRONT SHOULDER LINE
(8a) Position a **front shoulder level 6 cm** from the left side of the rectangle.

(8b) Construct the **shoulder line 17.5 cm** from the front neck point **(fr NP)** to intersect the shoulder level line.

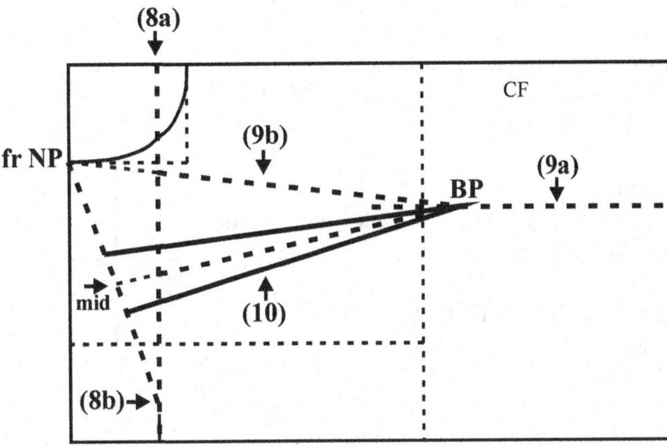

FRONT SHOULDER DART
(9a) Construct a line parallel to the CF between the bust and waist levels, the **bust prominence width 9.5 cm**.

(9b) Draw a **front length to bust point line 27 cm** from the fr NP to intersect bust prominence width line for the bust point **(BP)**.

(10) Connect a line from a mid point on the shoulder line to BP. Insert a **shoulder dart 4.5 cm wide with the apex finishing at BP**.

BACK SHOULDER LINE
(11a) Position a **back shoulder level line 4 cm** from the left side of the rectangle.

(11b) Construct a line from the back neck point **(bk NP) 14.5 cm long** (the shoulder length plus 1.5 cm dart) to intersect the shoulder level line.

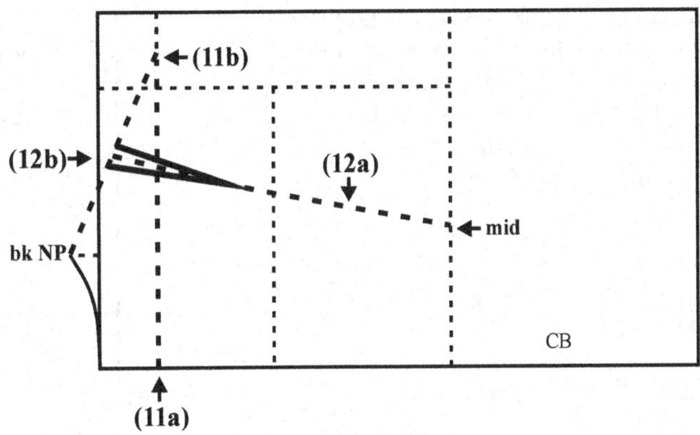

BACK SHOULDER DART
(12a) For the back shoulder dart position mark a point midway on the depth of the armhole level between the CB and back armhole **9.25 cm**. Connect this point for a central line to the middle of the shoulder line.

(12b) Insert a **shoulder dart 1.5 cm wide and 8 cm long** along the central line.

ARMHOLE CONSTRUCTION

(13a) Curve the back armhole line from the shoulder end, touching the across back line to the side seam. (A 3 cm bisect of the right angle gives a guide for the curve.)

(13b) Curve the front armhole from the shoulder end, touching a point at two-thirds down the front armhole depth to the side seam. (A 2.5 cm bisect of the right angle gives a guide for the curve.)

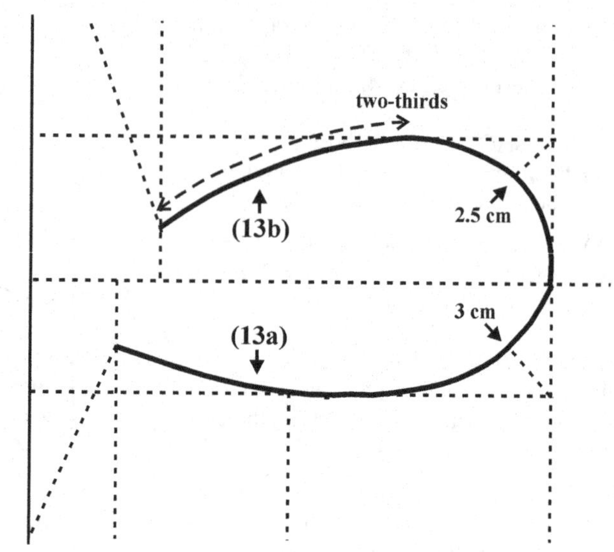

Check that the armhole measures approximately **45 cm (back and front equally 22.5 cm)**.

WAISTLINE SUPPRESSION

Calculate the waist suppression by subtracting half the bodice waist from the rectangle width **(48.5 – 37.0 = 11.5 cm)**. The size of the darts can vary according to the figure shape. The darts suggested are for average proportions, back and front waist 4 cm, side seam 3.5 cm.

FRONT WAIST DART

(14a) Extend from the BP the line parallel to the CF beyond the right side of the rectangle. The **front bodice length measures 44 cm from the fr NP over the bust prominence BP to the front waist level**. At this level construct a line parallel to the right side of the rectangle that intersects an extension of the CF line.

(14b) At the intersection of the front length line and waist level insert **a waist dart 4 cm wide and apex finishing 3 cm from BP**.

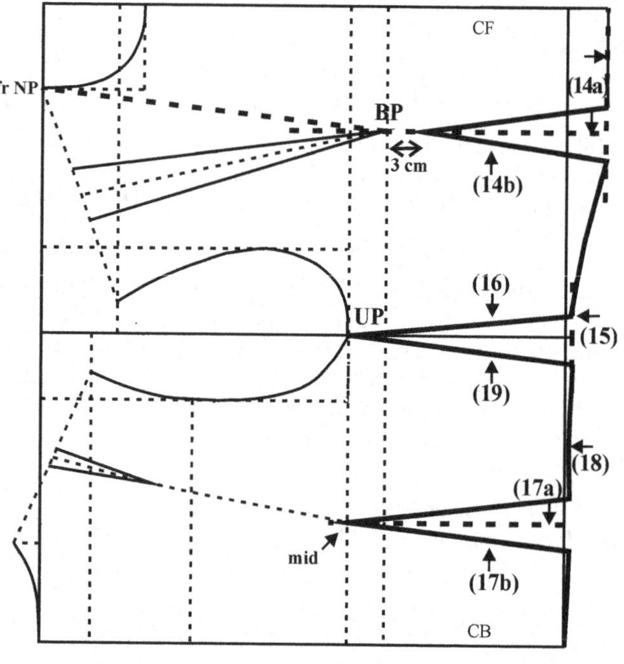

FRONT WAIST LINE AND SIDE SEAM

(15) For the side waist level construct a line **0.5 cm** parallel to the right of the rectangle. Complete the waistline by connecting a line of 12.0 cm from the side of the dart to intersect the side waist level **(front waist measure 19.5 cm from the CF to side seam excluding the dart)**.

(16) For the **front side seam** connect a line from the **side waist to underarm point UP**.

BACK WAIST DART

(17a) Construct a parallel line to the CB from the previously marked mid point (9.25 cm) on the depth of armhole level to the waist level.

(17b) At the intersection of this line and the waist level insert a **waist dart 4 cm wide and apex finishing at the depth of armhole level.**

BACK WAISTLINE AND SIDE SEAM

(18) Complete the back waistline by connecting a line of 10.25 cm from the side of the dart to intersect the side seam level **(back waist measures 17.5 cm from the CB to the side seam excluding the dart).**

(19) For the side back seam connect a line from the side waist to the underarm point UP.

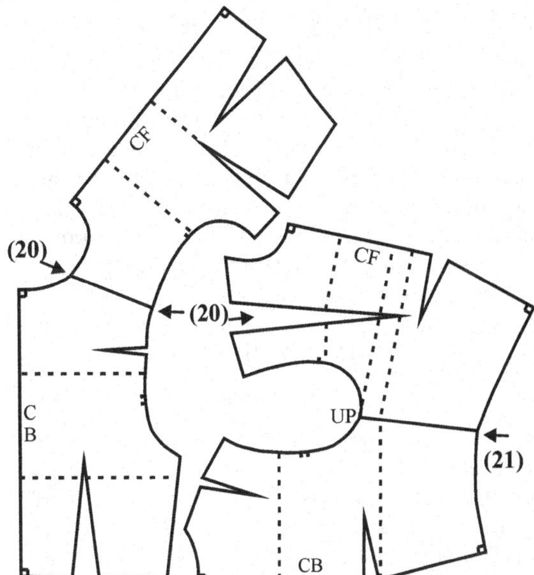

PATTERN COMPLETION

The final pattern has to be traced off the draft and the shoulder and waist seams completed. All the darts must have equal stitching lines and the adjoining seams must be of equal length. The curves have to run smoothly at the seam junctions. To achieve this the block patterns have to be pivoted to check the matching of the seams and smooth run of the curves. The adjoining seams to the CF and CB should be at right angles for the first centimetre.

(20) The shoulder seams of this bodice block are straight. To achieve this the back and front shoulder darts have to be temporarily transferred. The **shoulder seam** can then be straightened and the **shoulder measurement of 13 cm** checked. The darts can then be returned, opening at the mid shoulder.

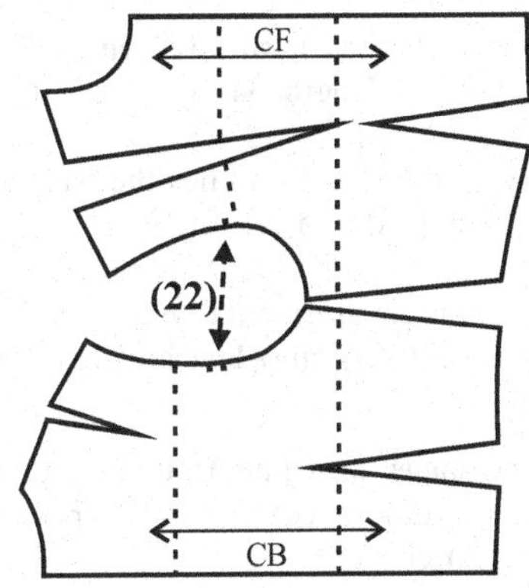

(21) The waist seams have to curve smoothly. To achieve this the waist darts are transferred temporarily. The **curve of the front waist measures 19.5 cm and the back waist 17.5 cm.** The darts can then be returned, opening at the original position.

(22) Balance marks are positioned for inserting the sleeve at the correct pitch. Position the **sleeve balance marks** a quarter of the armhole measurement from the side seam **(11 cm).** One notch identifies the front and two identify the back.

Straight sleeve block size 12

This sleeve hangs straight from the upper arm to the wrist and has no shaping for the contour of the arm. The height and shape of the sleeve head are calculated from the circumference of the bodice armhole. Therefore this sleeve can only be constructed after the bodice block has been completed. The balance marks positioned on the sleeve head have to match those on the armhole to retain the correct pitch of the sleeve in harmony with the arm.

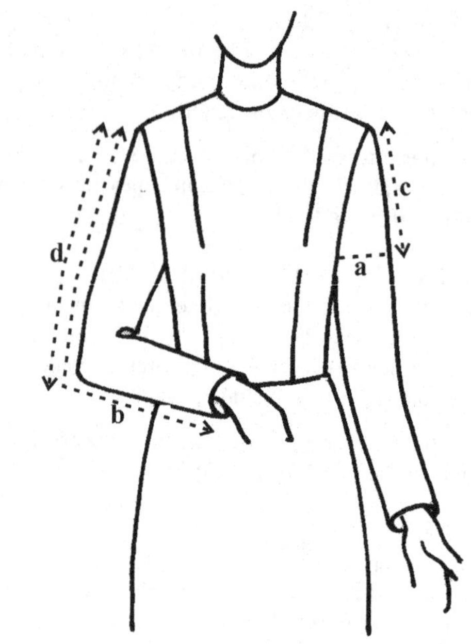

MEASUREMENTS REQUIRED

	Body	*Ease*	*Sleeve*
(a) Upper arm girth	28.0	+6.0	34.0
(b) Full length	—	—	59.0
(c) Depth of sleeve head (approx. $\frac{1}{3}$ armhole)	—	—	15.0
(d) Top of sleeve head to back arm at elbow	—	—	35.0

The measurements for this draft are for size 12 quoted in Chart 1.3. Other size charts or measurements of an individual can be used.

SLEEVE DRAFT CONSTRUCTION LINES
Construct a **basic rectangle, horizontal length 59 cm** (sleeve length), **vertical length 34 cm** (sleeve upper arm.)

For the following construction lines the basic rectangle lines can be copied and off-set.

(1) Divide the rectangle horizontally into four to give **forearm line**, **centre (grain) line** and **back arm line**.

(2) Construct the **upper arm line 15.0 cm** ($\frac{1}{3}$ of the bodice armhole circumference) vertically from the left side of the rectangle.

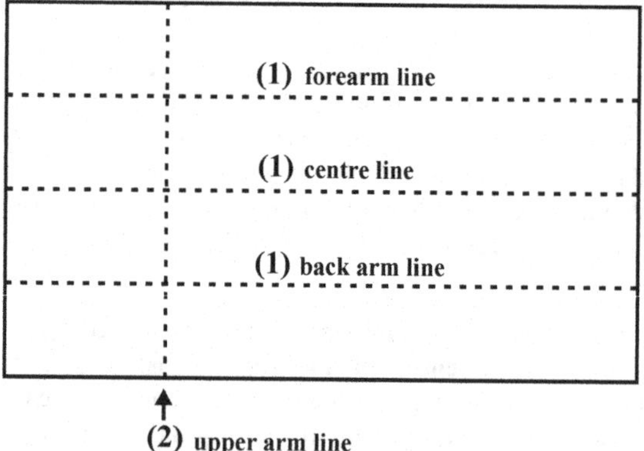

SLEEVE HEAD SHAPING

(3) The sleeve head height on the **back arm line** is **9.0 cm** from the upper arm line ($\frac{1}{6}$ of bodice armhole circumference + 1.5 cm). This can be obtained by reducing the back arm line by 6.0 cm from the left side of the rectangle at **A**.

(4) The sleeve head height on the **forearm line** is **7.5 cm** from the upper arm line ($\frac{1}{6}$ bodice armhole circumference). Reduce the forearm line 7.5 cm from the left side of the rectangle at **B**.

(5) Construct a line from the back underarm seam, at the upper arm line, to point (3) on the back arm line, from point (3) to the end of the centre line, and continue to point (4) on the forearm line and another to the underarm on the upper arm line.

(6) Shape the sleeve head by a curved line from the marked points and the middle of the connecting lines as shown in the diagram (approximately 1.25 cm at the back mid points and 1.75 cm at the front mid points). The curve from the underarm seam to the first mid point can be similar to the back and front bodice armholes.

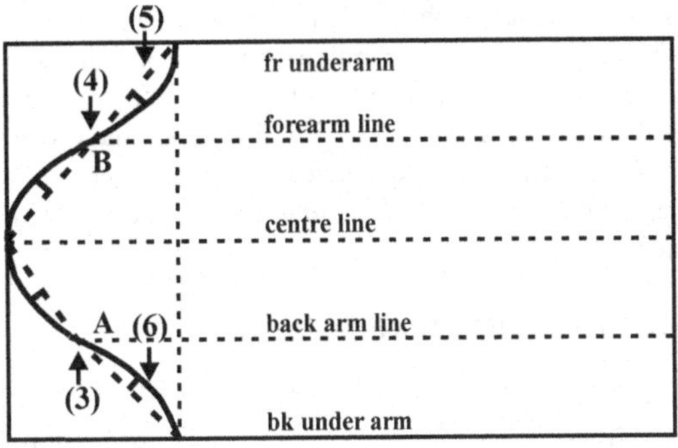

ELBOW LEVEL AND WRIST LINE

(7) Draw a diagonal line **35 cm** long from the sleeve head at the end of the central line to intersect the back arm line. At this intersection draw the **elbow line** parallel to and 19 cm below the upper arm line.

(8) If required the wrist can be shaped by shortening the length on the forearm line by approx. 1 cm and shortening the length on the underarm seam by 0.5 cm. Connect these points to the end of the back arm line.

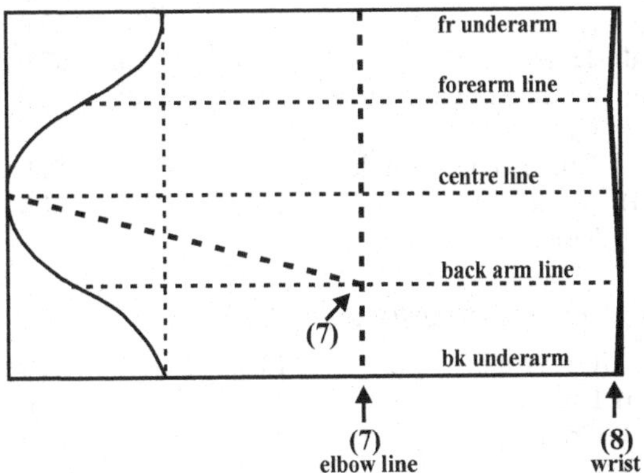

BALANCE MARKS

(9) The balance marks on the front and back of the sleeve head have to match those on the bodice, starting from the underarm seam. Position one notch to identify the front sleeve and two for the back. This sleeve head has little or no ease allowance between the underarm seam and notches. There should be some easing in allowed at the crown of the sleeve to mould around the top of the arm, approximately 2.0 to 3.0 cm. A notch at the top of the sleeve head should be positioned to indicate the shoulder seam and distribute the crown ease evenly.

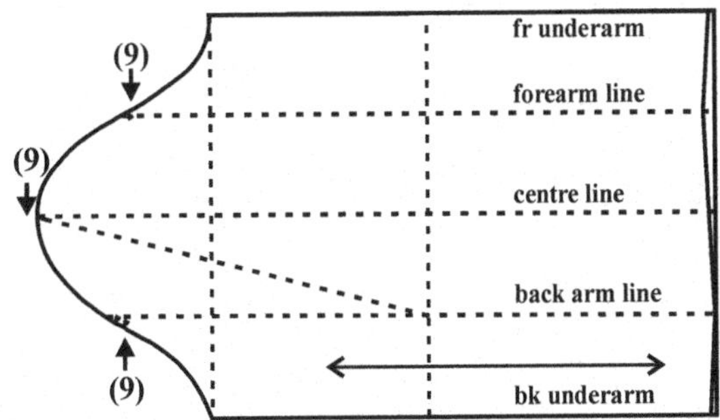

Basic trouser block size 12

The measurement for this draft are for size 12 quoted in Chart 1.3. Other size charts or measurements for an individual can be used. Some trouser measurements are difficult to obtain by measuring the body. Therefore these have been calculated by using a **proportional measurement system of drafting**. This system calculates shorter measurements as a proportion of a major measurement. Sometimes it requires adjustment by adding or subtracting a constant measurement. Proportional systems have been developed through experience and trial and error over many years by the skilled cutters of the tailoring trade.

MEASUREMENTS REQUIRED

Girth measurements	*Body*	*Ease*	*Trouser*
(a) Waist	70.0	+4.0	74.0
Waist band	70.0	+2.0	72.0
(b) Upper hip	90.0	+4.0	94.0
(c) Hip	96.0	+4.0	100.0
(d) Thigh (at crutch level)	57.0	+10.0	67.0
(e) Knee (straight trouser leg)	37.0	+15.0	52.0
Knee (slim trouser leg)	37.0	+9.0	46.0
(f) Ankle (slim trouser leg)	25.0	+9.0	34.0

Length Measurements (from waist)			
(g) Upper hip	10.0	—	10.0
(h) Hip	20.0	—	20.0
(i) Crutch	28.0	+1.0	29.0
(j) Knee	60.0	—	60.0
(k) Outside leg to ankle	100.0	—	100.0
(l) Inside leg to ankle	72.0	−1.0	71.0

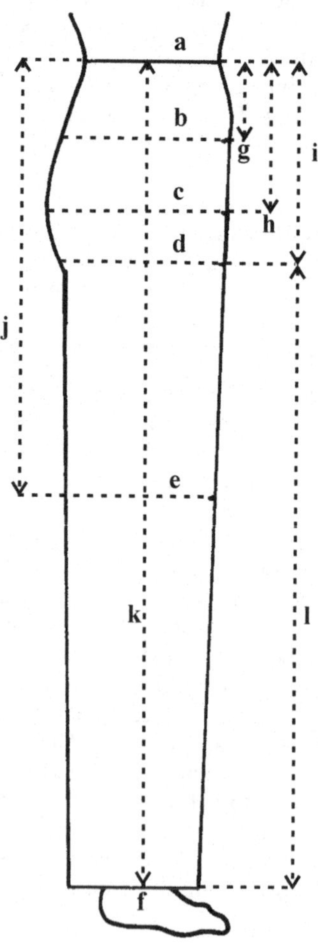

FRONT TROUSER CONSTRUCTION LINES

Construct a **basic rectangle**, **horizontal length 100 cm** (outside leg to ankle), **vertical length 29.5 cm** (quarter trouser hip minus 1 cm plus front crutch fork, calculated as $\frac{1}{20}$ of trouser hip, plus 0.5 cm, e.g. (25 – 1) + (5 + 0.5) = 29.5 cm).

The left side of the rectangle is the waist level and the right side the ankle. The inside leg will be towards the lower side and the outside leg towards the top side. For the following construction lines the basic rectangle lines can be copied and off-set.

(1) Construct the **upper hip level vertically at 10 cm** from the waist level.

(2) Construct the **hip level vertically at 20 cm** from the waist level.

(3) Construct the **crutch level vertically at 29 cm** from the waist level.

(4) Construct the **knee level vertically at 60 cm** from the waist level.

(5) A **centre front construction line** is positioned **horizontally at 24 cm** from the side seam (quarter of trouser hip minus 1.0 cm) to the crutch.

(6) A **trouser crease line** is positioned **horizontally 10 cm** from the CF ($\frac{1}{10}$ trouser hip girth towards the side seam).

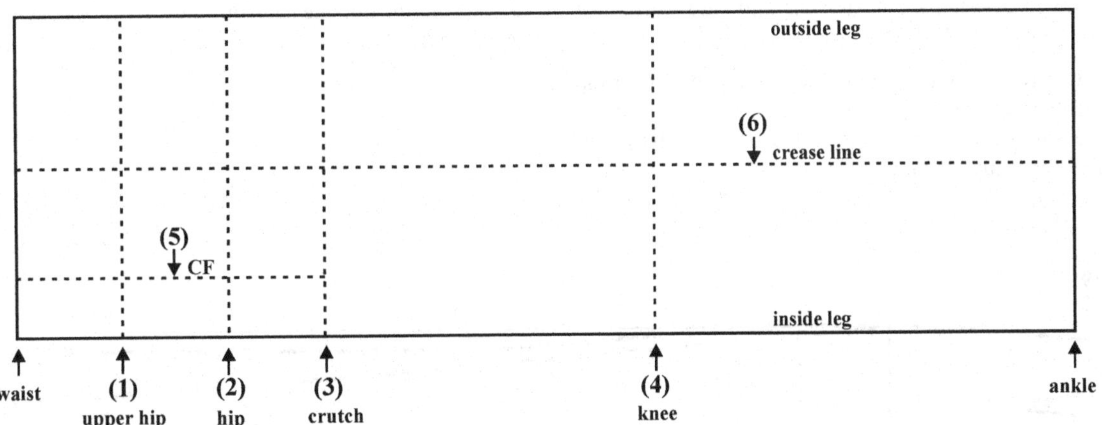

FRONT CENTRE SEAM AND CRUTCH SHAPING

(7) At the waist level construct a **CF seam line 0.5 cm** towards the crease line, then connect it to the hip level at the CF construction line.

(8) Continue this CF seam in a curve to shape the crutch from the hip level to the crutch level at the lower side of the rectangle (right angle bisect of 2.5 cm gives a guide for the curve).

FRONT WAIST SHAPING

(9) From the CF waist level mark the **side waist at 22.5 cm** (a quarter of the trouser waist measurement plus 4.0 cm dart suppression).

(10) Locate two **waist darts of 2 cm wide and 10 cm long**, one at the top of the crease line, the other midway between the crease line and side seam. The centre of each dart is square to the waistline.

(11) From the CF upper hip level mark the **upper hip width at 23.5 cm** (a quarter of the trouser upper hip measurement). Join the side seam between the waist, the upper hip and hip levels.

FRONT LEG SHAPING

(12) Between the knee and hem levels construct the **front leg width 24 cm** by positioning two lines equally at **12 cm** either side of the crease (a quarter of the trouser knee measurement minus 1 cm).

(13) Join the inside leg from the knee to the crutch level by slightly curving in 0.5 cm midway between the knee and crutch. Join the outside leg from the knee to the hip or crutch levels.

(14) For a slimmer leg mark **7.5 cm** at the ankle level (a quarter of the trouser ankle measurement minus 1 cm) either side of the crease line. At the knee level mark **10.5 cm** (a quarter of the slim leg trouser knee measurement minus 1.0 cm) either side of the crease line. For the inside leg join a line from the hem to the knee finishing midway between the knee and crutch levels. Join the outside leg from the knee to the hip or crutch levels.

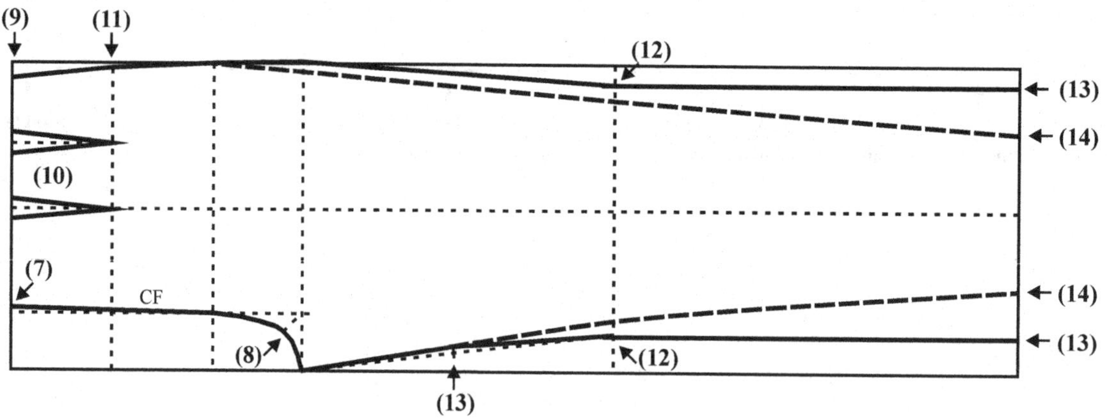

BACK TROUSER CONSTRUCTION LINES

Construct a **basic rectangle**, **horizontal length 100 cm** (outside leg to ankle), **vertical length 37.5 cm** (quarter trouser hip plus 1.0 cm plus back crutch fork calculated as $\frac{1}{10}$ of trouser hip plus 1.5 cm, e.g. (25 + 1) + (10 + 1.5) = 37.5 cm.

The left side of the rectangle is the waist level and the right side the ankle. The inside leg will be towards the lower side and the outside leg towards the top side. For the following construction lines the basic rectangle lines can be copied and off-set.

(15) Construct the **upper hip level vertically at 10 cm** from the waist level.

(16) Construct the **hip level vertically at 20 cm** from the waist level

(17) Construct the **crutch level vertically at 29 cm** from the waist level.

(18) Construct the **knee level vertically at 60 cm** from the waist level.

(19) A **centre back construction line** is positioned **horizontally at 26 cm** from the side seam (a quarter of trouser hip plus 1 cm) to the crutch.

(20) A **trouser crease line** is positioned **horizontally 9 cm** towards the side seam from the CB ($\frac{1}{10}$ trouser hip girth minus 1 cm).

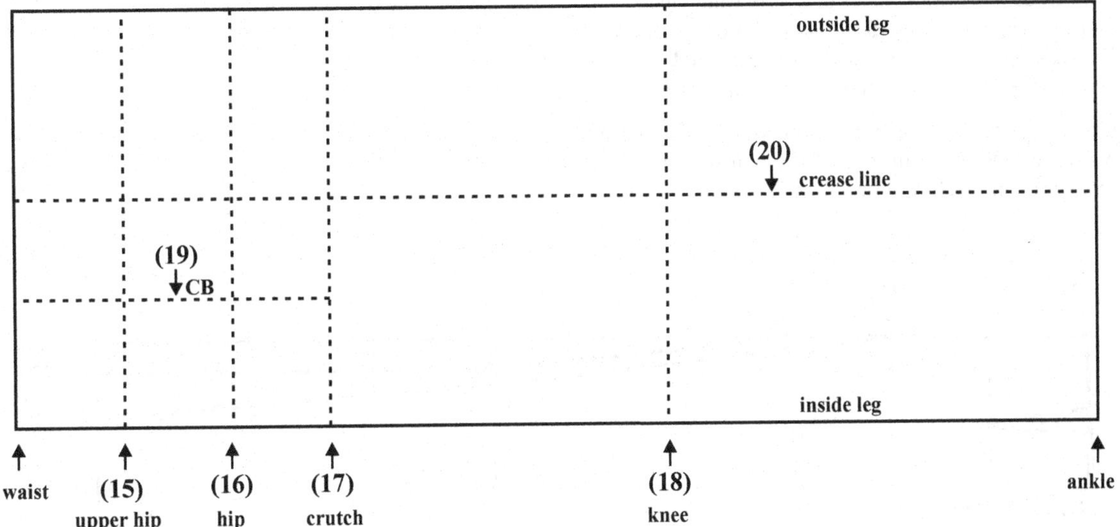

BACK CENTRE SEAM AND CRUTCH SHAPING

(21) From the CB construction line mark a point **2 cm** towards the crease line, then square to the left **1 cm** from the waist level. Connect this to the hip level at the CB construction line.

(22) Continue this CB seam in a curve to shape the crutch from the hip level to a point at the mid back fork. Then **1.0 cm** to the right of the crutch level at the lower side of the rectangle (right angle bisect of 3.0 cm gives a guide for the curve).

BACK WAIST SHAPING

(23) From the CB waistline mark the **side waist at 23.5 cm** (a quarter of the trouser waist measurement plus 5.0 cm dart suppression).

(24) Locate two **waist darts of 2.5 cm wide**, one at the top of the crease line **15 cm long**, the other midway between the crease line and side seam **13 cm long**. The centre of each dart is square to the waistline.

(25) From the CB upper hip level, mark the **upper hip width at 23.5 cm** (a quarter of the trouser upper hip measurement excluding the darts).

LEG SHAPING

(26) Between the knee and hem levels construct the **back leg width 28 cm** by positioning two lines equally at **14 cm** either side of the crease (a quarter of the trouser knee measurement plus 1 cm).

(27) Join the inside leg from the knee to the crutch level by curving in 1.5 cm midway between the knee and crutch. Join the outside leg from the knee to hip or crutch levels.

(28) For a slimmer leg mark **9.5 cm** at the ankle level (a quarter of the trouser ankle measurement plus 1.0 cm) either side of the crease line. At the knee level mark **12.5 cm** (a quarter of the slim leg trouser knee measurement plus 1 cm). For the inside leg join a line from the hem to the knee finishing midway between the knee and crutch levels. Join the outside leg from the knee to the hip or crutch levels.

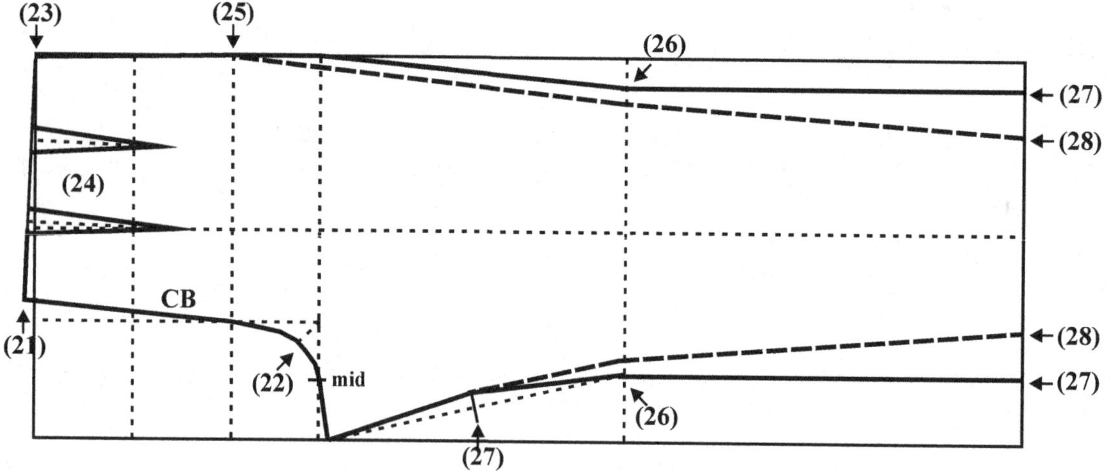

(29) The back and front waistlines should be curved smoothly by pivoting the darts away temporarily and correcting. The run of the side seam can also be checked, and each back and front waist to measure **18.5 cm**. Finally, the darts can be returned to their original position.

(30) Position balance mark notches on the back and front side seam at the hip and knee levels. On the inside leg seam position the notches at the knee level.

(31) The waist seam of this trouser block finishes at the waist level and 'eases in' 2 cm on to a band. This easing is distributed 0.5 cm in each quarter waist to give moulding over the upper hip area. The waist-band is constructed to finish **72 cm** plus an extension for fastening at the CF. The waist band is notched at equal distances of **18 cm** for the CB, side seams and CF. This band extends above the natural waist level.

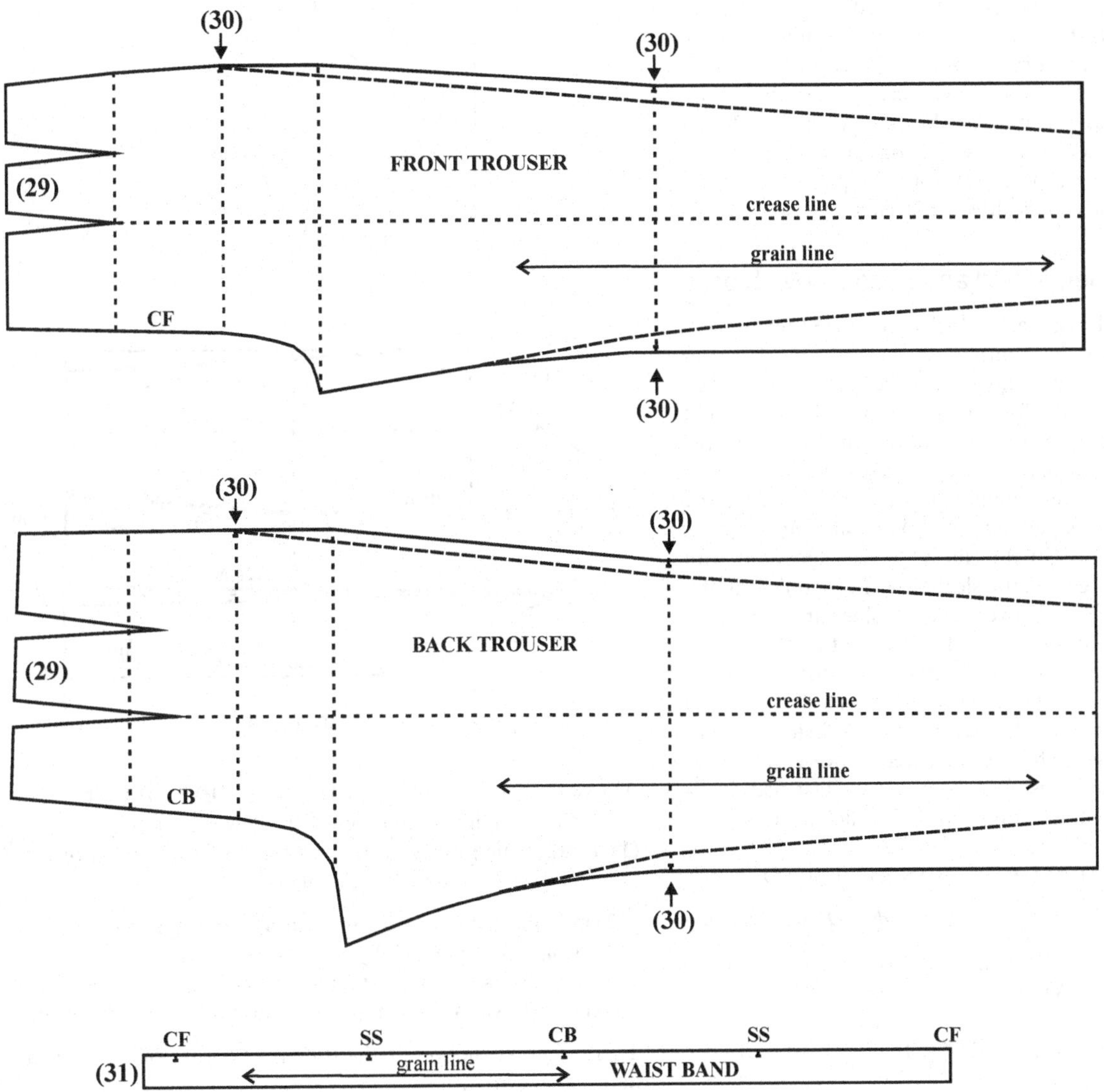

CONSTRUCTION OF SECONDARY BLOCK PATTERNS

Secondary block patterns are needed for garments that vary from the fit and silhouette of the primary block patterns. The examples explained below are for semi-fitted and close fitted sleeves, garments without waist seams such as one-piece dresses, close fitting tops and blouses. The silhouette and fit of these blocks are to some extent determined by the current fashion and, therefore, require revision from time to time. The secondary block patterns are developed by the adaptation of the primary block patterns. Different blocks can also be constructed for various types of fabric, for example woven, non-woven, knitted or stretch.

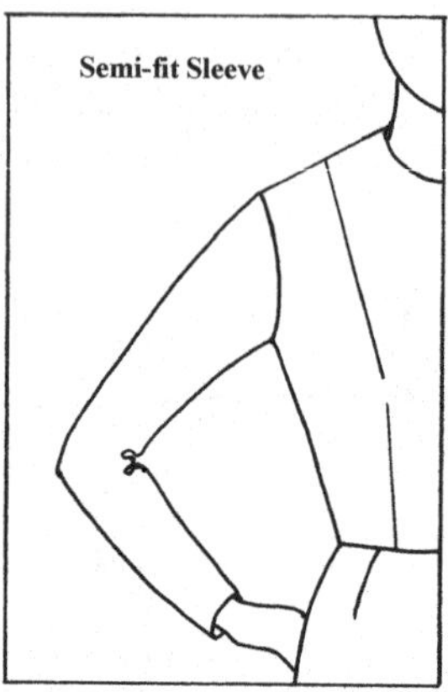

Semi-fitted and fitted sleeve blocks

These sleeve blocks are shaped to conform to the contour of the arm which hangs straight down to the elbow and then slightly forward at the wrist. They must not restrict the arm movement but when the arm hangs in a relaxed position these sleeves should hang smoothly without any folds or puckers. This shaping requires darts at the elbow. When compared with the straight sleeve the fit is only slightly closer between the shoulder and the elbow but more fitted between the elbow and wrist. However, the wrist opening for the semi-fit sleeve is large enough for the hand to pass through without a fastened opening, whereas the fitted sleeve is tighter at the wrist and so requires a fastened opening, generally along the back arm line.

MEASUREMENTS REQUIRED

	Body	*Ease*	*Sleeve*
(a) Elbow girth	26.0	+5.0	31.0
(b) Wrist girth (semi-fit)	16.0	+8.0	24.0
(c) Wrist girth (fitted)	16.0	+3.0	19.0

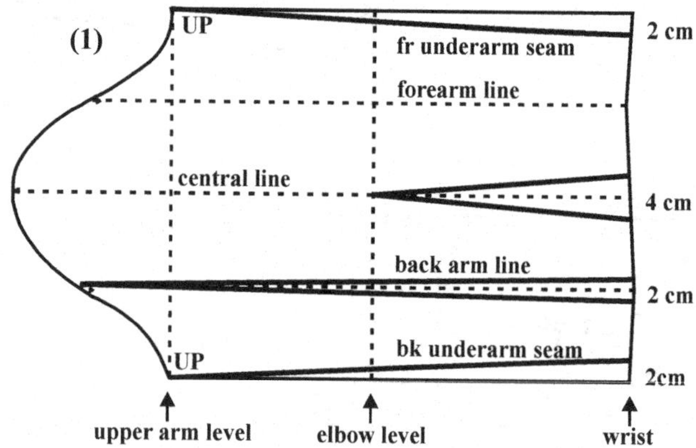

ADAPTATION FOR A SEMI-FIT SLEEVE BLOCK
Copy the straight sleeve block with the wrist shaping.
(1) Reduce the fullness at the wrist to **24 cm** (the positions are indicated on the diagrammatic plan):

2 cm at the front and back underarm seams tapering to nothing at the under point (UP)
2 cm at the back arm line tapering to nothing at the sleeve head
4 cm at the central line tapering to nothing at the elbow level.

There is no reduction at the forearm because of the forward hang of the wrist.

(2) Transfer the 4.0 cm dart at the wrist to the back elbow line.

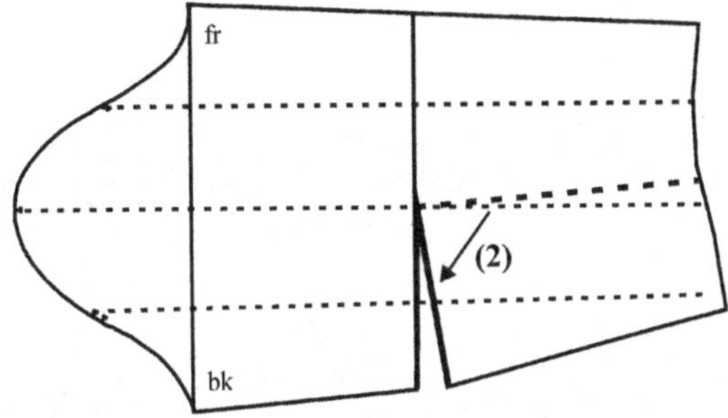

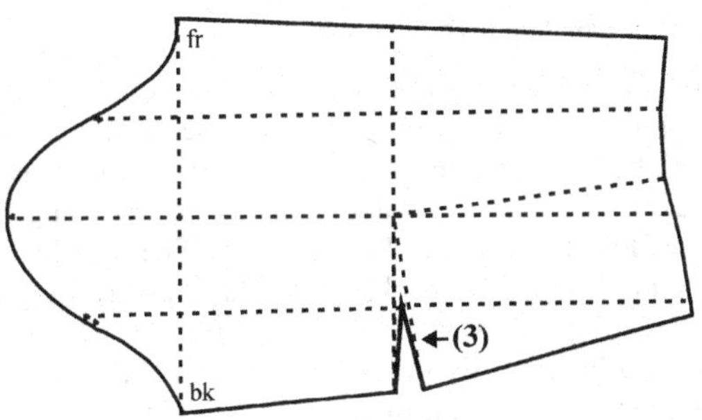

(3) Reduce the elbow dart length to the back arm line.

(4) The back arm line, central line and forearm line are redefined by dividing the wrist and elbow girth measurements by four. These points are then connected to the original lines at the sleeve head and are no longer straight. The new central line no longer conforms to the grain line. The original grain line has to be retained to maintain the correct forward hang of the sleeve.

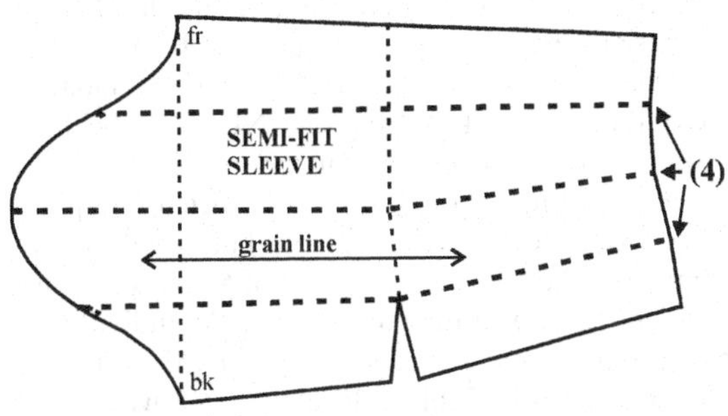

ADAPTATION FOR A FITTED SLEEVE BLOCK

Copy the semi-fit sleeve.

(1) Position a dart at the wrist 2.5 cm either side of the back arm line finishing approximately 5 cm from the elbow line **(total dart width 5 cm)**.

(2) The back arm wrist dart can be stitched to within 6 or 7 cm of the wrist for a fastened opening.

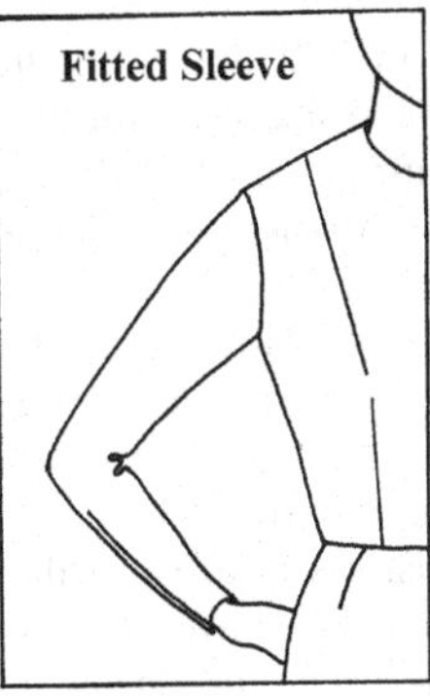

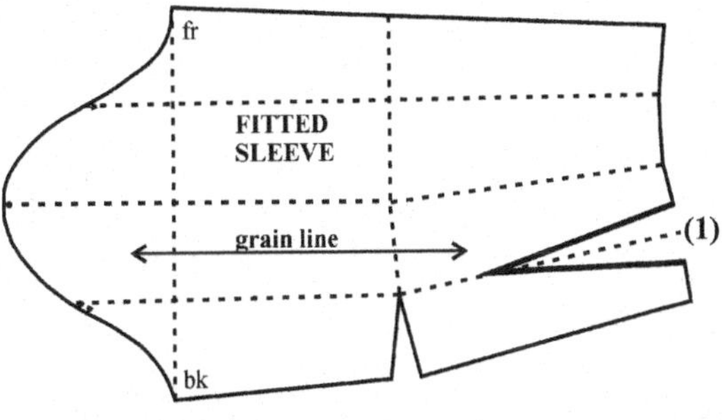

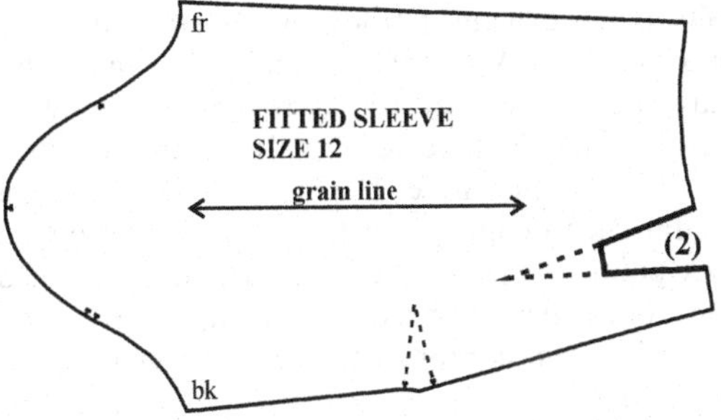

One-piece dress blocks

The fitting of one-piece dress blocks can be categorised into three sections: fitted, semi-fitted and straight loose fitting. The balance and ease allowances of the garments will be influenced according to how close the garment fits the body. The lengths of fitted dresses at the back, front and sides are longer when compared with those that hang straight and loose. Illustrated is the silhouette of a fitted dress compared with semi-fitted and straight one-piece dresses.

The back balance measurement is taken from the neck point NP (the highest point of the shoulder at the base of the neck) over the shoulder blades to the hip hem or ground levels (**A**). The front balance measurement is taken from the NP over the bust prominence to the hip hem or ground levels (**B**). In the case of this average size 12 the fitted one-piece dress has an equal back and front balance measurement from the NP to the hip level of 64 cm, the semi-fit 63 cm and the straight dress 62 cm. An individual's balance measurements may vary according to her posture. If it is difficult to locate the NP when constructing a pattern for an individual an overall measurement can be taken of the back length to the NP and continued down the front. This measurement is then divided in half for estimating the NP. The measurement at the side seam for this average size 12 from the underarm point UP to the hip level for a fitted one-piece dress is 38 cm, semi-fit 37.5 cm and straight dress 37 cm.

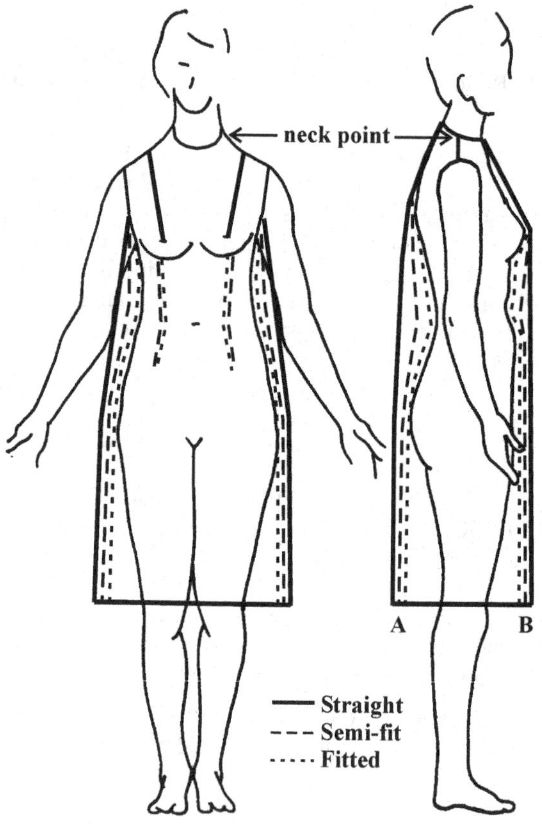

Theoretically these blocks are achieved by joining the fitted bodice and skirt blocks together. However, some adjustments have to be made as the waist seams do not always match in width and shape.

When the waist seams are joined, in some cases it will be found that the centre backs and fronts of the bodice and skirts are not in a straight line. Therefore, the exact matching of the waist seams is only suitable for panelled garments with seams at the centre back and front. The construction for a fitted one-piece dress block explained below has a straight centre back and front. When the CB and CF of the bodice and skirt are positioned in a straight line the waist seams do not fit together exactly. Therefore, there has to be a compromise. The bodice back waist seam is adjusted to conform to the skirt but the overall neck point to hip balance measurement is retained. The front bodice suppression is manipulated so that the waist seam conforms to the angle of the front skirt waist.

Fitted one-piece dress block

The fit around the waist area is slightly looser than the fitted dress with a waist seam. This is because a close fitted waistline can cause the one-piece dress to ride up and make the hip and hem levels uneven. This pattern is for the torso between the shoulders and hip line. For a complete dress the remainder of the skirt length can be added from the hip level.

BASIC RECTANGLE
Construct a rectangle with a **horizontal measurement of 64 cm** (NP to hip level) and a **vertical measurement of 50 cm** (half skirt hip girth). The lower side is the CB and the top side is the CF, the right side the hip level and the left side for positioning the neck points.

(1) Construct parallel to the hip level an **upper hip level at 10 cm and a waist level at 20 cm**.

(2) Construct a line midway between the CB and CF with a **length of 38 cm for armhole** depth from the hip to the underarm point (UP).

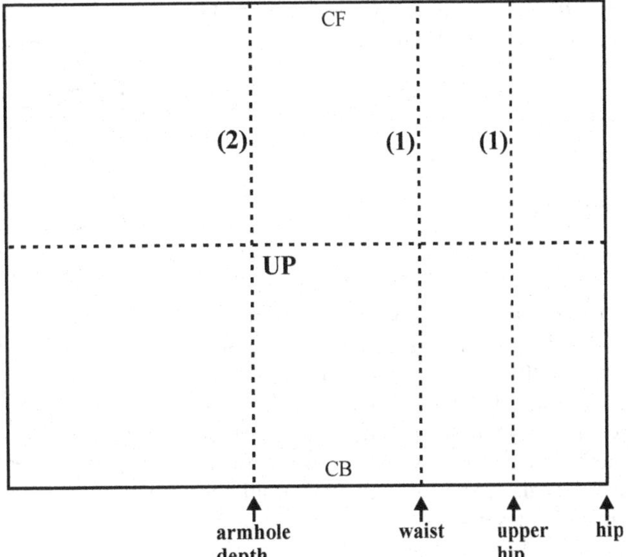

BACK DRESS
(3) Outline the back skirt from the hip level to waist seam and position within the rectangle to the CB and hip level (some computer systems require the relevant lines to be merged and then copied for transferring to the rectangle).

(4) Outline the back bodice and position within the rectangle conforming to the CB with the NP meeting the left side of the rectangle.

(5) Widen the back bodice side seam to match the skirt waist width and raise at UP to match the underarm level (this releases some of the waist suppression that is concealed between the back armhole and waist).

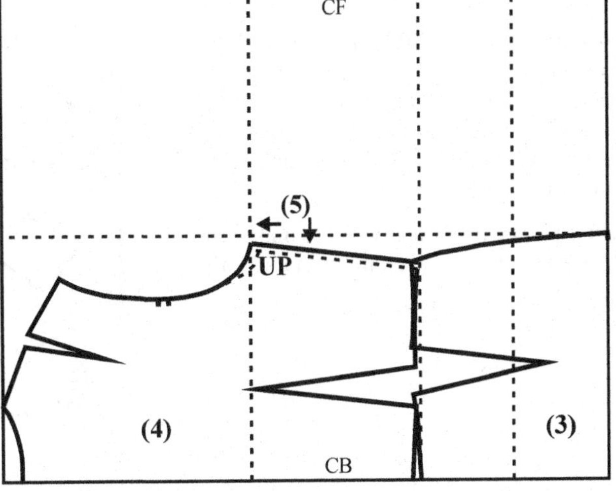

(6) Position a back waist dart by constructing a central line parallel to the CB from the top of the bodice waist dart to where the skirt dart originally finished. Draw the dart 3 cm wide at the skirt waist level.

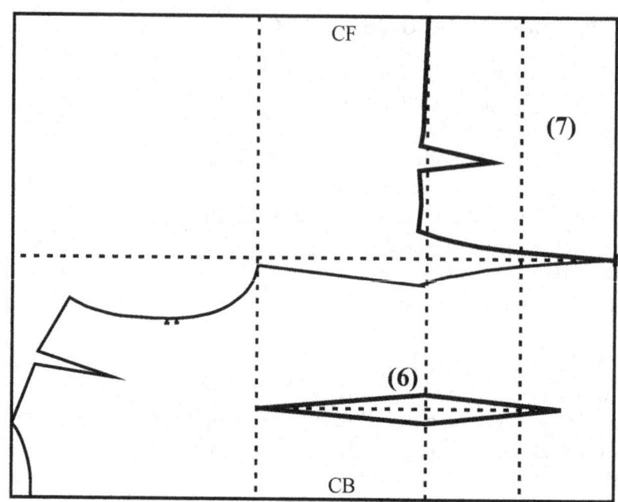

FRONT DRESS

(7) Outline the front skirt from the hip level to waist seam and position within the rectangle to the CF and hip level.

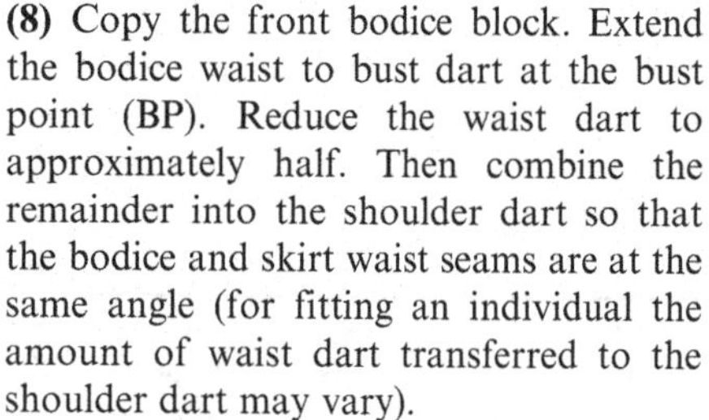

(8) Copy the front bodice block. Extend the bodice waist to bust dart at the bust point (BP). Reduce the waist dart to approximately half. Then combine the remainder into the shoulder dart so that the bodice and skirt waist seams are at the same angle (for fitting an individual the amount of waist dart transferred to the shoulder dart may vary).

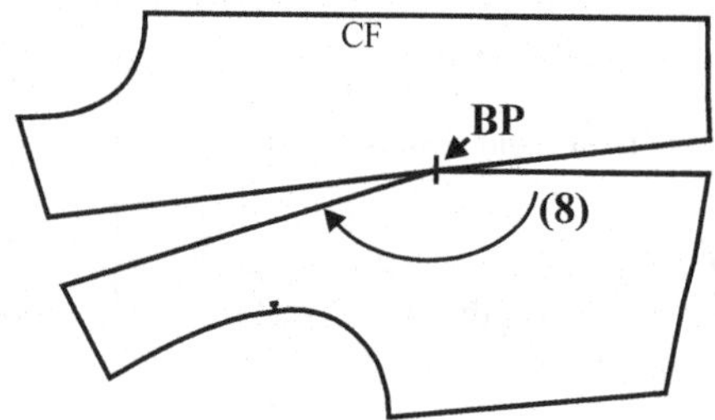

(9) Outline the modified bodice and position within the rectangle to the CF and the neck point NP touching the left side of the rectangle. The bodice and skirt waist seams should meet at the CF and at the side of the skirt waist level.

(10) Increase the bodice waist to match the skirt waist at the side seam and join to the UP.

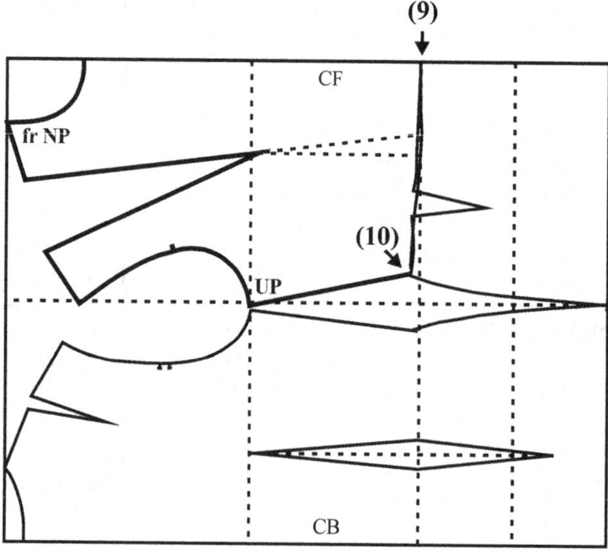

(11) Position the front waist dart by constructing a central line between the BP and upper hip that is parallel to the CF line. Draw a dart 3 cm from the BP, 2 cm wide at the skirt waist level (or the remainder of the bodice waist dart), to finish at the upper hip level.

(12) Curve the side seam smoothly at the waist.

(13) Trace off the new block patterns. A skirt length of 40 cm can be added from the hip level to the knee level or as required.

(14) Position balance notches in the middle of the armholes, one to indicate the front sleeve and two for the back. Matching notches can also be positioned on the side seams at the waist and hip levels. Position the grain line and label the pattern.

The straight, semi-fitted and fitted sleeve blocks can be used with this fitted one-piece dress block as the armhole has remained the same size as the original bodice block. The position of the original balance notches has to be retained to match the sleeves.

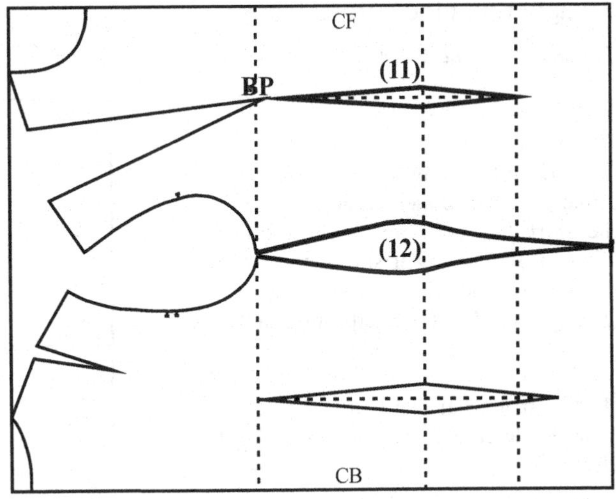

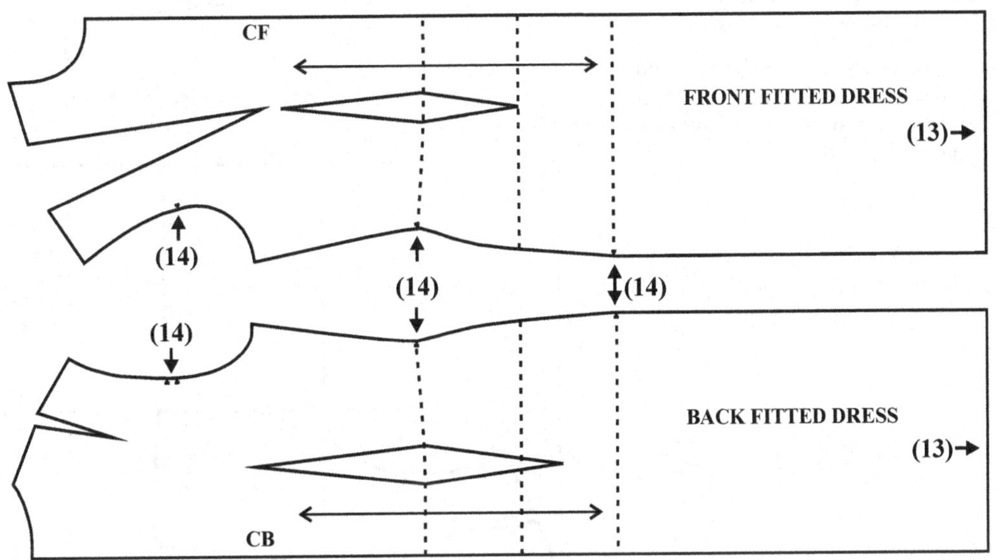

Straight one-piece dress block

This block has no waist suppression.

BASIC RECTANGLE
Construct a rectangle **horizontal measurement 62 cm** (the length from neck point to hip level), **vertical measurement 51 cm** (half the skirt hip girth plus 1 cm extra ease allowance). The lower side is the CB and the top side the CF, the right side the hip level and the left for positioning the neck points.

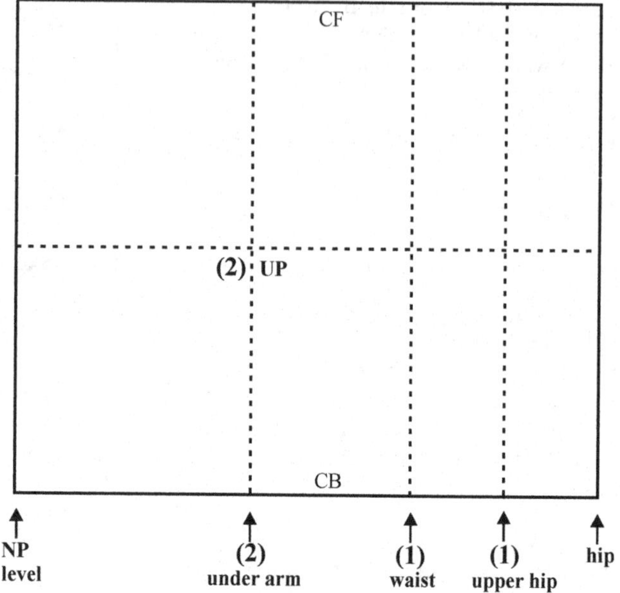

(1) Construct parallel to the hip level an **upper hip level at 10 cm and a waist level at 20 cm**.

(2) Construct a side seam line midway between the CB and CF, with a **length of 37 cm** (the measurement from hip to underarm point (UP)).

BACK DRESS
(3) Copy the back bodice and construct two lines parallel to the side seam one at 3 cm the other at 6 cm. These are the positions for releasing the concealed suppression between the armhole and waist.

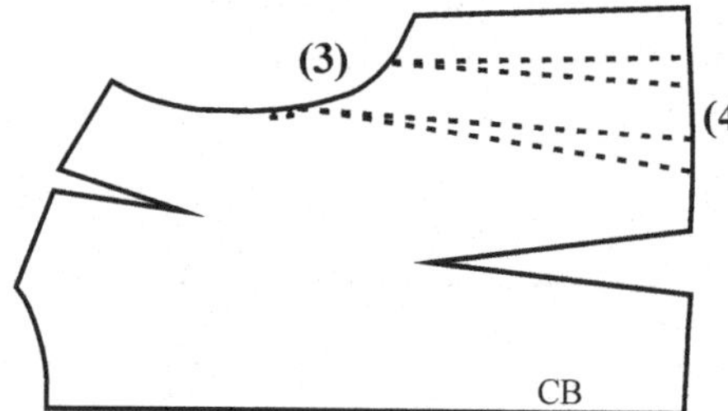

(4) Using the armhole end of these lines as pivot point add fullness of 2 cm at the waist end of each line (or an appropriate amount to straighten the bodice back waist).

(5) Outline the adapted back bodice and position to the CB line with the neck point NP meeting the left side of the rectangle and the side seam length at UP.

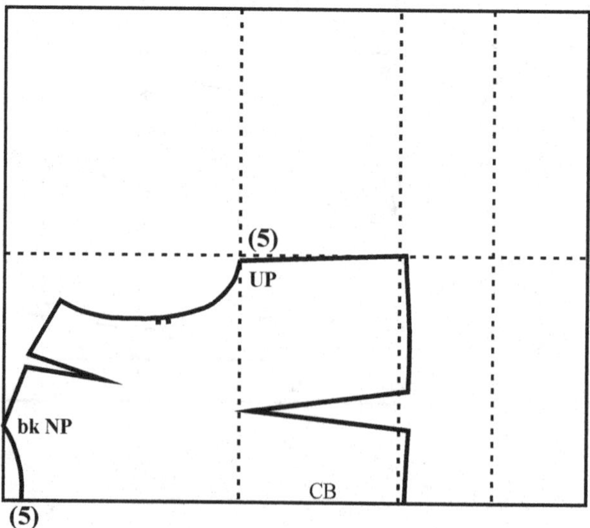

FRONT DRESS

(6) Copy the front bodice block. Connect the end of the waist dart to the bust point (BP), then pivot one-third of the waist dart into the shoulder dart (or sufficient to lower the armhole to match the side seam length at UP).

(7) Outline the adapted front bodice and position to the CF line with the neck point NP meeting the left side of the rectangle and the side seam length at UP.

(8) Connect the back and front armhole with a smooth continuous line to meet the side seam at the underarm point UP.

(9) Trace off the new block patterns with a straight side seam midway between and parallel to the CB and CF. This increases the bust girth measurement to the same as the hip. Check the armhole measurement. For a looser easier fitting garment this can be lowered slightly. The corresponding sleeve would also have to be modified. A skirt length of 40 cm can be added from the hip level to the knee level or as required.

(10) Position balance notches in the middle of each armhole. One notch to indicate the front and two notches for the back. Matching notches can also be positioned on the side seams at the waist and hip levels.

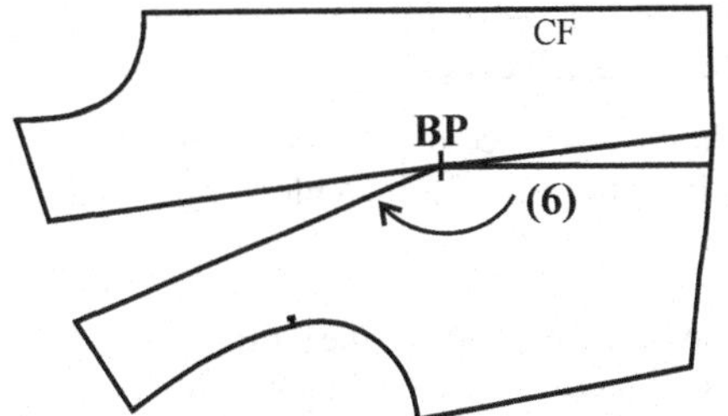

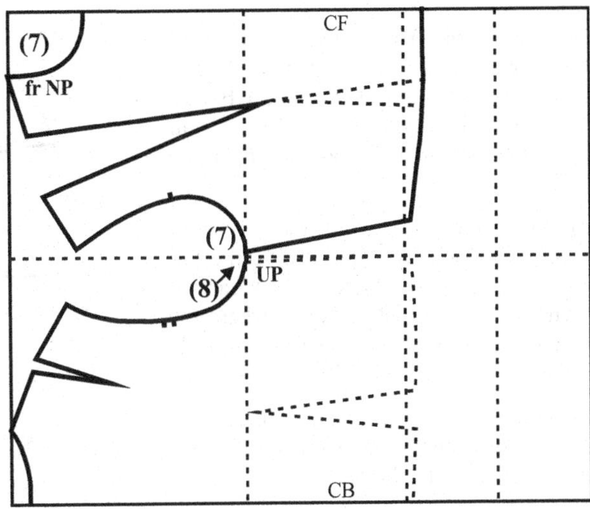

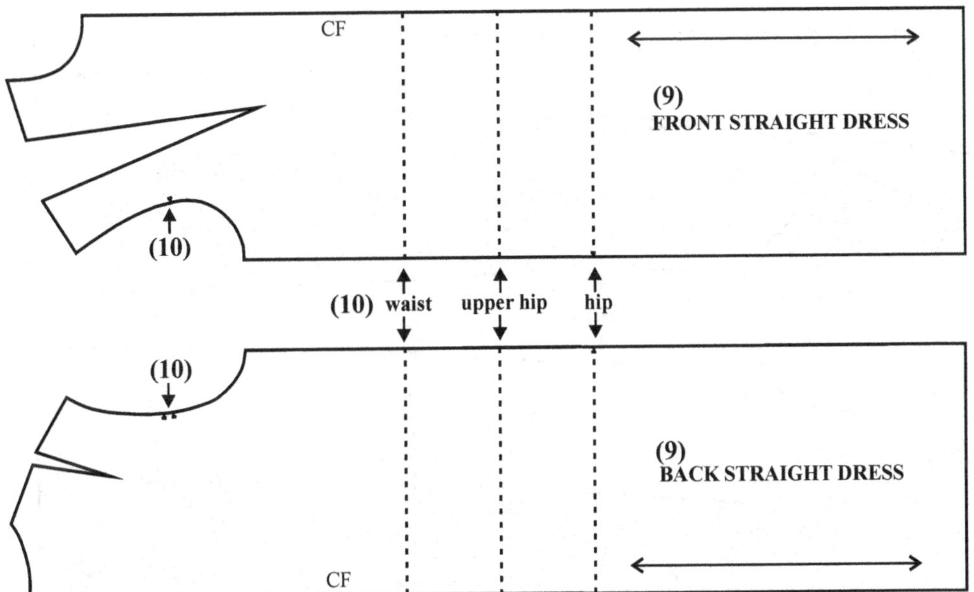

Adaptation for a semi-fitted one-piece dress block

This semi-fitted one-piece dress block can be adapted from the straight dress block by adding waist suppression.

(1) Copy the back and front of the straight dress block. Lengthen these blocks by lowering the waist, upper hip and hip levels 1 cm.

(2) Position the front waist dart by constructing a central line between the BP and upper hip that is parallel to the CF line. Draw a dart 3 cm from the BP, 2 cm wide at the waist level, to finish at the upper hip level.

(3) Position a back waist dart by constructing a central line parallel to the CB from the top of the bodice waist dart to 14 cm below the waistline, or to where the skirt dart finished. Draw the dart 3 cm wide at the waist level.

(4) Shape the side seam by reducing the width at the waist 1.5 cm from the straight side seam for both the back and front.

(5) A skirt length of 40 cm can be added from the hip level to the knee level or as required.

(6) Position balance notches matching the side seams at the waist and hip levels.

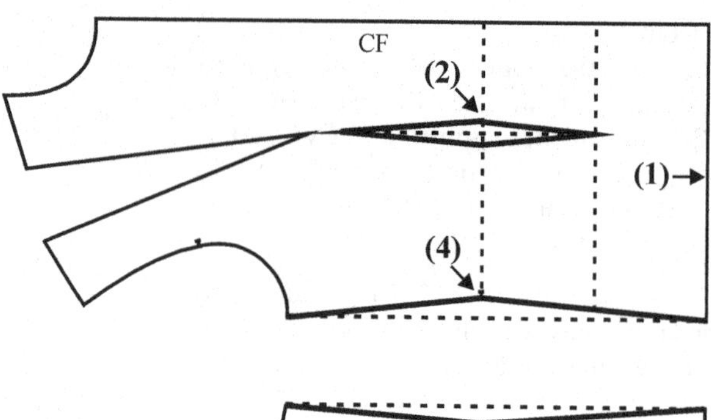

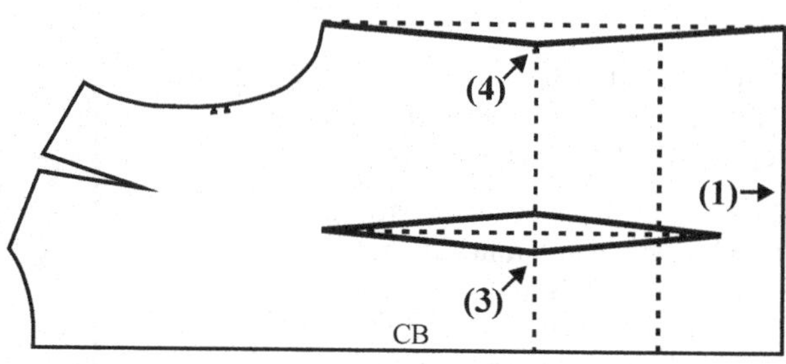

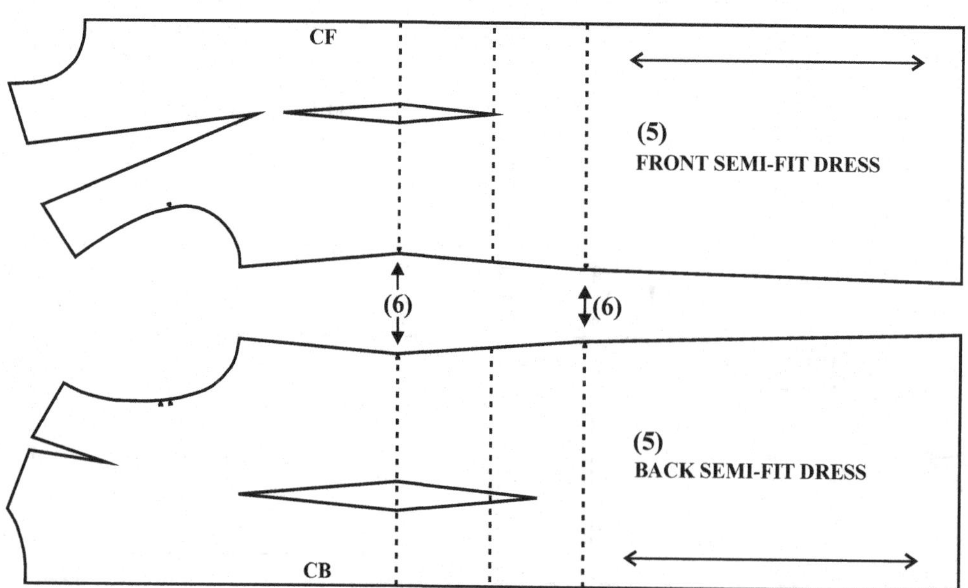

Dartless block tops and blouses

These blocks are adapted from the one-piece dress blocks. Two types of dartless blocks are explained below. The first is for a close fitting top suitable for two-way stretch knitted fabric. The second is for a loose fitting blouse in woven fabric or a firm knit.

ADAPTATION FOR A FITTED DARTLESS TOP FOR STRETCH FABRIC

The amount of stretch in knitted fabric varies greatly so this block for two-way stretch can only be an estimate. This pattern adaptation uses the actual body measurements for the bust and hips girths, also for the torso length from neck point to hip level and the sleeve length. Styled patterns based on these blocks can then be adjusted larger or smaller according to the fabric stretch and recovery.

Two pattern construction methods are given. The first is more theoretical to explain how this shape is created from the fitted one-piece dress block. It was developed from the comparison of two modelled fitted torsos, one in woven fabric, the other in two-way stretch. The suppression of the woven fabric was in the form of darts, whereas some of the suppression for the knitted fabric was incorporated into the stretch and some transferred into the armholes. The second method of construction is a simplified draft from measurements based on the shape of the first method. Dartless tops can also be created by modelling on a workroom stand in the actual stretch fabric and then digitising into the computer.

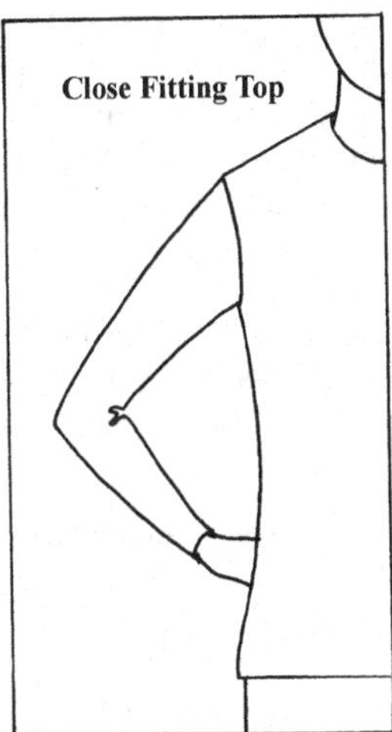

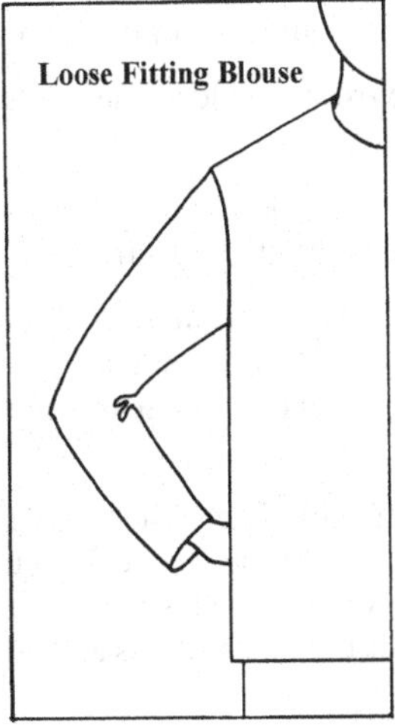

MEASUREMENT REQUIRED

	Body	*Ease*	*Pattern*
(a) Bust girth	88.0	—	88.0
(b) Waist girth	70.0	+10.0	80.0
(c) Hip girth	96.0	—	96.0
(d) Shoulder	13.0	—	13.0
(e) Across back	35.0	+2.0	37.0
(f) Across front	32.0	+1.0	33.0
(g) Neck point to hip level	64.0	—	64.0
(h) Neck point to waist level	44.0	—	44.0
(i) Armhole depth	21.0	+1.0	22.0
(j) Sleeve length	59.0	—	59.0
(k) Sleeve head depth	14.0 ($\frac{1}{3}$ of armhole)		
(l) Upper arm girth	28.0	+2.0	30.0
(m) Wrist girth	16.0	+2.0	18.0

METHOD 1: BACK AND FRONT TOP ADAPTATION

Copy the back and front fitted dress block from the neck points to the hip level.

Reduce the back and front widths at the side seams:

(1) Bust width to 22 cm ($\frac{1}{4}$ body bust) and raise to the new **underarm depth of 22 cm** (CB nape of neck to armhole depth plus 1.0 cm) (21 + 1 = 22).

(2) Waist width to 20 cm ($\frac{1}{4}$ body waist plus 2.5 cm ease).

(3) Hip width to 24 cm ($\frac{1}{4}$ body hip).

(4) Omit the back and front darts.

(5) Transfer the front shoulder dart into the armhole at approximately $\frac{1}{3}$ of the armhole from the underarm point (UP).

(6) Lower the front shoulder line 1.5 cm at the armhole end.

(7) Transfer the back shoulder dart into the armhole at right angles to the CB.

(8) Blend the back and front armholes into a smooth curve with the underarm point UP at the same level (the side seams should match in length).

(9) From the front UP keep the bust width of 22 cm parallel with the CF for 7 cm then curve to the waist level. This is to accommodate some of the bust fullness and curve the side seam.

(10) From the back UP join to the waist with a straight line. Then shape as the front side seam from the waist to the upper hip and hip.

(11) If the neck has no fastened opening for the head to pass through the neck, measurements may have to be increased. For a 3 or 4 cm wide rib the neckline can be lowered by 1.5 or 2 cm.

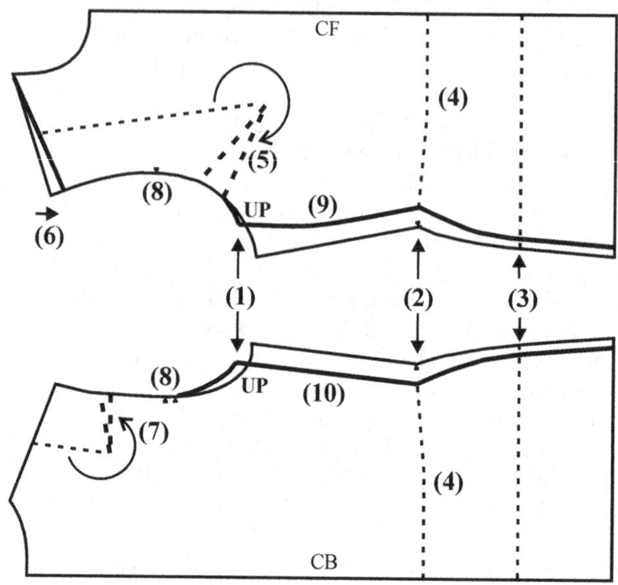

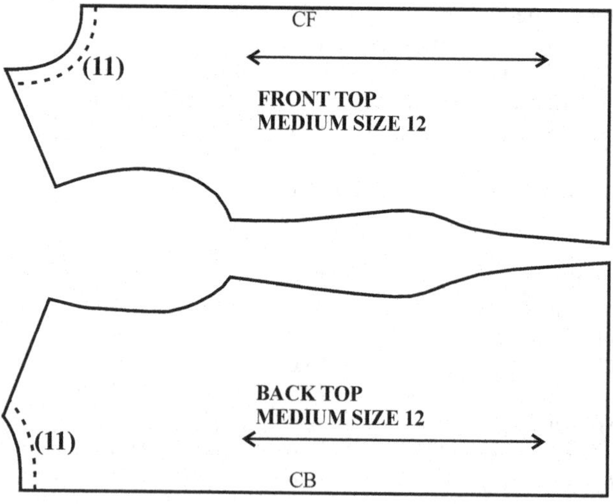

Top sleeve adaptation
Copy the back half of the straight sleeve block. For two-way stretch fabric the back and front of the sleeve can be identical.

(12) Reduce the **wrist measurement to 9 cm** ($\frac{1}{2}$ the sleeve wrist). Remove 2 cm at the centre line from the wrist tapering to nothing to the top of the sleeve head. Remove 2 cm from the underarm seam tapering from the wrist to nothing at the underarm point UP. On the back arm line remove 4 cm at the wrist tapering to nothing at the sleeve head (the diagram illustrates this in the form of darts). Using a computer the fullness can be reduced at these specific points at the wrist.

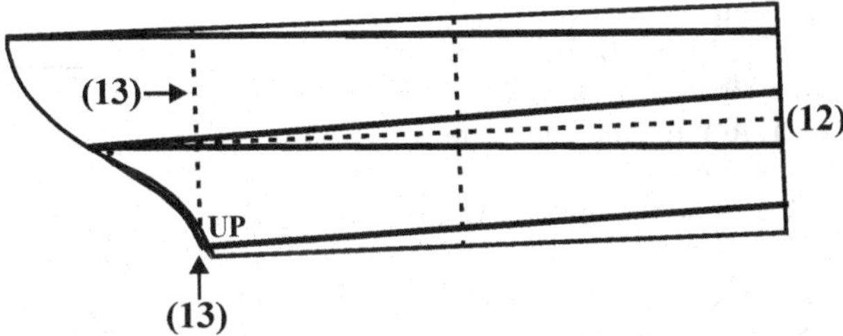

(13) Construct a **new upper arm level 14 cm** ($\frac{1}{3}$ of armhole) from the top of the sleeve head and square out from the central line **15 cm** ($\frac{1}{2}$ upper arm width). Join the new underarm point UP to the wrist.

(14) Blend the sleeve head into a smooth curve. The sleeve head should measure the same or 1 or 2 cm more than the armhole. The shape of the lower $\frac{1}{3}$ of the sleeve head from the underarm seam can be similar to the armhole from UP.

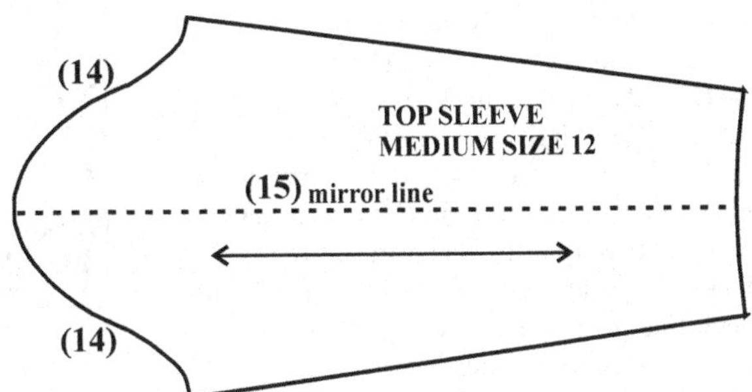

(15) Trace the new sleeve block patterns. The central lines can be mirror lines. There are no notches on these patterns as it is difficult to notch knitted fabric accurately.

METHOD 2: DRAFT FOR A FITTED DARTLESS TOP

This method is a direct draft from measurements positioned within a rectangle. The back and front of the torso from the neck point to the hip level are identical except for the neckline. This draft is based on the shape developed from method 1. When the back and front were superimposed they were very similar in shape. With a slight adjustment to the neck width and armhole shape they could be made identical. This simplified method does not provide the same subtleties of shape as method 1, nor does it explain how the shape has been developed.

Sleeve draft
Construct a rectangle:
A–B = 59 cm (sleeve length)
A–C = 15 cm ($\frac{1}{2}$ upper arm girth)
A–D = 14 cm (sleeve head depth $\frac{1}{3}$ armhole approximately)
E squared from **D** for underarm point
B–F = 9 cm ($\frac{1}{2}$ wrist girth)
Join **E to F** for the underarm seam
D–G = 8 cm ($\frac{1}{16}$ armhole plus 1 cm)
G–H = 8 cm the same as D–G

For sleeve head shaping join A to H then to E. Square a mid point line 1 cm on these lines to guide the curve from A to H to E. Shape the curve from E similar to the lower armhole. Trace out the new sleeve shape and mirror from line A–B for the total sleeve.

Top draft
Construct a rectangle:
A–B = 64 cm (neck point to hip level)
B–C = 24 cm ($\frac{1}{4}$ body hip girth)
Label the lower edge **CB and CF**
A–D = 2 cm (back neck rise)
A–E = 7 cm (half neck width)
D–E = square from CB then curve **back neck line to measure 8 cm** ($\frac{1}{2}$ back neck)
A–F = 8 cm (front neck rise)
F–E[1] = square from CF then curve **front neck line to measure 12 cm** ($\frac{1}{2}$ front neck)
A–G = 5 cm (shoulder level)
G–H = 19.5 cm (shoulder width squared from G)
A–I = 16 cm (across front level at mid CF neck to armhole level)
I–J = 17.5 cm (half across back and across front squared from I)
A–K = 24 cm (armhole depth)
K–L = 22 cm ($\frac{1}{4}$ body bust measurement squared from K)
H–J–L curve the **armhole to finish 20.5 cm**
A–M = 44 cm (waist level)
M–N = 20 cm ($\frac{1}{4}$ body waist girth plus 2.5 cm squared from M)
L–N–C connect these points for the side seam

Trace off from these drafts a separate back and front top and sleeve.

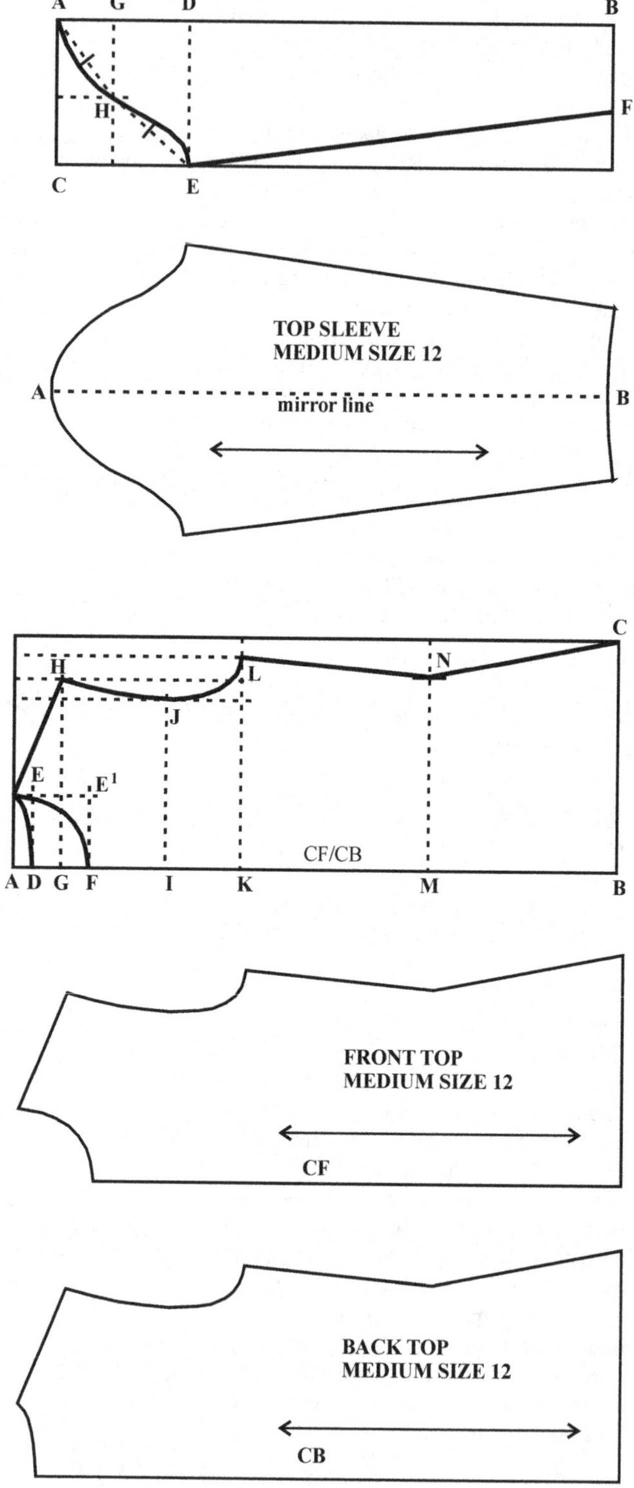

ADAPTATION FOR A DARTLESS BLOUSE BLOCK FOR WOVEN FABRIC OR FIRM KNIT

This is a loose fitting garment based on the straight one-piece dress block. The same adaptation can be applied to the semi-fit dress block for a more fitted blouse without increasing the garment width or lowering the armhole.

Front blouse adaptation

Copy the front straight dress block from the neck point to the hip level.

(1) Reduce the front shoulder dart width by 1.5 cm (which was the width of the original back shoulder dart). Pivot the remainder of the shoulder dart into the armhole at approximately $\frac{1}{3}$ of the armhole from the underarm point UP.

(2) Widen the front 0.5 cm from the shoulder to the hem (in the position indicated in the diagram). The new shoulder width is 15.0 to match the back.

(3) Lower the underarm point UP **1.5 cm** (or between 1 and 2 cm).

(4) Blend the armhole curve to UP. Position one balance notch at the mid armhole.

Back blouse adaptation

Copy the back straight dress block from the neck point to the hip level.

(5) Pivot the shoulder dart into the armhole. The new shoulder length is 15.5 cm (shoulder length plus 2 cm). The back shoulder seam eases slightly on to the front shoulder seam.

(6) Widen the back 0.5 cm from the shoulder to the hem (in the position indicated in the diagram).

(7) Lower the underarm point UP **1.5 cm** (or between 1 and 2 cm).

(8) Blend the armhole curve to UP. Position two balance notches at the mid armhole.

Trace the new back and front blouse blocks.

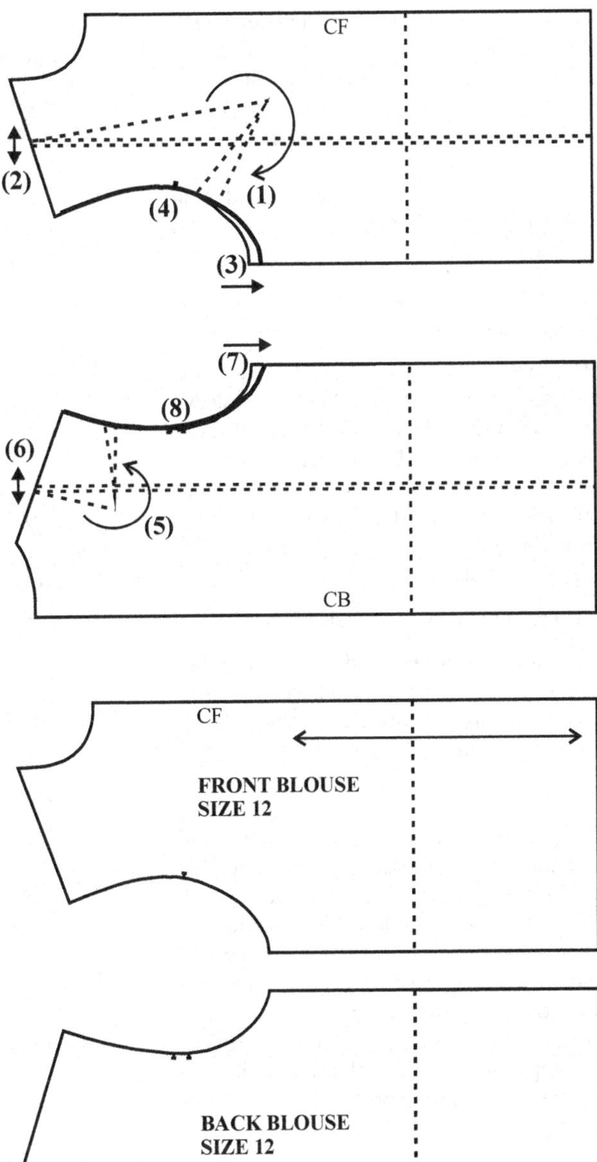

Blouse sleeve adaptation

The sleeve head height has to be reduced by the amount that the blouse shoulder seam has been extended to retain the same overall length measurement. Likewise the sleeve head at the underarm seam has to be lowered the same amount as the blouse. This more casual style of garment does not require much ease allowance of the sleeve head, approximately 1 to 2 cm. However, when the sleeve head is flattened it can reduce the measurement of the sleeve head seam. To compensate for this the upper arm width can be increased. This can increase the allowance for arm movement.

Copy the straight sleeve block.

(1) Lower the top of the sleeve head **2 cm** and blend the curve to the forearm and back arm lines.

(2) Lower the top of the underarm seam at UP **1.5 cm** (or between 1 and 2 cm).

(3) Measure the sleeve head seam and compare it with the armhole (the sleeve head seam should measure 1 to 2 cm more than the armhole). To compensate for any loss of measurement the width of the upper arm has been extended **2.5 cm** at the back and front side seam tapering to nothing at the wrist or elbow.

(4) Trace the new blouse sleeve block.

(5) Match the sleeve head balance notches with those of the armhole. There is no need for any ease allowance between the underarm seam and the notches.

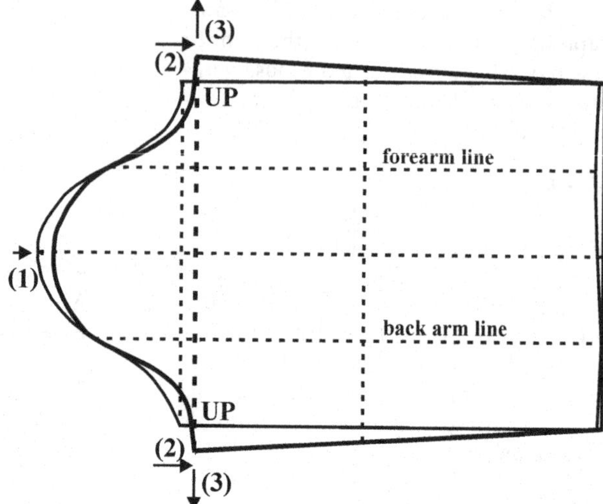

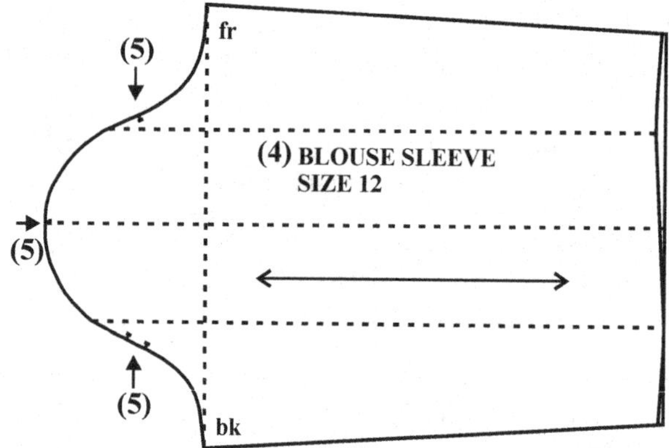

The understanding of how the primary and secondary block patterns are developed is important before grading patterns and creating garment styles. The position of each body measurement has to be known and how much ease allowance and suppression is required for the correct fit and balance. Even if the reduced size patterns in Appendix II are enlarged it is still worthwhile reading how they are developed. All new basic block patterns must be made into sample garments and tested for fit and size before they can be adapted into patterns for garment styles. In the next section the pattern grading of these blocks is described prior to pattern designing. This is because when using computers the grading from the block patterns is passed on to the new patterns.

Part 2
Computer pattern grading

Part 2 focuses entirely on pattern grading and:

- Introduces various computer grading techniques
- Explains grading increments, grading point movement, grade rules and grade rule tables
- Gives examples of grading primary block patterns
- Gives examples of grading secondary block patterns
- Explains computer pattern digitising and pattern verification.

PATTERN GRADING

Pattern grading is a process of producing patterns of difference sizes from a master pattern according to a specified size chart. The amount and direction that the patterns increase or decrease has to be determined. At the same time the correct proportions of the garments have to be maintained without distorting the style features.

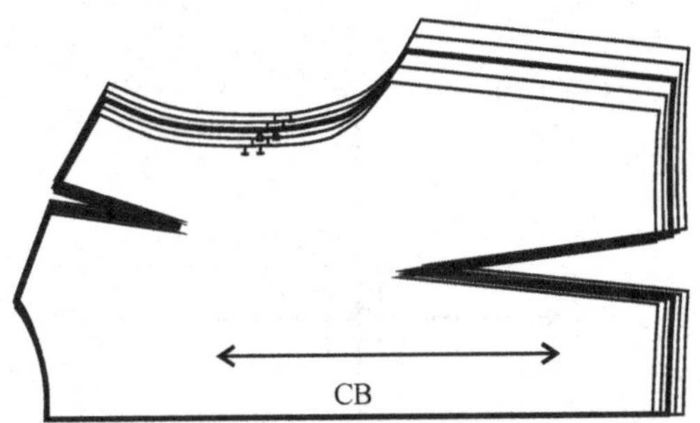

Figure 2.1 Stack or nest of graded patterns

The **master pattern**, sample pattern or base size pattern can be either a block pattern or a production pattern of a style from which the other sizes are graded. It is generally the centre of the size range for grading accuracy (Figure 2.1). This pattern can be produced manually or by computer. For digitising, a manually produced pattern can be either in card or drawn on tracing paper. Computer generated patterns can be graded within the computer program.

Grading increments

A **grading increment** is the difference in measurement between two sizes, either in a size chart or a specific point on a pattern. When looking at a set of stacked or nested graded patterns (Figure 2.1) it appears that only the perimeter of the patterns alters in size. This, however, is not the case as the changes in body size take place three-dimensionally. Therefore the increase or decrease in size takes place within the pattern area. The full girth and length grading increments are found in most size charts. The **sectional increments** which are a proportion or section of the full increments also have to be known, e.g. across back, across front, shoulder, depth and width of armhole. These can be found in some comprehensive size charts but are not always quoted. However, the increments can be calculated from any reliable drafting system by calculating two sizes then subtracting the smaller from the larger. The sectional increment can also be calculated if the measurement was derived as a proportion of a full major girth measurement. Examples of this are given in the explanation for grading the bodice, sleeve and trouser blocks.

There are two basic principles which should be adhered to for successful grading:

- All the sectional increments must add up to the full girth or length increments
- All the increments must maintain the matching of adjoining seams and notches.

The first two diagrams of Figure 2.2 show the position of the sectional increments on a three-dimensional figure for a straight skirt. This is followed by the position of the sectional increment within the two-dimensional pattern pieces.

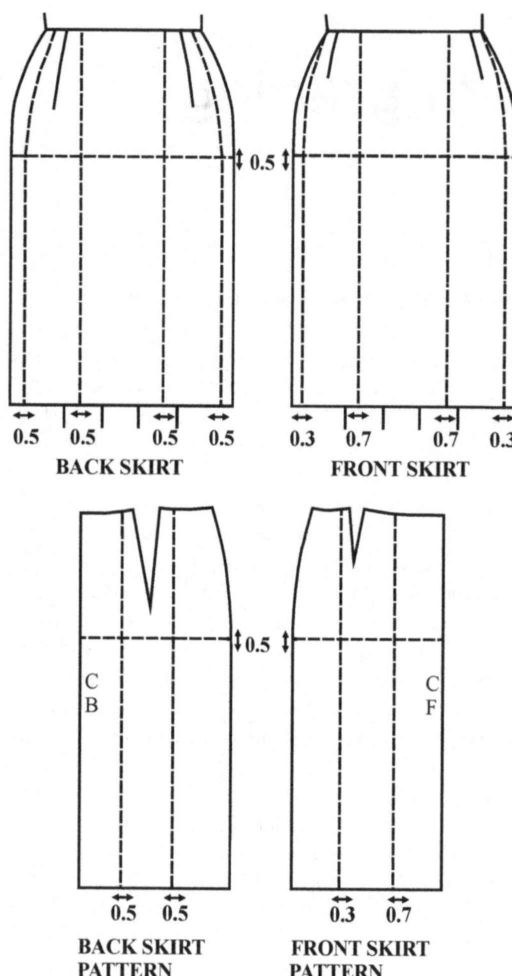

Figure 2.2 Position of sectional grading increments

Grade point movement

Grade points are positioned at the cardinal points of a pattern where the measurement to another size takes place. These can be perimeter or internal. This movement is in an X (horizontal) and Y (vertical) direction similar to the grid of manual grading. The distance and direction of the grade point movement is recorded on an X-axis and Y-axis from a **zero point** that is stationary at the junction of the axis. The measurements are marked from the zero point in one of the four directions of +X, −X, +Y, −Y. The X-axis generally represents the straight warp grain of the fabric and is generally used as the **grade reference line**.

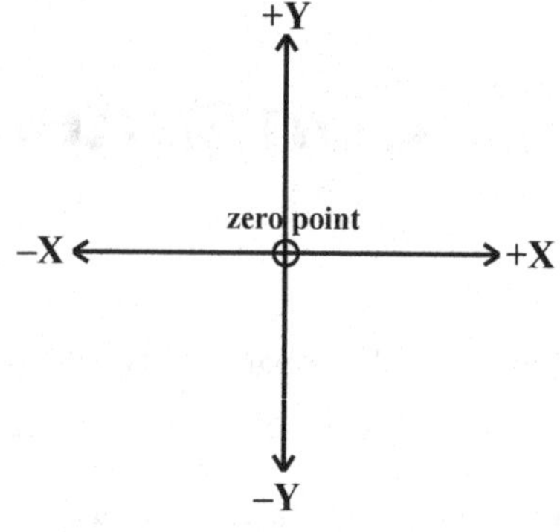

- The stationary zero point represents the position where no movement takes place in the X or Y direction. It is from this point that all the grading increments are calculated.
- The X-axis is horizontal. The movement to the right from the zero point is positive and to the left negative.
- The Y-axis is vertical. The movement up from the zero point is positive and down negative.

X AND Y CO-ORDINATES

An **X and Y co-ordinate** is a combination of a horizontal X movement and a vertical movement when neither X or Y are at zero. These diagonal movements can be in any of the four directions.

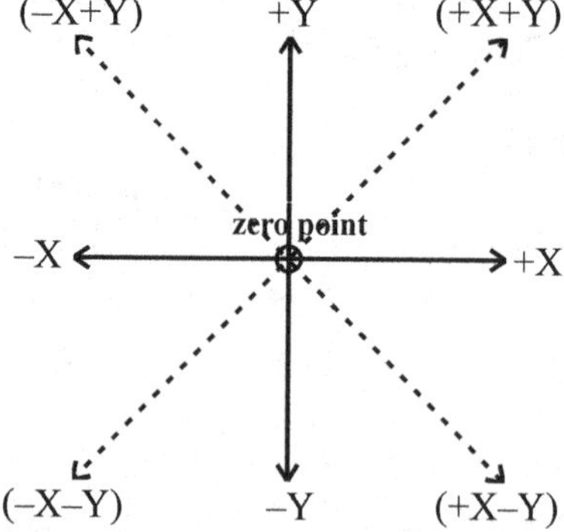

Methods of recording incremental growth

There are basically three methods of recording the growth of the pattern.

Method 1 records the growth from the smallest size to the largest incrementally.

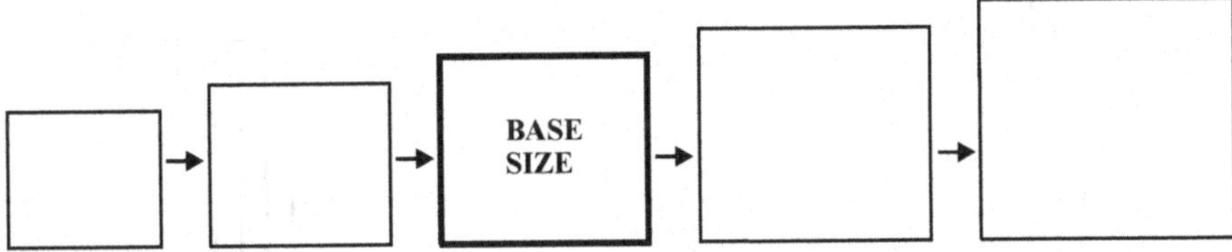

Method 2 records the growth from the base size down to the smallest and up to the largest incrementally.

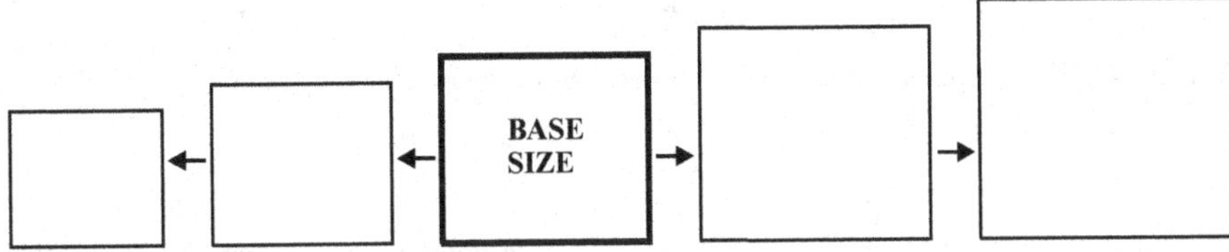

Method 3 records the growth of each pattern from the base size down to the smallest then up to the largest cumulatively.

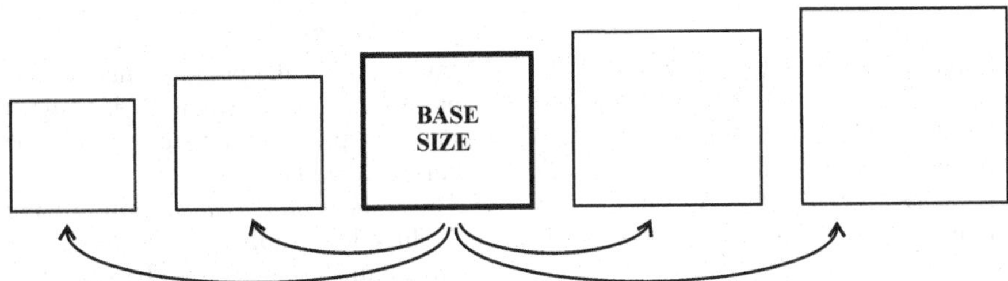

Grade rules

For computer grading each movement of a graded point in both X and Y direction is defined by a numbered grade rule. The grade rules are listed in a grade rule table. The same rules can be assigned to any point of any other pattern piece within the same size range. Once a grade point number is assigned to a point on the pattern the computer automatically redraws the shape of the piece by connecting the grade points.

Figure 2.3 is an example of a graded rectangle and is for the first method of recording the growth of the pattern from the smallest to the largest. It covers five women's sizes, 8, 10, 12, 14 and 16, with size 12 as the base size. There are four points that can be graded. The fifth point A is a notch with zero growth. The following is an explanation of the movement of each point graded:

- Point A does not move, therefore is 0 for zero growth
- Point B moves left on the X-axis 1.0 cm but does not move on the Y-axis
- Point C moves left on the X-axis 1.0 cm and up on the Y-axis 0.5 cm
- Point D moves right on the X-axis 1.0 cm and up on the Y-axis 0.5 cm
- Point E moves right on the X-axis 1.0 cm but does not move on the Y-axis.

GRADE RULE NUMBERS

A **grade rule** defines the movement of the graded points. Before setting up a grade rule table for the

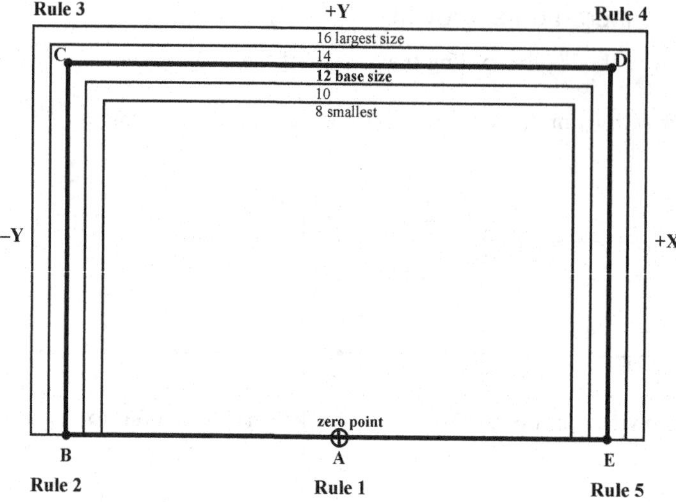

Figure 2.3 Positions of graded points and movement (Method 1 Small to large incremental)

pattern in Figure 2.3 the following X and Y values to points A to E would be assigned a grade rule number and a movement. Most computer systems consider all numbers as a positive so there is no need for a plus (+) sign only the minus (−) would have to be assigned.

- Rule 1 for Point A: X = 0.0 Y = 0.0 (zero growth)
- Rule 2 for Point B: X = −0.5 Y = 0.0
- Rule 3 for Point C: X = −0.5 Y = 0.5
- Rule 4 for Point D: X = 0.5 Y = 0.5
- Rule 5 for Point E: X = 0.5 Y = 0.0

Constructing a grade rule table

The **grade rule table** lists and defines the numbered grade rules. It is more convenient to enter the grade rule tables before digitising or using the pattern design systems. Rules can be altered or added later when the pattern piece is being verified. Each computer system has its own format but most require the following information before inputting the rules:

- A *name* has to be assigned to the rule table
- *Size names* have to be chosen, either numerical (10, 12, 14, 16) or in words (small, medium, large)
- A *base size* has to be stated that is the size upon which all grade rules are applied (a *size step* is often used for numerical sizes only to indicate the sizes are for step 1 (10, 11, 12, 13, 14) or step 2 (10, 12, 14, 16)
- *Rule numbers* are entered sequentially, often followed by a name of the point (generally the first rule number indicates a zero point as X = 0 and Y = 0)
- *Attributes* is a permanently assigned characteristic to a point but this can be optional (e.g. for Gerber Accumark System 'N' is assigned to indicate no smoothing of the point so that it always remains an angle. If the attribute 'S' is assigned the computer is programmed to smooth the point by curving the line)
- The *grade point movement* is inserted in the X and Y directions (only when a movement is in the minus direction need (−) be inserted).

The three examples of grade rule tables below (created using Gerber Accumark system) are for the graded rectangle in Figure 2.3, illustrating the three basic methods of recording the pattern growth in five sizes. These methods cannot be mixed within the same grade rule table. The recording of each method is different but the results of the gradings remain the same.

METHOD 1 Small to large incremental

NAME: RECTANGLE GRADING METHOD: SMALL TO LARGE INCREMENTAL

RULE NUMBER:		1		2		3	
COMMENT:		POINT A		POINT B		POINT C	
POINT ATTRIBUTE:		N		N		N	
SIZE BREAKS		X	Y	X	Y	X	Y
8	10	0.00	0.00	−0.50	0.00	−0.50	0.50
10	12	0.00	0.00	−0.50	0.00	−0.50	0.50
12	14	0.00	0.00	−0.50	0.00	−0.50	0.50
14	16	0.00	0.00	−0.50	0.00	−0.50	0.50

RULE NUMBER:		4		5			
COMMENT:		POINT D		POINT E			
POINT ATTRIBUTE:		N		N		N	
SIZE BREAKS		X	Y	X	Y	X	Y
8	10	0.50	0.50	0.50	0.00		
10	12	0.50	0.50	0.50	0.00		
12	14	0.50	0.50	0.50	0.00		
14	16	0.50	0.50	0.50	0.00		

METHOD 2 Base size down to the smallest and up to the largest incrementally

NAME: RECTANGLE GRADING METHOD: UP-DOWN INCREMENTAL

RULE NUMBER:		1		2		3	
COMMENT:		POINT A		POINT B		POINT C	
POINT ATTRIBUTE:		N		N		N	
SIZE BREAKS		X	Y	X	Y	X	Y
10	8	0.00	0.00	0.50	0.00	0.50	−0.50
12	10	0.00	0.00	0.50	0.00	0.50	−0.50
12	14	0.00	0.00	−0.50	0.00	−0.50	0.50
14	16	0.00	0.00	−0.50	0.00	−0.50	0.50

RULE NUMBER:		4		5			
COMMENT:		POINT D		POINT E			
POINT ATTRIBUTE:		N		N		N	
SIZE BREAKS		X	Y	X	Y	X	Y
10	8	−0.50	−0.50	−0.50	0.00		
12	10	−0.50	−0.50	−0.50	0.00		
12	14	0.50	0.50	0.50	0.00		
14	16	0.50	0.50	0.50	0.00		

METHOD 3 Base size down to the smallest and up to the largest cumulatively

NAME: RECTANGLE GRADING METHOD: UP-DOWN CUMULATIVE

RULE NUMBER:		1		2		3	
COMMENT:		POINT A		POINT B		POINT C	
POINT ATTRIBUTE:		N		N		N	
SIZE BREAKS		X	Y	X	Y	X	Y
12	8	0.00	0.00	1.00	0.00	1.00	−1.00
12	10	0.00	0.00	0.50	0.00	0.50	−0.50
12	14	0.00	0.00	−0.50	0.00	−0.50	0.50
12	16	0.00	0.00	−1.00	0.00	−1.00	1.00

RULE NUMBER:		4		5			
COMMENT:		POINT D		POINT E			
POINT ATTRIBUTE:		N		N		N	
SIZE BREAKS		X	Y	X	Y	X	Y
12	8	−1.00	−1.00	−1.00	0.00		
12	10	−0.50	−0.50	−0.50	0.00		
12	14	0.50	0.50	0.50	0.00		
12	16	1.00	1.00	1.00	0.00		

Variations in positioning the zero point

The five nests of graded squares in Figure 2.4 have the zero points in different positions. A has the zero point in the left lower corner and B at the lower right corner. The zero point C is central on the lower side and for D it is central. E has the zero point outside the pattern piece. This illustrates how the grading increments and grade rules at each corner vary according to the position of the zero point. However, the grading results are identical.

Computer grading techniques

GRADING STRAIGHT, HORIZONTAL AND VERTICAL LINES

Example 1 of Figure 2.5 illustrates that equal increments in both X and Y directions maintain the proportions of a square when graded. Example 2 illustrates that varying increments in X and Y directions change the proportion of the squares into rectangles for the larger and smaller sizes. The pattern technologist has to be aware of this variation in proportion when calculating and applying grading increments.

The grade points of a nest of grades that have equal increments between the sizes should lie in a straight line if the patterns have been correctly graded.

GRADING DIAGONAL LINES

Example 1 of Figure 2.6 illustrates that the angle of the diagonal line is altered when the pattern is graded in one direction only. The computer will connect the graded points automatically. Example 2 illustrates that the angle can be maintained if the pattern is graded in both the X and Y directions. In this example the increment in the X direction depends on both the angle and length of line. This may need correcting when the nest of grades as verified. Some computer systems enable a diagonal or curved perimeter line to be graded perpendicular to that line and not on the X and Y axis of the original grain line. If equal grades are given to each end of the line the graded sizes will become parallel, as illustrated in Example 3.

GRADING CURVED LINES

In Figure 2.7 Example 1 illustrates the grading of a curved line that maintains a similar parallel curve to the original. This is achieved by grading equal increments at the beginning and end of the line. Example 2 illustrates the grading of a curved line that is altered in shape. This is achieved by grading unequal increments at the beginning and end of the curved line. The computer will try to connect the best fitting line but if this is distorted intermediate graded points will have to be added.

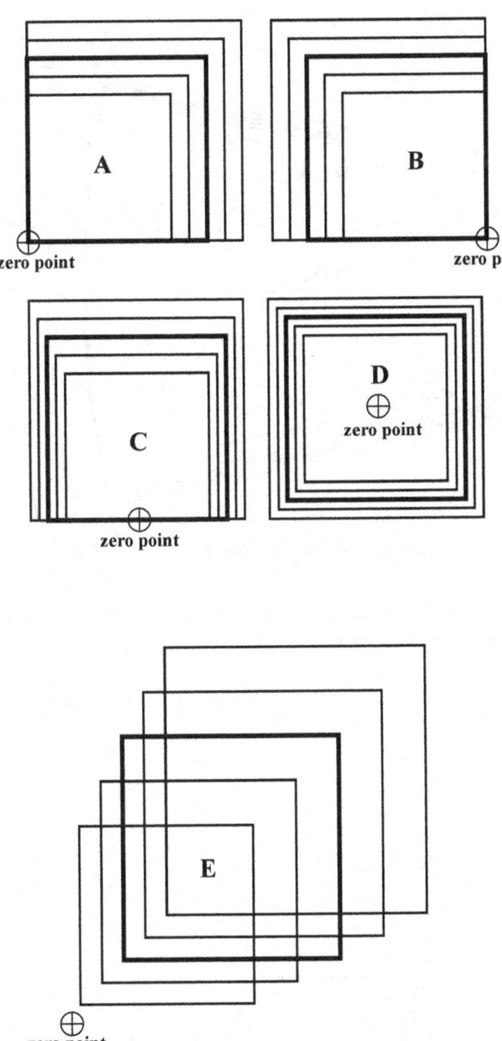

Figure 2.4 Different positions of the zero point

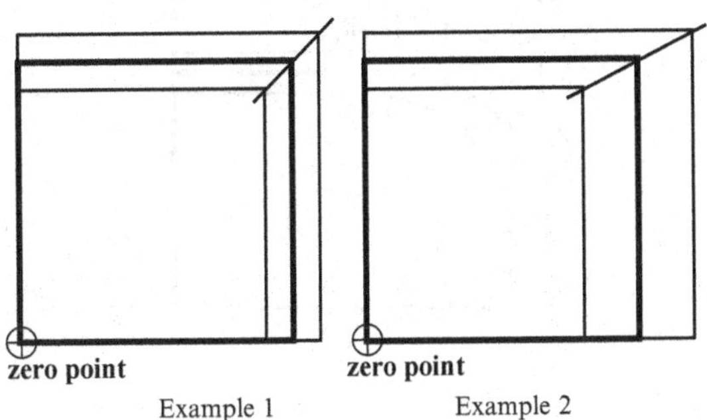

Figure 2.5 Maintaining and changing the pattern proportion

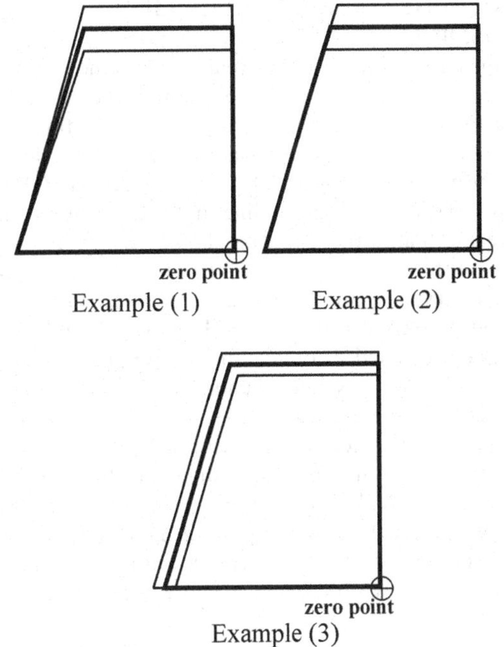

Figure 2.6 Grading diagonal lines

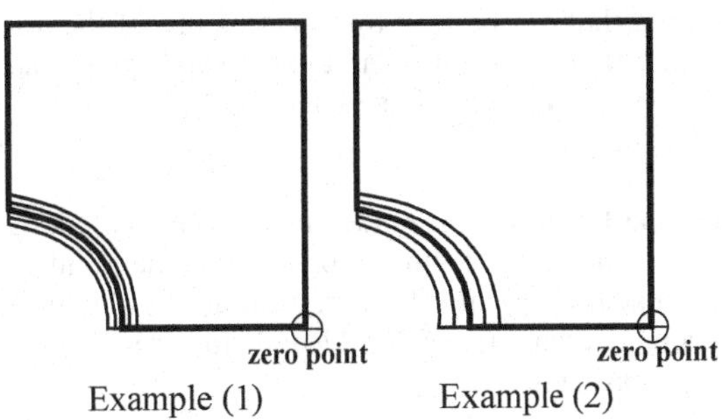

Figure 2.7 Grading curved lines

MULTI-GRADE REFERENCE LINE GRADING

Some patterns are drafted circular or become circular by pattern adaptation. When grading, the curve of the circle has to be maintained. To achieve this the pattern is graded from **alternate grade reference lines**. In the example in Figure 2.8 the first grain line (GRL 0) is the original grain line indicating the warp threads, that is the direction the pattern should be laid on the fabric. The +X in the centre of the right-hand curve is graded from this grain line (GRL 0). From the second grain line (GRL 1) the +Y and +X will be orientated. From the third grain line (GRL 2) the −Y and +X increments will be orientated. Generally these grain lines have to be digitised in a numerical sequence starting with the one representing the warp threads. When digitising the pattern perimeter each alternative grain line and number has to be selected from the digitising menu before the related grade points are recorded. Some computer programs allow up to eight or more grade reference lines.

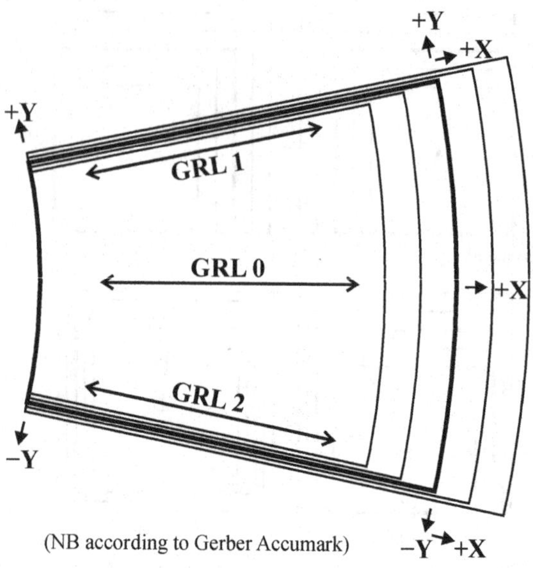

Figure 2.8 Alternate grade reference line grading (according to Gerber Accumark)

GRADING MATCHING NOTCHES ON ADJOINING SEAMS

The grading of matching notches on the pattern perimeter can be a problem. There are four alternatives that can be used by assigning one of the following but these may not be available on all computer pattern grading systems.

Method 1 averages the graded notch increment along the perimeter line from the previous grade point to the next grade point. This is best employed where the adjoining seams are of equal length and there is little or no easing in.

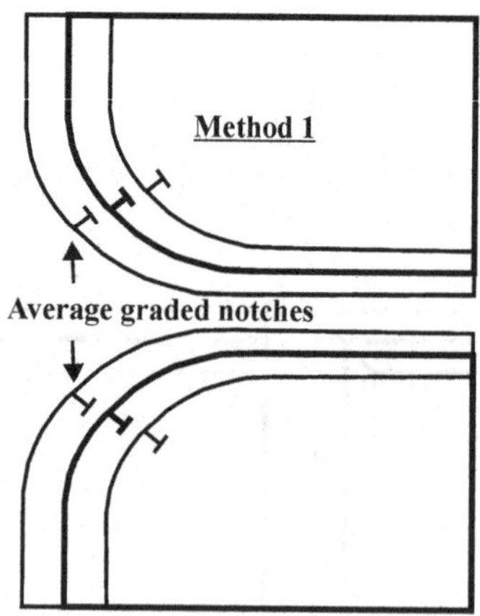

Method 2 fixes the graded notch to the previous grade point so that the distance between them remains the same for all sizes along the perimeter line.

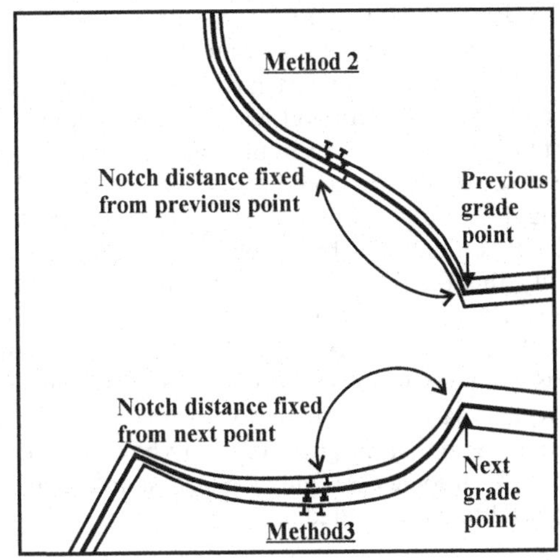

Method 3 fixes the graded notch to the next grade point so that the distance between them remains the same for all sizes along the perimeter line.

Method 4 slides the notch along a curved line a specific amount without affecting the curve of the perimeter line. This can be useful for adjoining seams of differing curves and grades as it maintains the same grade of the matching notches from specific matching reference points on the adjoining seams.

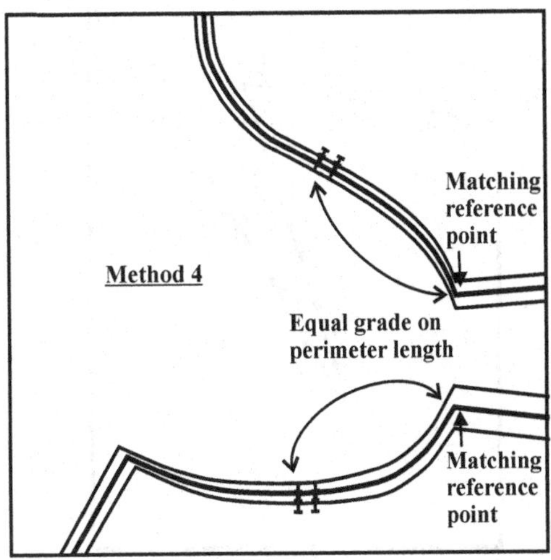

Care has to be taken when grading the sleeve balance notches to maintain the correct amount of easing-in of the crown into the armhole for all the sizes.

GRADING DART SUPPRESSION

When the proportion of the garment has to be altered for a different size, the dart suppression may require grading in both width and length. The angle of the dart may also alter. In Example 1 in Figure 2.9, when the base size dart is closed the seam line A to B is straight. A problem arises after the darts are graded, to maintain the straight line A to B on all sizes. To rectify this the darts of each size have to be temporarily transferred to another position and the line A to B straightened, as in Example 2. After this the darts can be returned to their original position. The corrected dart is compared with the original in Example 3 so that the grade rules can be corrected at C and D. In some cases only the largest and smallest size need to be compared.

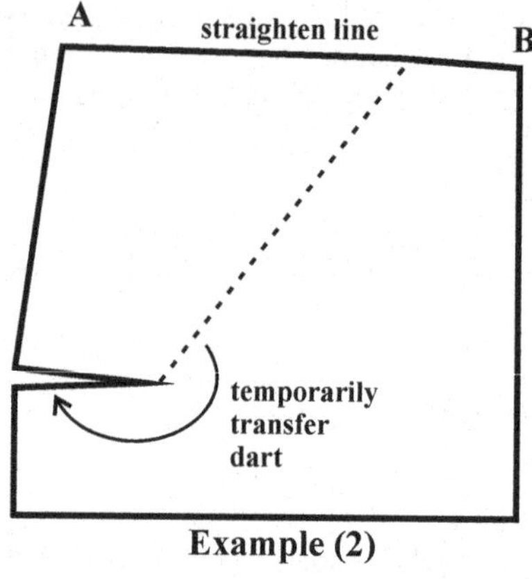

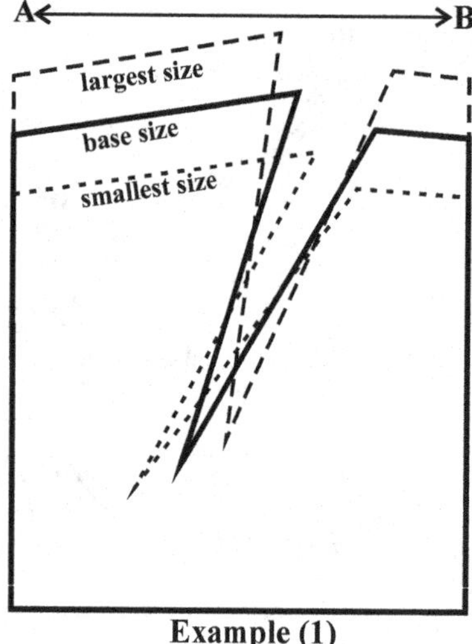

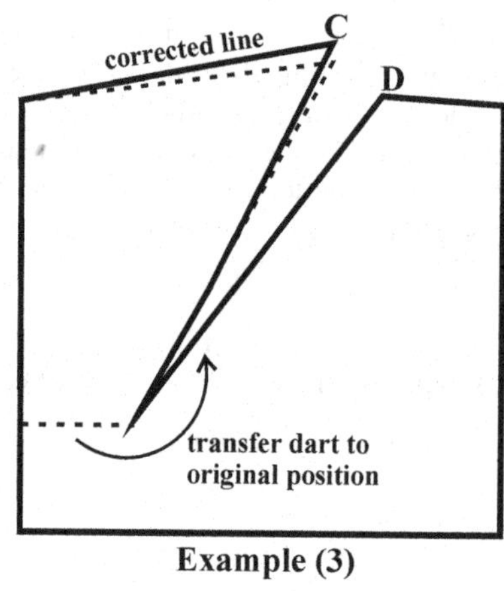

Figure 2.9 Grading dart suppression

The pattern grading details explained in this section are important for maintaining accuracy between the sizes and to ensure that all adjoining seams and notches match. To verify the graded pattern a nest of grades can be displayed on the visual display unit. The grade rules can be checked and edited if revision is required. Some computer systems have the facility to measure the adjoining graded perimeter lines. However, it is the matching of the stitching lines that is important and not always the pattern's perimeter edge. This can be checked more easily by a plotted full scale nest of graded patterns which can be viewed and checked on a light box to ensure that all adjoining seams, stitching lines and notches match and all the grading increments are correct.

The focus of this section has been on the understanding of the importance of grading increments, grade points and the construction of grade rule tables prior to grading. This is preliminary to grading block patterns and styles.

GRADING PRIMARY BLOCK PATTERNS

It is an advantage to store graded block patterns within the computer memory because the grades are automatically transferred to the patterns adapted from these blocks. This eliminates the necessity of grading completed patterns for a style as a separate task. The calculation and positioning of the grading increments suggested in this section are based on the block drafting methods described in Part 1. This should result in both the drafted and graded block patterns being identical.

The three size charts from which the grading increments are calculated are those that were developed in Part 1 Size chart formulation. However, any reliable size chart can be used providing the grading principles are adhered to. Three distributions of grading increments are suggested based on varying girth width for a medium height:

- 4.0 cm increment for bust, waist and hips for sizes 8 to 16 covering five sizes
- 5.0 cm increment for bust, waist and hips for sizes 8 to 16 covering five sizes
- Incremental variation giving proportional change for sizes 8 to 20 covering seven sizes.

The first two ranges for 4.0 cm and 5.0 cm (Part 1, Chart 1.2 and Chart 1.3) retain the same body proportion and balance for all the sizes. This is because the dart suppression is not graded as there is no proportional variation. Therefore this distribution of grading increments should be restricted to generally five sizes because beyond this the garments become disproportionate. The third size range (Part 1, Chart 1.8 and Chart 1.9) is suitable for a larger size range of seven sizes or more due to the dart suppression being graded. This consequently provides for the proportion changes needed for smaller and larger women. It is advisable to have the master or base size pattern approximately the central size in the range. This increases the accuracy of the pattern shape compared with grading from the smallest or largest size. For symmetrical styles only half the pattern need be considered as the computer will mirror the other side or create a pair.

You will notice that the examples of graded patterns have been positioned in the same direction with the warp grain line running horizontally, the top of the pattern to the left and the CB and CF towards the lower edge. If the patterns are lying face side up this will display a right back and left front. This is to enable the same grade rules to be applied to both patterns where identical movement is required. This aids accuracy, reduces the number of grade rules and keeps them simple and concise.

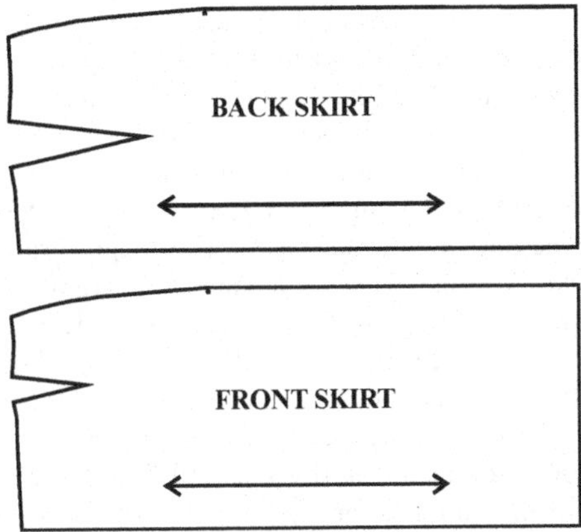

Examples of pattern grading will be given below for:

Straight skirt	Semi-fitted one-piece dress
Fitted bodice	Knitted top
Semi-fitted sleeve	Blouse
Trouser	

When preparing each pattern piece for grading certain decisions have to be made:

- The position and amount of the grading increments within the pattern piece
- The position of the X axis, which is the grading reference line
- The location of the zero point from which all the increase or decrease of the other sizes is calculated
- The identification of all the grade points
- The amount each grade point will move calculated from the zero point
- The formulation of a grade rule table

To explain how these points are implemented a detailed example of grading a straight skirt will be described.

Straight skirt grading that retains the same proportion

Illustrated below the positioning and distribution of the grading increments for a straight skirt and waistband with a 4 cm waist and hip girth grade for sizes 8 to 16, according to Size Chart 1.3 in Part 1. In this style the side seam is central between the CB and CF. Therefore a quarter of the girth grade of 1 cm is distributed to each half of the back and front skirts. The back waist to hip dart is central between the CB and side seam. To maintain this proportion the 1.0 cm increment is divided equally 0.5 cm either side of the dart. The front dart suppression is approximately two-thirds of the waist measurement from the CF as the dart apex finishes at the bone of the iliac crest. Therefore, the grade of 1 cm increment is proportioned 0.7 cm from the CF to dart and 0.3 cm from the dart to the side seam. The length grade is optional. This 0.5 cm is not a change in height but in the contour of hip level. This length grade has to match at the side seam. When the skirt is attached to a bodice the increments at the waist seam have to match. Separate skirts often have a waistband. This example is for a total skirt waistband that has a back opening and notches for the CB, CF and side seams. The grading increments of 1.0 cm are positioned centrally for each quarter of the waistband to match the skirt.

0.5
0.5
BACK SKIRT
0.5
CB

0.5
0.7
FRONT SKIRT
0.3
CF

1.0 1.0 1.0 1.0
CB CF CB
WAIST BAND

LOCATION OF THE GRADING REFERENCE LINE

The **grading reference line (GRL) is the X-axis** and generally located on the warp grain line. Alternatively the CB or CF can be used when it is a mirror line, as in the last example below. This has to be indicated on each pattern piece.

SELECTION OF THE ZERO POINT POSITION

The most usual positioning of the **zero points** on skirt patterns is at the centre front or back, at **(A)** the waist level, **(B)** the hip level or **(C)** the hem level, as illustrated below. In this example of skirt grading the CBs are stationary as these lines are often a mirror line when the left and right sides of the skirt are cut as one piece.

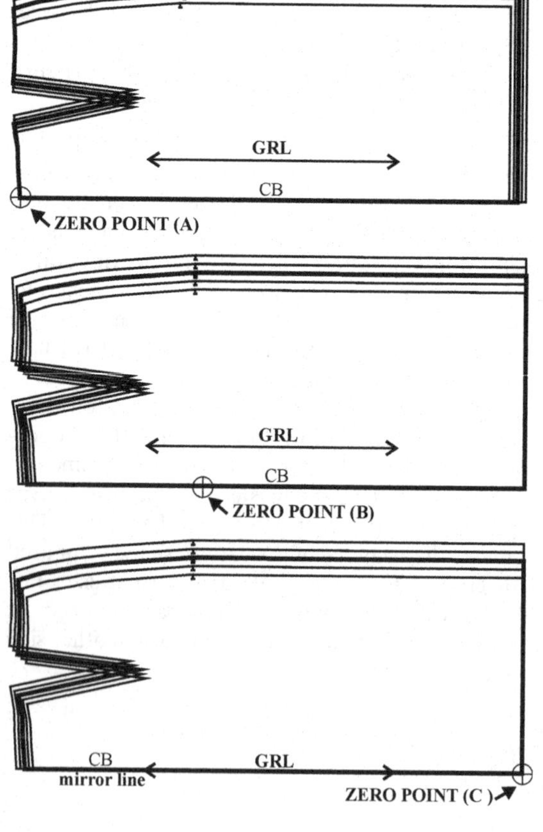

POSITION AND MOVEMENT OF THE SKIRT GRADE POINTS

Illustrated here are the positions and movement of the **grade points when the zero point is at position (A)**. The increments quoted are for the increase of a size 12 to size 14 for a skirt back, front and waistband for a 4.0 cm waist and hip girth grade.

CALCULATION OF THE SKIRT GRADE RULES

The **grade rules have to be calculated from the zero point** by adding up all the sectional increments. For example *Rule 3* is calculated by adding the CB waist to dart increment of 0.5 cm to the dart to side seam increment of 0.5 cm, which equals 1.0 cm. It will be noticed that most of the grade rules are common to both the back and front skirts where the pattern movement is identical. It is only the movement of the back dart *Rule 2* and the front dart *Rule 7* which have to be different to maintain the same proportion for each size.

The waistband is graded from the CB the full waist increment of 4 cm, divided into 1 cm for each quarter of the skirt.

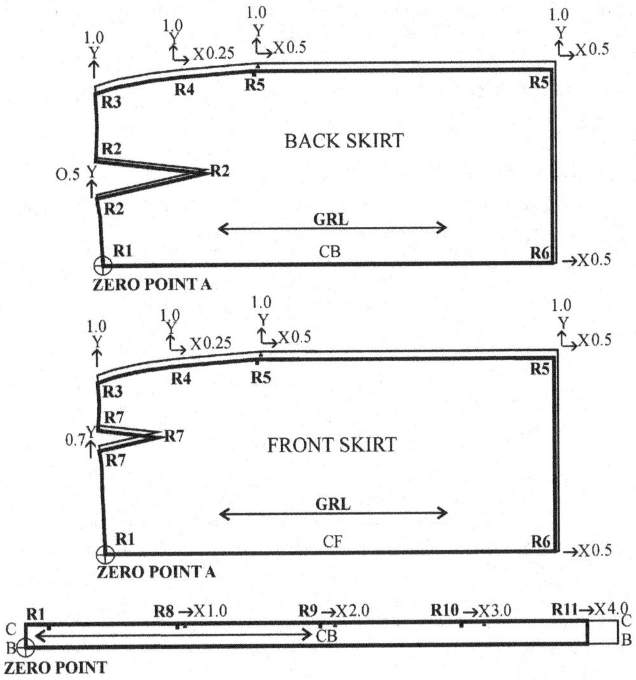

SKIRT GRADE RULE TABLES FOR 4 cm GIRTH GRADE

The **grade rules are input into the computer as a grade rule table**. This is generally prior to pattern digitising, although, the grade points can be edited later if the pattern is already within the computer system. Table 2.1 is an example of a grade rule table using the method of recording the incremental growth from the smallest to the largest size.

Table 2.1 Grade rule table for a straight skirt 4.0 cm girth grade for sizes 8 to 16

RULE TABLE: Skirt 4.0 cm — RULE METHOD: Incremental Small to Large
SIZE RANGE: 8–16 — SIZE NAMES: Numeric
SAMPLE SIZE: 12 — NOTION: Metric

RULE NUMBER:	1		2		3		4		5		6	
COMMENT:	zero		back dart		side waist		side upper hip		side hip/hem		CB CF hem	
POINT ATTRIBUTE:	N		N		N		N		N		N	
BREAKS	X	Y	X	Y	X	Y	X	Y	X	Y	X	Y
8 to 16	0.0	0.0	0.0	0.5	0.0	1.0	0.25	1.0	0.5	1.0	0.5	0.0
RULE NUMBER:	7		8		9		10		11			
COMMENT:	fr dart		waist band rt side		waist band CF		waist band left side		waist band CB			
POINT ATTRIBUTE	N		N		N		N		N			
BREAKS	X	Y	X	Y	X	Y	X	Y	X	Y		
8 to 16	0.0	0.7	1.0	0.0	2.0	0.0	3.0	0.0	4.0	0.0		

NB POINT ATTRIBUTE N means no smoothing at the grade point

Table 2.2 Grade rule table for a straight skirt 5.0 cm girth grade for sizes 8 to 16

RULE TABLE: Skirt 5.0 cm		RULE METHOD: Incremental Small to Large
SIZE RANGE: 8–16		SIZE NAMES: Numeric
SAMPLE SIZE: 12		NOTION: Metric

RULE NUMBER:	1		2		3		4		5		6	
COMMENT:	zero		back dart		side waist		side upper hip		side hip/hem		CB CF hem	
POINT ATTRIBUTE:	N		N		N		N		N		N	
BREAKS	X	Y	X	Y	X	Y	X	Y	X	Y	X	Y
8 to 16	0.0	0.0	0.0	0.6	0.0	1.25	0.25	1.25	0.5	1.25	0.5	0.0

RULE NUMBER:	7		8		9		10		11	
COMMENT:	fr dart		waist band rt side		waist band CF		waist band left side		waist band CB	
POINT ATTRIBUTE	N		N		N		N		N	
BREAKS	X	Y	X	Y	X	Y	X	Y	X	Y
8 to 16	0.0	0.8	1.25	0.0	2.5	0.0	3.75	0.0	5.0	0.0

NB POINT ATTRIBUTE N means no smoothing at the grade point

SKIRT GRADE RULE TABLE FOR 5 cm GIRTH GRADE

There is no need to grade a new set of patterns if there is only a difference in the increments between the sizes. For example, in Table 2.1 for 4 cm and Table 2.2 for 5 cm increments in the major girth measurements, the rule numbers apply to the same grade point movement. Only the amount of movement varies. By editing the rule table name *Skirt 4 cm* and replacing it by *Skirt 5 cm* the same skirt patterns can be used for both size ranges.

Straight skirt grading that changes the proportion

Larger size ranges extending above five sizes require a proportional change that accommodates the variation in the body shape. These changes occur in the flesh areas. Not only do they expand outwards but also can become lower. This change in proportion has to be considered three-dimensionally by changing the grading increments between the sizes and grading the amount of suppression. For the larger sizes above size 12 there is a need to increase the ease allowance. This is illustrated by the grading of the straight skirt seven sizes from size 8 to 20 according to the Size Chart 1.11 in Part 1. In this example the full waist size increase is

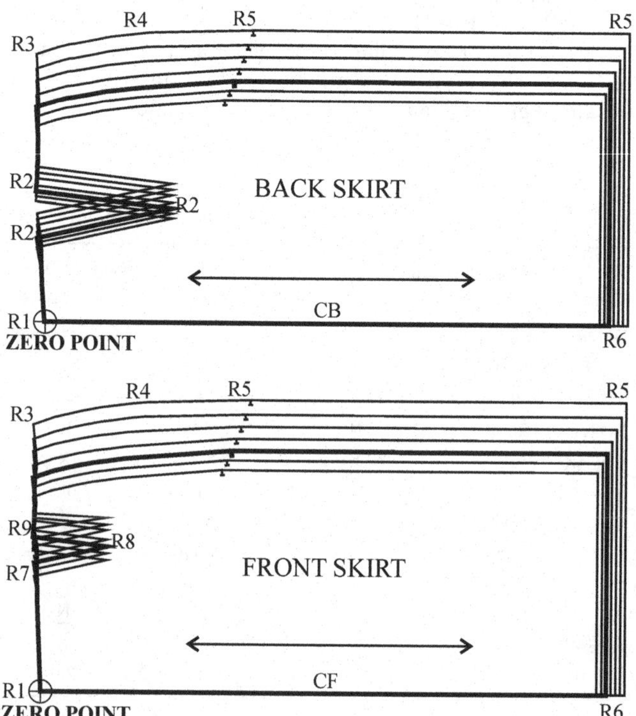

4.0 cm below size 12, 5.0 cm for size 14 and 6.0 cm for size 16 and above. For size 16 and above the front waist increases more than the back by grading a reduction in the front dart suppression. The full grade for the hip and hem girths are 4.0 cm below size 12 and 5.0 cm above size 12. The length grade of 0.5 cm is optional as it is not for change in height but in contour of hip level. Table 2.3 shows an example of a grade rule table using the method of recording the incremental growth from the smallest to the largest size. (The other methods of recording the growth can also be applied as explained in the earlier section 'Methods of recording incremental growth'). The waistband is graded from the CB the full waist increment, divided according to each quarter of the skirt grade.

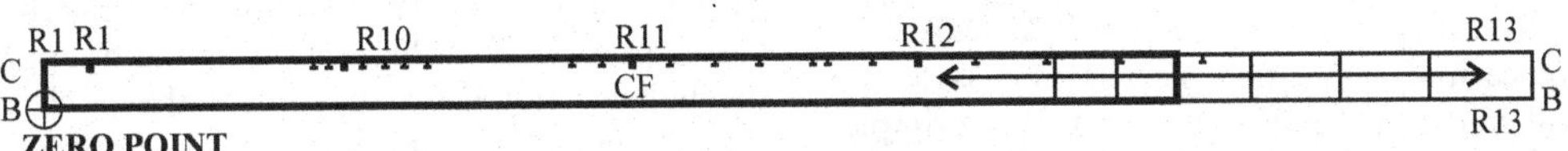

Table 2.3 Grade rule table for a straight skirt with a proportional grade for sizes 8 to 20

RULE TABLE: SKIRT PROPORTIONAL
SIZE RANGE: 8–20
SAMPLE SIZE: 12

RULE METHOD: Incremental Small to Large
SIZE NAMES: Numeric
NOTION: Metric

RULE NUMBER:	1		2		3		4		5	
COMMENT:	zero		back dart		side waist		side upper hip		side hip/hem	
POINT ATTRIBUTE:	N		N		N		N		N	
BREAKS	X	Y	X	Y	X	Y	X	Y	X	Y
8 to 10	0.0	0.0	0.0	0.5	0.0	1.0	0.25	1.0	0.5	1.0
10 to 12	0.0	0.0	0.0	0.5	0.0	1.0	0.25	1.0	0.5	1.0
12 to 14	0.0	0.0	0.0	0.6	0.0	1.25	0.25	1.25	0.5	1.25
14 to 16	0.0	0.0	0.0	0.75	0.0	1.5	0.25	1.5	0.5	1.5
16 to 18	0.0	0.0	0.0	0.6	0.0	1.25	0.25	1.5	0.5	1.25
18 to 20	0.0	0.0	0.0	0.6	0.0	1.5	0.25	1.5	0.5	1.5

RULE NUMBER:	6		7		8		9		10	
COMMENT:	CB CF hem		CF to fr dart		fr dart apex		fr dart side		waist band right side	
POINT ATTRIBUTE:	N		N		N		N		N	
BREAKS	X	Y	X	Y	X	Y	X	Y	X	Y
8 to 10	0.5	0.0	0.0	0.7	0.0	0.7	0.0	0.7	1.0	0.0
10 to 12	0.5	0.0	0.0	0.7	0.0	0.7	0.0	0.7	1.0	0.0
12 to 14	0.5	0.0	0.0	0.7	0.0	0.7	0.0	0.7	1.25	0.0
14 to 16	0.5	0.0	0.0	1.0	0.0	1.0	0.0	1.0	1.5	0.0
16 to 18	0.5	0.0	0.0	0.8	0.0	0.55	0.0	0.3	1.25	0.0
18 to 20	0.5	0.0	0.0	1.0	0.0	0.75	0.0	0.5	1.5	0.0

RULE NUMBER:	11		12		13	
COMMENT:	CF waistband		waist band left side		waist band left CB	
POINT ATTRIBUTE:	N		N		N	
BREAKS	X	Y	X	Y	X	Y
8 to 10	2.0	0.0	3.0	0.0	4.0	0.0
10 to 12	2.0	0.0	3.0	0.0	4.0	0.0
12 to 14	2.5	0.0	3.75	0.0	5.0	0.0
14 to 16	3.0	0.0	4.75	0.0	6.0	0.0
16 to 18	3.0	0.0	4.75	0.0	6.0	0.0
18 to 20	3.5	0.0	5.5	0.0	7.0	0.0

NB POINT ATTRIBUTE N means no smoothing at the grade point

Fitted bodice and semi-fitted sleeve grading that retains the same proportion

The bodice and sleeve grade are illustrated together as there is a relationship between the grading increments of the armhole and sleeve head. Where the bodice is attached to a skirt the grade of the waist seams must be equal. Likewise the grade of the bodice neckline will influence that of an attached collar. Illustrated below are the positioning and distribution of the grading increments for the back and front bodice and semi-fitted sleeve. This is for a 4.0 cm full grade for the bust and waist. It is a simple grade with no proportional change and so is limited to five sizes, 8 to 16.

The full 1.0 cm neck grade is divided between the half back neck of 0.2 cm and the larger half front neck of 0.3 cm, giving the total of 0.5 cm for the half neck. Both the width and rise of the front neck are graded 0.2 cm to retain an equal curve to the neck edge for each size, whereas only the width of the back neck is graded 0.2 cm. This conforms to the proportional calculation for the neck construction of the bodice in Part 1. This proportion was $\frac{1}{5}$ of the bodice neck; therefore $\frac{1}{5}$ of the 1.0 cm full neck grade is 0.2 cm. Grading increments can be calculated from any reliable drafting system when the sectional increment is not found in the size chart.

None of the dart suppression is graded, only the length of the front dart changes. The position of the darts is determined by the bust and shoulder blade prominence. There is only a slight increase in width of 0.2 cm from the centre back and centre front to these prominences. The greater part of the bust and waist grading takes place three-dimensionally at the side seam. There is a slight increase in length from the shoulder to the depth of armhole of 0.5 cm. This is to accommodate the change in the bust girth, not the height of the wearer. The underarm seams of both the bodice and sleeve remain the same length for all the sizes. Only the depth of the sleeve head increases 0.5 cm for each size to match the armhole.

The sleeve head depth is calculated as one-third of the armhole circumference according to the sleeve draft on p. 32. The grade per size of this bodice armhole is 1.5 cm and therefore the depth of sleeve head grade is 0.5 cm. Where the shape of the armhole varies greatly due to the style it is worth measuring the grade of the full armhole for each size. The full wrist grade is only 0.5 cm as this is a comparatively bony area when compared with the fleshy upper arm with a grade of 1.0 cm. The grading of notches discussed previously in this section should be applied to the balance notches that control the pitch of the sleeve head when attached to the armhole. The example illustrated retains the same distance from the underarm seam to the balance notches for all the sizes. This allows any discrepancies in the graded measurements to be accommodated in the 'easing in' of the sleeve head.

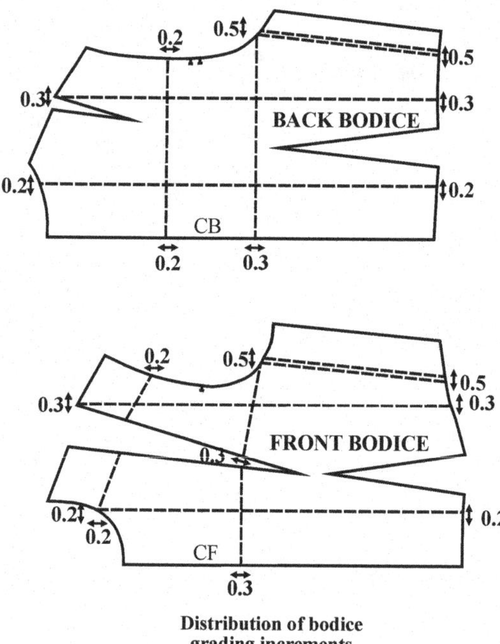

Distribution of bodice grading increments

POSITION AND MOVEMENT OF THE BODICE AND SLEEVE GRADE POINTS

The increments quoted are for the increase to a size 14 from a size 12. Each movement is calculated from the zero point positioned at the neck of the bodice and the top of the sleeve head. The grading reference line is the grain line. The sleeve reference line is positioned on the centre grain line. The shoulder *Grade rule 3* has a slight grade in the X direction of 0.15 cm. This may seem small but it helps to retain a straight shoulder seam for each size when the dart is closed.

BODICE AND SEMI-FIT SLEEVE GRADE RULE TABLES

The two grade rule tables in Tables 2.4 and 2.5 are for the grading method of a small to large incremental. Table 2.4 is for a full bust and waist grade of 4 cm and Table 2.5 for 5 cm. It will be noticed that most grade rules are common to the back and front bodice when they are graded positioned in the same direction. This assists accuracy and reduces the number of rules. Working from very large rule tables can become very confusing. The sleeve with the central grade reference line has the front sleeve graded in the positive Y-axis and the back sleeve in the negative Y-axis. The straight sleeve block can be graded in a similar way except the wrist can be graded the same width as the upper arm.

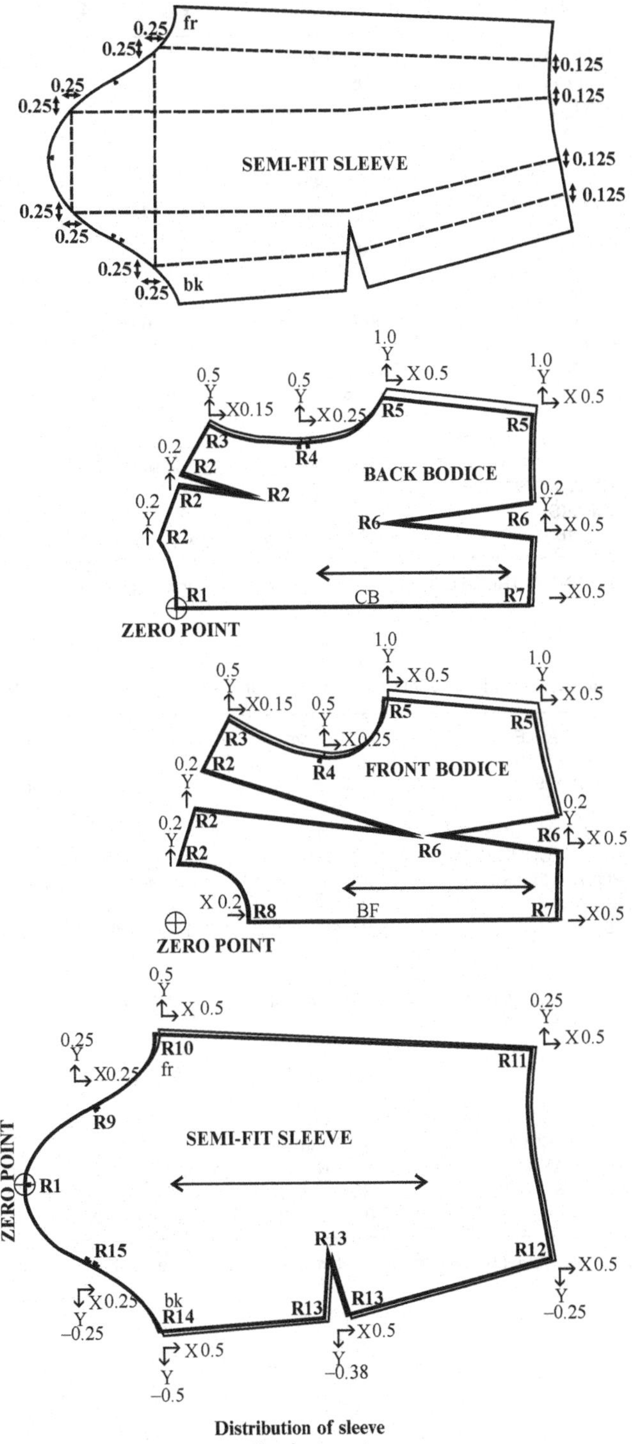

Distribution of sleeve grading increments

Table 2.4 Grade rule table for a bodice and sleeve with a 4 cm girth grade for sizes 8 to 16

RULE TABLE:	BODICE 4 cm	RULE METHOD:	Incremental Small to Large
SIZE RANGE:	8–16	SIZE NAMES:	Numeric
SAMPLE SIZE:	12	NOTION:	Metric

RULE NUMBER	COMMENT	POINT ATTRIBUTE	BREAKS	X	Y
1	zero	N	8 to 16	0.0	0.0
2	neck width	N	8 to 16	0.0	0.2
3	shoulder width	N	8 to 16	0.15	0.5
4	Xback Xfront	N	8 to 16	0.25	0.5
5	bust waist width	N	8 to 16	0.5	1.0
6	Fr/Bk dart width	N	8 to 16	0.5	0.2
7	CF CB length	N	8 to 16	0.5	0.0
8	CF neck	N	8 to 16	0.2	0.0
9	Fr slv crown width	N	8 to 16	0.25	0.25
10	Fr upper arm width	N	8 to 16	0.5	0.5
11	Fr slv wrist	N	8 to 16	0.5	0.25
12	Bk slv wrist	N	8 to 16	0.5	−0.25
13	Bk elbow	N	8 to 16	0.5	−0.38
14	Bk slv width	N	8 to 16	0.5	−0.5
15	Bk slv crown	N	8 to 16	0.25	−0.25

NB POINT ATTRIBUTE N means no smoothing at the grade point.

Table 2.5 Grade rule table for a bodice and sleeve with a 5 cm girth grade for sizes 8 to 16

RULE TABLE:	BODICE 5 cm	RULE METHOD:	Incremental Small to Large
SIZE RANGE:	8–16	SIZE NAMES:	Numeric
SAMPLE SIZE:	12	NOTION:	Metric

RULE NUMBER	COMMENT	POINT ATTRIBUTE	BREAKS	X	Y
1	zero	N	8 to 16	0.0	0.0
2	neck width	N	8 to 16	0.0	0.2
3	shoulder width	N	8 to 16	0.15	0.6
4	Xback Xfront	N	8 to 16	0.25	0.6
5	bust waist width	N	8 to 16	0.5	1.25
6	Fr Bk dart width	N	8 to 16	0.5	0.2
7	CF CB length	N	8 to 16	0.5	0.0
8	CF neck	N	8 to 16	0.2	0.0
9	Fr slv crown width	N	8 to 16	0.25	0.4
10	Fr upper arm width	N	8 to 16	0.5	0.8
11	Fr slv wrist	N	8 to 16	0.5	0.25
12	Bk slv wrist	N	8 to 16	0.5	−0.25
13	Bk elbow	N	8 to 16	0.5	−0.4
14	Bk slv width	N	8 to 16	0.5	−0.8
15	Bk slv crown	N	8 to 16	0.25	−0.4

NB POINT ATTRIBUTE N means no smoothing at the grade point.

Fitted bodice and semi-fitted sleeve grading that changes the proportion

For a larger size range of seven sizes 8 to 20 there has to be a more complex proportional grade. The grade rule table in Table 2.6 for a bodice and Table 2.7 for a sleeve for sizes 8 to 20 are based on the Size Chart 1.10 in Part 1. The bodice waist grade has to match the skirt waist grade. However, the bust grade is 5.0 cm from size 8 to 14, then 7.0 cm to size 16 as 2.0 cm extra ease allowance is added. The increase is then 6.0 cm for sizes 18 and 20. There is also a greater increase in length from the front neck point to bust and waist for sizes 14 to 20. To accommodate this increase in front bust and waist girths the dart suppression has to be graded. There has to be an adjustment in the X-axis at the top of the shoulder dart so that the shoulder finishes in a straight line when the dart is closed. There is also a greater increase in the upper arm girth and depth of armhole for size 14 to 20.

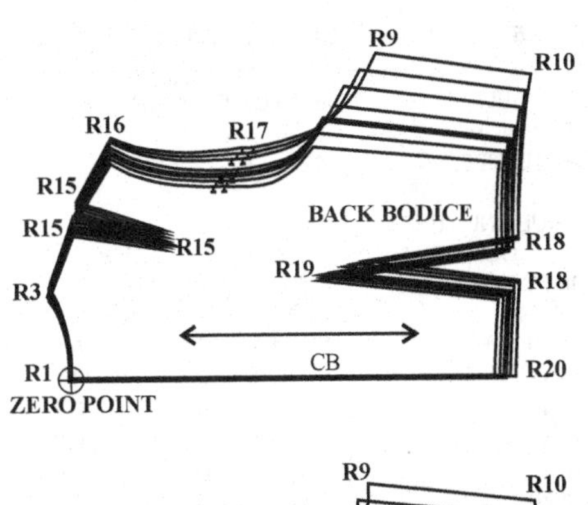

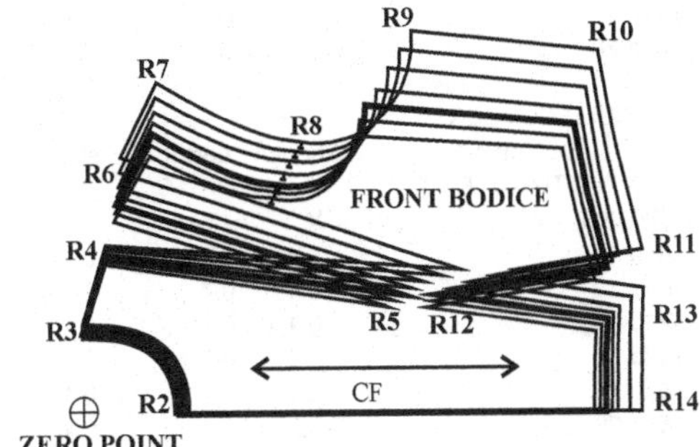

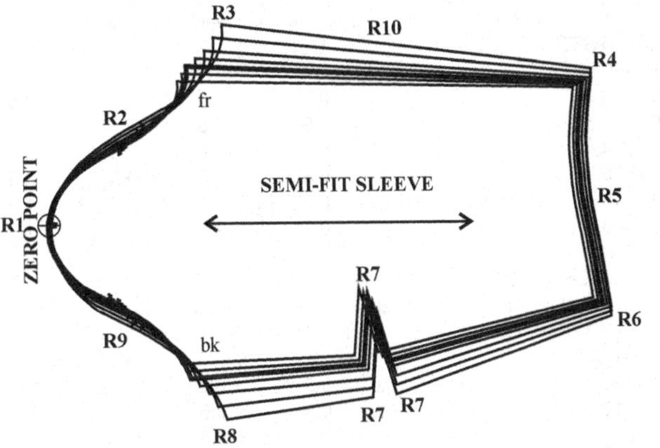

Grade rules applied to a nest of graded bodice and semi-fit sleeve blocks for sizes 8 to 20

Table 2.6 Grade rule table for a bodice with a proportional grade for sizes 8 to 20

RULE TABLE:	BODICE PROPORTIONAL	RULE METHOD:	Incremental Small to Large
SIZE RANGE:	8–20	SIZE NAMES:	Numeric
SAMPLE SIZE:	12	NOTION:	Metric

RULE NUMBER:	1		2		3		4		5	
COMMENT:	zero		CF neck		neck point		fr shoulder dart 1		fr shoulder dart apex	
POINT ATTRIBUTE:	N		N		N		N		N	
BREAKS	X	Y	X	Y	X	Y	X	Y	X	Y
8 to 10	0.0	0.0	0.2	0.0	0.0	0.2	0.0	0.3	0.5	0.5
10 to 12	0.0	0.0	0.2	0.0	0.0	0.2	0.0	0.3	0.5	0.5
12 to 14	0.0	0.0	0.2	0.0	0.0	0.2	0.0	0.3	1.0	0.5
14 to 16	0.0	0.0	0.2	0.0	0.0	0.2	0.0	0.3	1.0	0.5
16 to 18	0.0	0.0	0.2	0.0	0.0	0.2	−0.1	0.2	1.5	0.5
18 to 20	0.0	0.0	0.2	0.0	0.0	0.2	−0.1	0.2	1.5	0.5

RULE NUMBER:	6		7		8		9		10	
COMMENT:	fr shoulder dart 2		shoulder width		across front		bust width		waist width	
POINT ATTRIBUTE:	N		N		N		N		N	
BREAKS	X	Y	X	Y	X	Y	X	Y	X	Y
8 to 10	0.1	0.7	0.1	0.9	0.25	0.5	0.5	1.25	0.5	1.0
10 to 12	0.1	0.7	0.1	0.9	0.25	0.5	0.5	1.25	0.5	1.0
12 to 14	0.1	0.7	0.1	0.9	0.5	1.0	1.0	1.25	0.5	1.25
14 to 16	0.1	0.7	0.1	0.9	0.5	1.0	1.0	1.75	0.5	1.5
16 to 18	−0.3	1.2	0.1	1.2	0.5	1.0	1.0	1.5	0.5	1.25
18 to 20	−0.3	1.2	0.1	1.2	0.5	1.0	1.0	1.5	0.5	1.5

RULE NUMBER:	11		12		13		14		15	
COMMENT:	waist dart side		waist dart apex		waist dart CF		CF waist		bk shoulder dart	
POINT ATTRIBUTE:	N		N		N		N		N	
BREAKS	X	Y	X	Y	X	Y	X	Y	X	Y
8 to 10	0.5	0.5	0.5	0.5	0.5	0.5	0.5	0.0	0.0	0.3
10 to 12	0.5	0.5	0.5	0.5	0.5	0.5	0.5	0.0	0.0	0.3
12 to 14	0.5	0.5	1.0	0.5	0.5	0.5	0.5	0.0	0.0	0.3
14 to 16	0.5	0.5	1.5	0.5	0.5	0.5	0.5	0.0	0.0	0.3
16 to 18	1.0	0.25	2.0	0.5	1.0	0.75	1.0	0.0	0.0	0.3
18 to 20	1.0	0.25	2.0	0.5	1.0	0.75	1.0	0.0	0.0	0.3

RULE NUMBER:	16		17		18		19		20	
COMMENT:	back shoulder		across back		back waist dart		bk waist dart apex		CB waist	
POINT ATTRIBUTE:	N		N		N		N		N	
BREAKS	X	Y	X	Y	X	Y	X	Y	X	Y
8 to 10	0.0	0.5	0.25	0.5	0.5	0.25	0.5	0.25	0.5	0.0
10 to 12	0.0	0.5	0.5	0.5	0.25	0.5	0.25	0.5	0.0	0.0
12 to 14	0.0	0.5	0.5	0.5	0.5	0.25	0.5	0.25	0.5	0.0
14 to 16	0.0	0.5	0.5	1.0	0.0	0.5	1.0	0.5	0.0	0.0
16 to 18	0.0	0.2	0.5	0.5	0.5	0.25	1.0	0.25	0.5	0.0
18 to 20	0.0	0.2	0.5	0.5	0.0	0.25	1.0	0.25	0.0	0.0

NB POINT ATTRIBUTE N means no smoothing at the grade point.

Table 2.7 Grade rule table for a sleeve with a proportional grade for sizes 8 to 20

RULE TABLE:	BODICE PROPORTIONAL	RULE METHOD:	Incremental Small to Large
SIZE RANGE:	8–20	SIZE NAMES:	Numeric
SAMPLE SIZE:	12	NOTION:	Metric

RULE NUMBER:	1		2		3		4		5	
COMMENT:	zero		fr sleeve head		fr upper arm		fr waist		mid wrist	
POINT ATTRIBUTE:	N		N		N		N		N	
BREAKS	X	Y	X	Y	X	Y	X	Y	X	Y
8 to 10	0.0	0.0	0.25	0.4	0.5	0.8	0.5	0.25	0.5	0.0
10 to 12	0.0	0.0	0.25	0.4	0.5	0.8	0.5	0.25	0.5	0.0
12 to 14	0.0	0.0	0.5	0.4	1.0	0.8	0.5	0.25	0.5	0.0
14 to 16	0.0	0.0	0.5	0.4	1.0	0.8	0.5	0.25	0.5	0.0
16 to 18	0.0	0.0	0.5	0.7	1.0	1.4	0.0	0.5	0.0	0.0
18 to 20	0.0	0.0	0.5	0.7	1.0	1.4	0.0	0.5	0.0	0.0

RULE NUMBER:	6		7		8		9		10	
COMMENT:	bk wrist		bk elbow		bk upper arm		bk sleeve head		fr elbow	
POINT ATTRIBUTE:	N		N		N		N		N	
BREAKS	X	Y	X	Y	X	Y	X	Y	X	Y
8 to 10	0.5	−0.25	0.5	−0.5	0.5	−0.8	0.25	−0.4	0.5	0.6
10 to 12	0.5	−0.25	0.5	−0.5	0.5	−0.8	0.25	−0.4	0.5	0.6
12 to 14	0.5	−0.25	0.5	−0.5	1.0	−0.8	0.5	−0.4	0.5	0.6
14 to 16	0.5	−0.25	0.5	−1.0	1.0	−0.8	0.5	−0.4	0.5	0.6
16 to 18	0.0	−0.5	0.0	−1.0	1.0	−1.4	0.5	−0.7	0.0	1.1
18 to 20	0.0	−0.5	0.0	−1.0	1.0	−1.4	0.5	−0.7	0.0	1.1

NB POINT ATTRIBUTE N means no smoothing at the grade point.

Trouser grading that retains the same proportion

Illustrated in Figure 2.10 are the positioning and distribution of the grading increments for five sizes 8 to 16 with equal increments between the sizes. The increments quoted are from Size Chart 1.3 in Part 1 with 4 cm at the waist and hips 5 cm (in parenthesis). There is a slight increase in length of 0.5 cm between the waist and crutch which is not for an increase in height but variation in girths. The girths' sectional increments are calculated as a proportion of the trouser hip girth. The following are based on the trouser draft in Part 1. From the CB and CF to the crease line is $\frac{1}{10}$ of the hip girth. Therefore $\frac{1}{10}$ of the 4.0 cm grade equals 0.4 cm. The width of the back crutch fork is also $\frac{1}{10}$. The front crutch width is only $\frac{1}{20}$ and therefore 0.2 cm. Half the back and front waist and hip widths are based on $\frac{1}{4}$ of these girths and therefore grade 1.0. The crease line to side seam increment is calculated by subtracting the CB or CF to crease line from the $\frac{1}{4}$ waist or hip grade, 1.0 cm minus 0.4 cm equals 0.6 cm. This is halved again to centralise the side dart. The knee sectional increment of 0.5 cm is based on $\frac{1}{4}$ of the 2.0 cm girth grade. For straight trousers this remains the same at the ankle. Figure 2.11 illustrates the grade point movement for a 4.0 cm full girth grade. The zero point is positioned at the junction of the crease line and crutch level.

TROUSER GRADE RULE TABLES

The grade rules for a 4 cm waist and hip girth are shown in Table 2.8. This can be substituted for a 5 cm girth grade quoted in grade rule Table 2.9, providing that the same movement and direction of the grade rule numbers are used. Both these tables are for small ranges of five sizes 8 to 16 that retain the same proportion for each size.

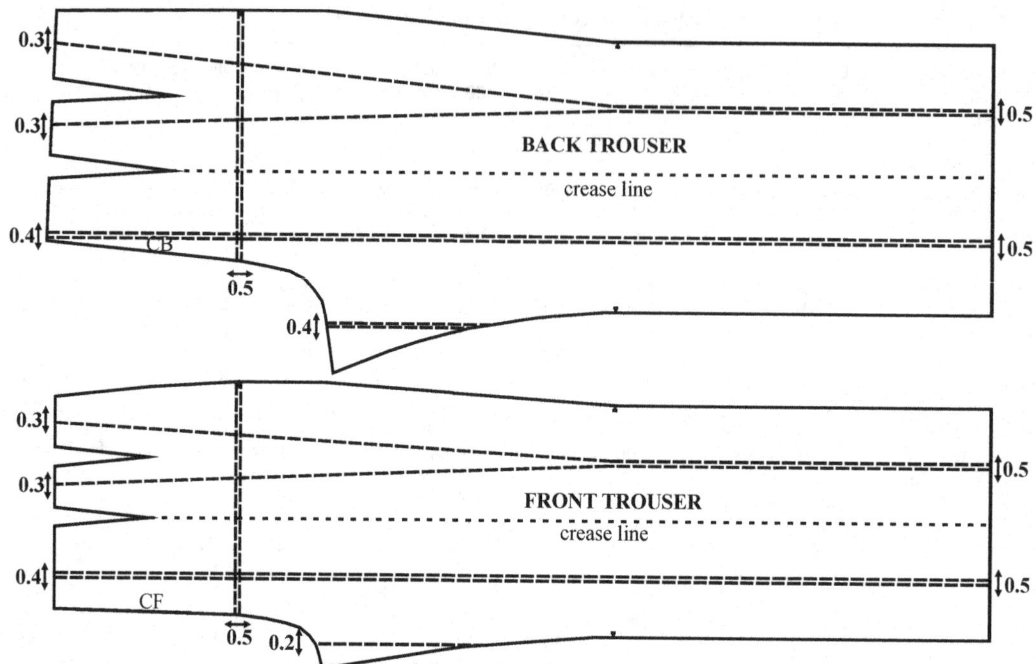

Figure 2.10 Distribution of trouser grading increments for 4 cm grade

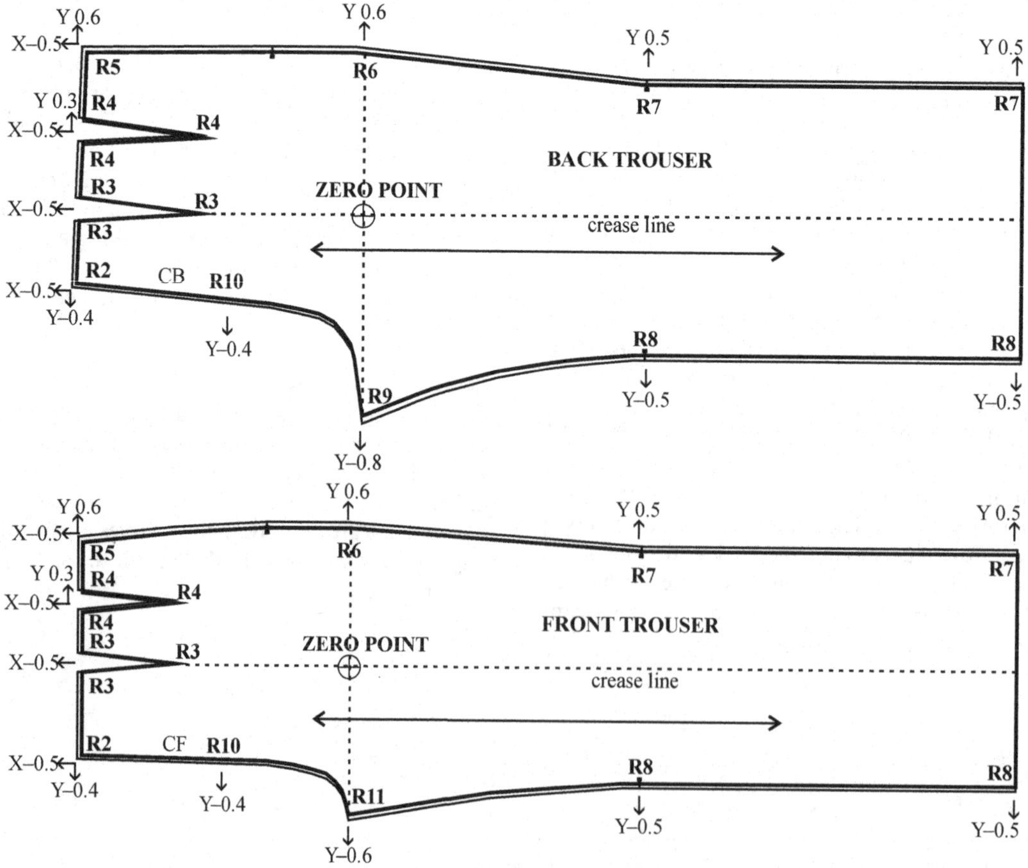

Y 0.6
X–0.5
Y 0.6
Y 0.5
Y 0.5
R5
R6
Y 0.3
R4
X–0.5
R4
R7
R7
R4
BACK TROUSER
R3
ZERO POINT
X–0.5
R3
R3
crease line
R2
CB
R10
X–0.5
Y–0.4
Y–0.4
R8
R8
R9
Y–0.5
Y–0.5
Y–0.8
Y 0.6
Y 0.6
Y 0.5
Y 0.5
X–0.5
R5
R6
Y 0.3
R4
R7
R7
X–0.5
R4
R4
FRONT TROUSER
R3
ZERO POINT
R3
X–0.5
R3
crease line
R2
CF
R10
X–0.5
R8
R8
Y–0.4
Y–0.4
R11
Y–0.5
Y–0.5
Y–0.6

Figure 2.11 Grade point movement and grade rules for a trouser with a 4 cm grade for sizes 12 to 14

Table 2.8 Grade rule table for a trouser with a 4.0 cm girth grade for sizes 8 to 16

RULE TABLE:	TROUSER 4 cm	RULE METHOD:	Incremental Small to Large
SIZE RANGE:	8–16	SIZE NAMES:	Numeric
SAMPLE SIZE:	12	NOTION:	Metric

RULE NUMBER:	1		2		3		4		5	
COMMENT:	zero		CB CF waist		centre waist dart		side waist dart		side waist	
POINT ATTRIBUTE:	N		N		N		N		N	
BREAKS	X	Y	X	Y	X	Y	X	Y	X	Y
8 to 16	0.0	0.0	−0.5	−0.4	−0.5	0.0	−0.5	0.3	−0.5	0.6

RULE NUMBER:	6		7		8		9		10	
COMMENT:	side hip		side knee/hem		inside knee/ankle		back crutch		centre hip	
POINT ATTRIBUTE:	N		N		N		N		N	
BREAKS	X	Y	X	Y	X	Y	X	Y	X	Y
8 to 16	0.0	0.6	0.0	0.5	0.0	−0.5	0.0	−0.8	0.0	−0.4

RULE NUMBER:	11	
COMMENT:	front crutch	
POINT ATTRIBUTE:	N	
BREAKS	X	Y
8 to 16	0.0	−0.6

NB POINT ATTRIBUTE N means no smoothing at the grade point.

Table 2.9 Grade rule table for a trouser with a 5.0 cm girth grade for sizes 8 to 16

RULE TABLE:	TROUSER 5 cm	RULE METHOD:	Incremental Small to Large
SIZE RANGE:	8–16	SIZE NAMES:	Numeric
SAMPLE SIZE:	12	NOTION:	Metric

RULE NUMBER:	1		2		3		4		5	
COMMENT:	zero		CB CF waist		centre waist dart		side waist dart		side waist	
POINT ATTRIBUTE:	N		N		N		N		N	
BREAKS	X	Y	X	Y	X	Y	X	Y	X	Y
8 to 16	0.0	0.0	−0.5	−0.5	−0.5	0.0	−0.5	0.35	−0.5	0.75

RULE NUMBER:	6		7		8		9		10	
COMMENT:	side hip		side knee/hem		inside knee/ankle		back crutch		centre hip	
POINT ATTRIBUTE:	N		N		N		N		N	
BREAKS	X	Y	X	Y	X	Y	X	Y	X	Y
8 to 16	0.0	0.75	0.0	0.5	0.0	−0.5	0.0	−1.0	0.0	−0.5

RULE NUMBER:	11	
COMMENT:	front crutch	
POINT ATTRIBUTE:	N	
BREAKS	X	Y
8 to 16	0.0	−0.75

NB POINT ATTRIBUTE N means no smoothing at the grade point.

Trouser grading that changes the proportion

A proportional trouser grade for an extended size range of seven sizes 8 to 20 can also be calculated from Size Chart 1.11 in Part 1 and the method of drafting trouser blocks. Figure 2.12 illustrates a nest of grades for this range including the grade rule numbers given in Table 2.10.

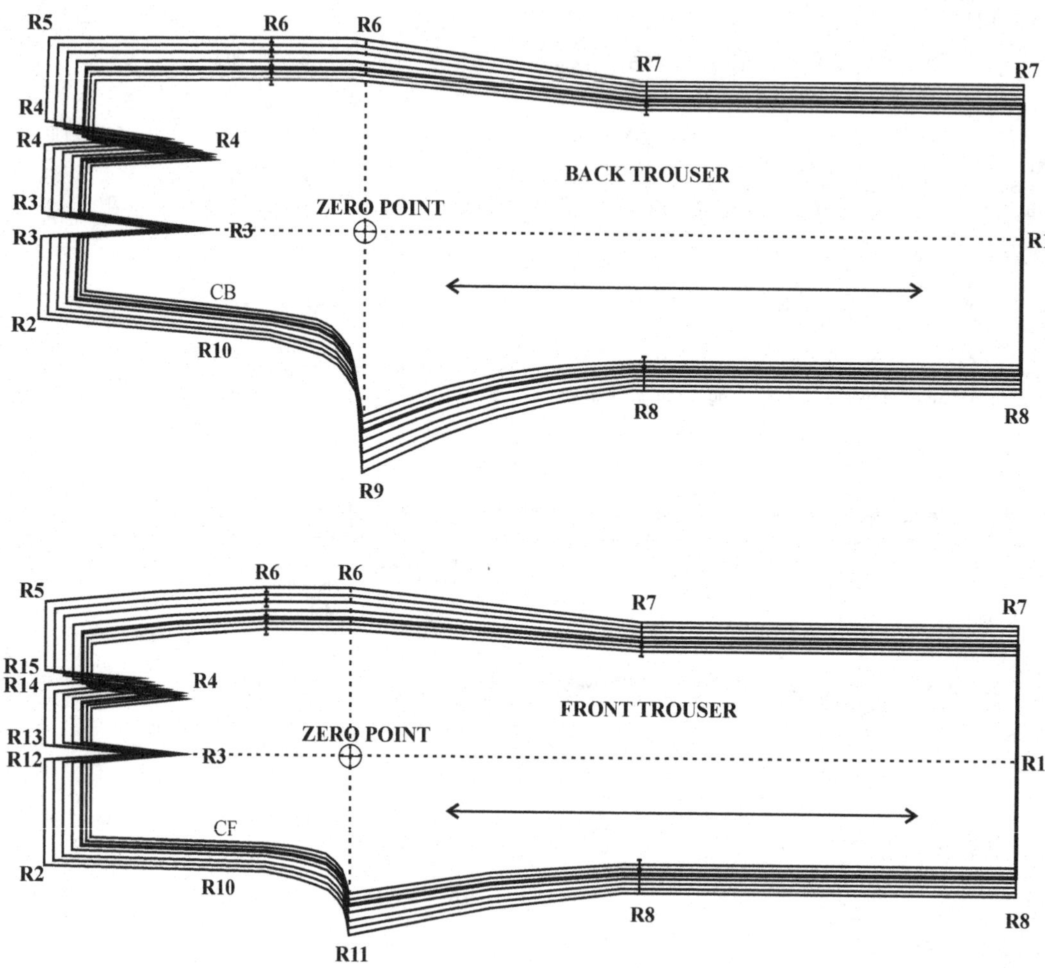

Figure 2.12 The graded point movement and grade rules for a trouser with a proportional grade for sizes 8 to 20

Table 2.10 Grade rule table for a trouser with a proportional grade for sizes 8 to 20

RULE TABLE:	TROUSER PROPORTIONAL	RULE METHOD:	Incremental Small to Large
SIZE RANGE:	8–20	SIZE NAMES:	Numeric
SAMPLE SIZE:	12	NOTION:	Metric

RULE NUMBER:	1		2		3		4		5	
COMMENT:	zero		CB CF waist		bk centre dart		bk side dart		side waist	
POINT ATTRIBUTE:	N		N		N		N		N	
BREAKS	X	Y	X	Y	X	Y	X	Y	X	Y
8 to 10	0.0	0.0	−0.5	−0.4	−0.5	0.0	−0.5	0.3	−0.5	0.6
10 to 12	0.0	0.0	−0.5	−0.4	−0.5	0.0	−0.5	0.3	−0.5	0.6
12 to 14	0.0	0.0	−1.0	−0.5	−1.0	0.0	−1.0	0.35	−1.0	0.75
14 to 16	0.0	0.0	−1.0	−0.6	−1.0	0.0	−1.0	0.4	−1.0	0.9
16 to 18	0.0	0.0	−1.0	−0.5	−1.0	0.0	−1.0	0.35	−1.0	0.75
18 to 20	0.0	0.0	−1.0	−0.5	−1.0	0.0	−1.0	0.4	−1.0	0.75

RULE NUMBER:	6		7		8		9		10	
COMMENT:	side hip/crutch		side knee/hem		inside knee/hem		bk crutch		CB CF hip	
POINT ATTRIBUTE:	N		N		N		N		N	
BREAKS	X	Y	X	Y	X	Y	X	Y	X	Y
8 to 10	0.0	0.6	0.0	0.5	0.0	−0.5	0.0	−0.8	0.0	−0.4
10 to 12	0.0	0.6	0.0	0.5	0.0	−0.5	0.0	−0.8	0.0	−0.4
12 to 14	0.0	0.75	0.0	0.5	0.0	−0.5	0.0	−1.0	0.0	−0.5
14 to 16	0.0	0.9	0.0	0.5	0.0	−0.5	0.0	−1.2	0.0	−0.6
16 to 18	0.0	0.75	0.0	0.5	0.0	−0.5	0.0	−1.0	0.0	−0.5
18 to 20	0.0	0.75	0.0	0.5	0.0	−0.5	0.0	−1.0	0.0	−0.5

RULE NUMBER:	11		12		13		14		15	
COMMENT:	fr crutch		fr centre dart 1		fr centre dart 2		fr side dart 1		fr side dart 2	
POINT ATTRIBUTE:	N		N		N		N		N	
BREAKS	X	Y	X	Y	X	Y	X	Y	X	Y
8 to 10	0.0	−0.6	−0.5	0.0	−0.5	0.0	−0.5	0.3	−0.5	0.3
10 to 12	0.0	−0.6	−0.5	0.0	−0.5	0.0	−0.5	0.3	−0.5	0.3
12 to 14	0.0	−0.75	−1.0	0.0	−1.0	0.0	−1.0	0.35	−1.0	0.35
14 to 16	0.0	−0.9	−1.0	0.0	−1.0	0.0	−1.0	0.40	−1.0	0.4
16 to 18	0.0	−0.75	−1.0	0.25	−1.0	−0.25	−1.0	0.35	−1.0	0.35
18 to 20	0.0	−0.75	−1.0	0.0	−1.0	−0.0	−1.0	0.65	−1.0	0.15

NB POINT ATTRIBUTE N means no smoothing at the grade point.

GRADING SECONDARY BLOCK PATTERNS

One-piece semi-fit dress grading

The grading of the one-piece semi-fit dress block is similar to that of the bodice and sleeve with the added extension of the skirt length. The waist darts grading increments are according to the bodice increments. This is illustrated in Figure 2.13 for the position of the increments for a 4.0 cm bust, waist and hip full girth grade. Figure 2.14 shows the grade point movement and grade rules for an increase in size from a 12 to a 14. The semi-fit dress sleeve grading is the same as the previous semi-fit sleeve. The grade rules are given in Table 2.11.

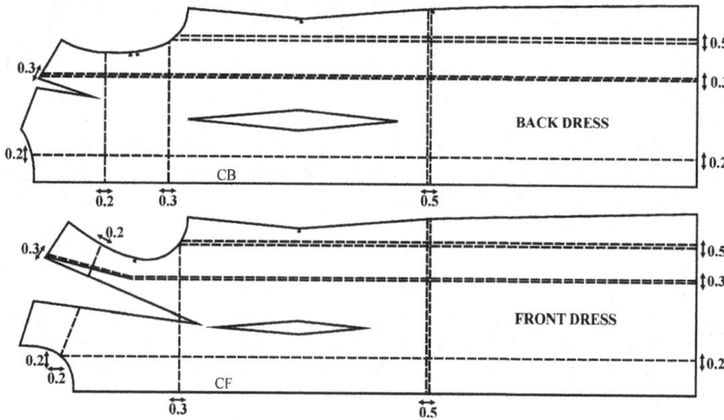

Figure 2.13 Grading increment positions and distribution for a semi-fit dress block 4 cm grade

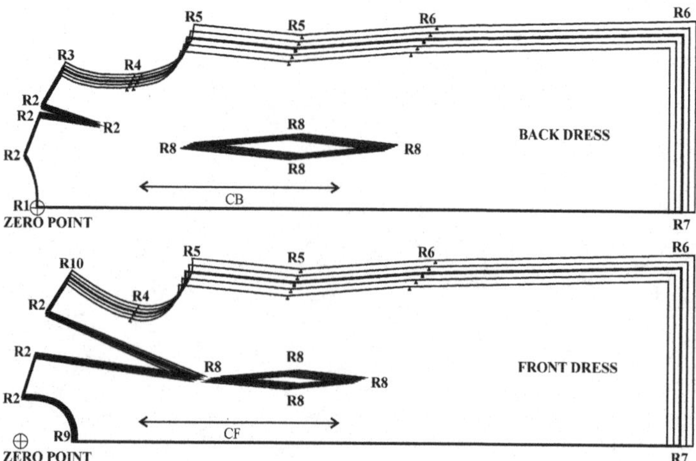

Figure 2.14 Grade point movement and grade rules for a one-piece semi-fit dress increasing size 12 to 14 for a 4 cm grade

Table 2.11 Grade rule table for a semi-fit one-piece dress with a 4.0 cm girth grade for sizes 8 to 16

RULE TABLE:	ONE-PIECE DRESS 4 cm	RULE METHOD:	Incremental Small to Large
SIZE RANGE:	8–16	SIZE NAMES:	Numeric
SAMPLE SIZE:	12	NOTION:	Metric

RULE NUMBER:	1		2		3		4		5	
COMMENT:	zero		bk neck width		shoulder width		Xback/Xfront		bust/waist width	
POINT ATTRIBUTE:	N		N		N		N		N	
BREAKS	X	Y	X	Y	X	Y	X	Y	X	Y
8 to 16	0.0	0.0	0.0	0.2	0.15	0.5	0.3	0.5	0.5	1.0

RULE NUMBER:	6		7		8		9		10	
COMMENT:	hip/hem width		CF CB length		fr bk waist dart		CF neck		fr shoulder width	
POINT ATTRIBUTE:	N		N		N		N		N	
BREAKS	X	Y	X	Y	X	Y	X	Y	X	Y
8 to 16	1.0	1.0	1.0	0.0	0.5	0.2	0.2	0.0	0.2	0.5

NB POINT ATTRIBUTE N means no smoothing at the grade point.

Knitted top grading

Stretch fabric often requires a larger girth grading increment compared with the woven fabric of the previous blocks. This can vary according to the extensibility of the knitted fabric. This example is for a size range labelled small, medium and large with a bust, waist and hip girth grade 8.0 cm. Figure 2.15 illustrates the position of the increments and distribution. Figure 2.16 shows a nest of sizes including the grade rules from Table 2.12. When a pattern is mirrored the grade rules 10* and 11* are automatically mirrored.

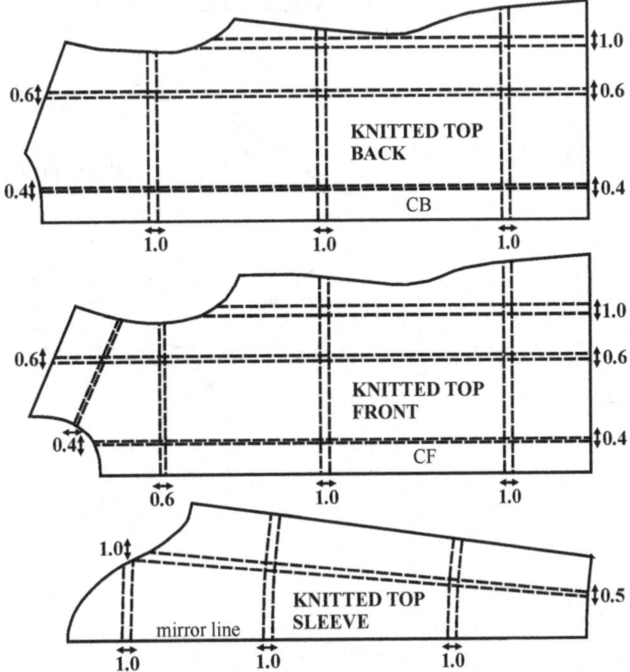

Figure 2.15 Grading increment positions and distribution for a knitted top 8 cm grade

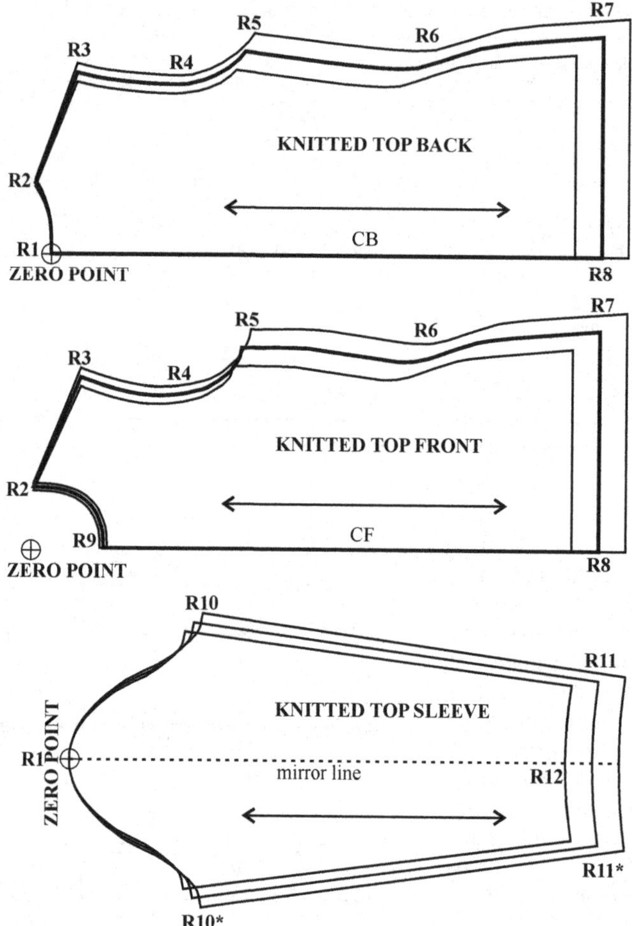

Figure 2.16 Grade point movement and grade rules for a knitted top increasing size S, M and L for an 8 cm grade

Table 2.12 Grade rule table for a knitted top with an 8.0 cm girth grade for sizes small, medium and large

RULE TABLE:	Top		RULE METHOD:	Incremental Small to Large
SIZE RANGE:	S, M, L		SIZE NAMES:	Alphabetical
SAMPLE SIZE:	M		NOTION:	Metric

RULE NUMBER:	1		2		3		4		5	
COMMENT:	zero		neck width		shoulder width		mid armhole		bust width	
POINT ATTRIBUTE:	N		N		N		N		N	
BREAKS	X	Y	X	Y	X	Y	X	Y	X	Y
S to M	0.0	0.0	0.0	0.4	0.0	1.0	0.5	1.0	1.0	2.0
M to L	0.0	0.0	0.0	0.4	0.0	1.0	0.5	1.0	1.0	2.0

RULE NUMBER:	6		7		8		9		10	
COMMENT:	waist width		hip/hem width		CB CF length		CF neck		upper arm width	
POINT ATTRIBUTE:	N		N		N		N		N	
BREAKS	X	Y	X	Y	X	Y	X	Y	X	Y
S to M	2.0	2.0	3.0	2.0	3.0	0.0	0.4	0.0	1.0	1.0
M to L	2.0	2.0	3.0	2.0	3.0	0.0	0.4	0.0	1.0	1.0

RULE NUMBER:	11		12	
COMMENT:	sleeve wrist width		wrist length	
POINT ATTRIBUTE:	N		N	
BREAKS	X	Y	X	Y
S to M	3.0	0.5	3.0	0.0
M to L	3.0	0.5	3.0	0.0

NB POINT ATTRIBUTE N means no smoothing at the grade point.

Blouse grading

The grading of the blouse block is also based on the bodice and sleeve grading. The only difference is that there is no dart suppression to grade. Positioning a length grading increment between the waist and hips is optional. This is illustrated in Figure 2.17 for the position of the increments for a 4.0 cm bust, waist and hip full girth grade, and in Figure 2.18 for the blouse sleeve. Figure 2.19 shows the grade point movement and grade rules for an increase in size from a 12 to a 14. The grade rules are given in Table 2.13.

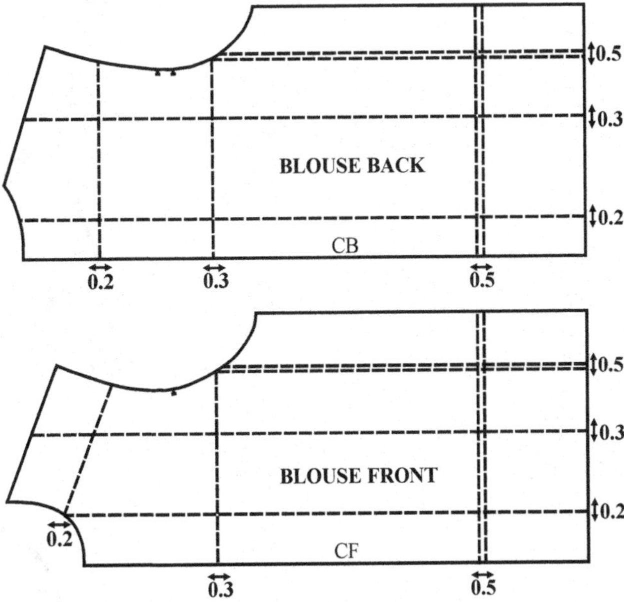

Figure 2.17 Grading increment positions and distribution for a blouse 4 cm grade

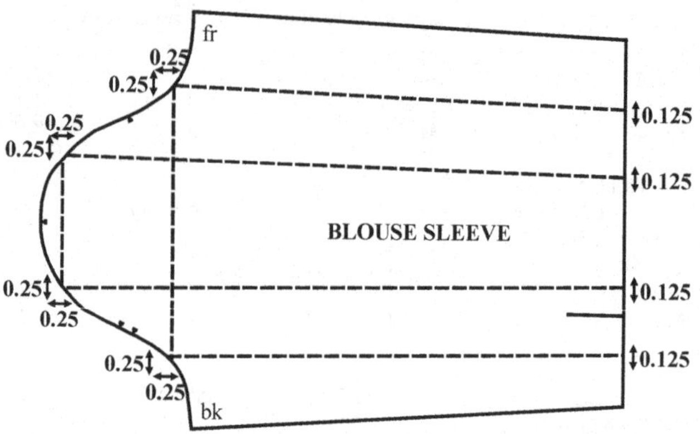

Figure 2.18 Grading increment positions and distributions for a blouse sleeve 4 cm grade

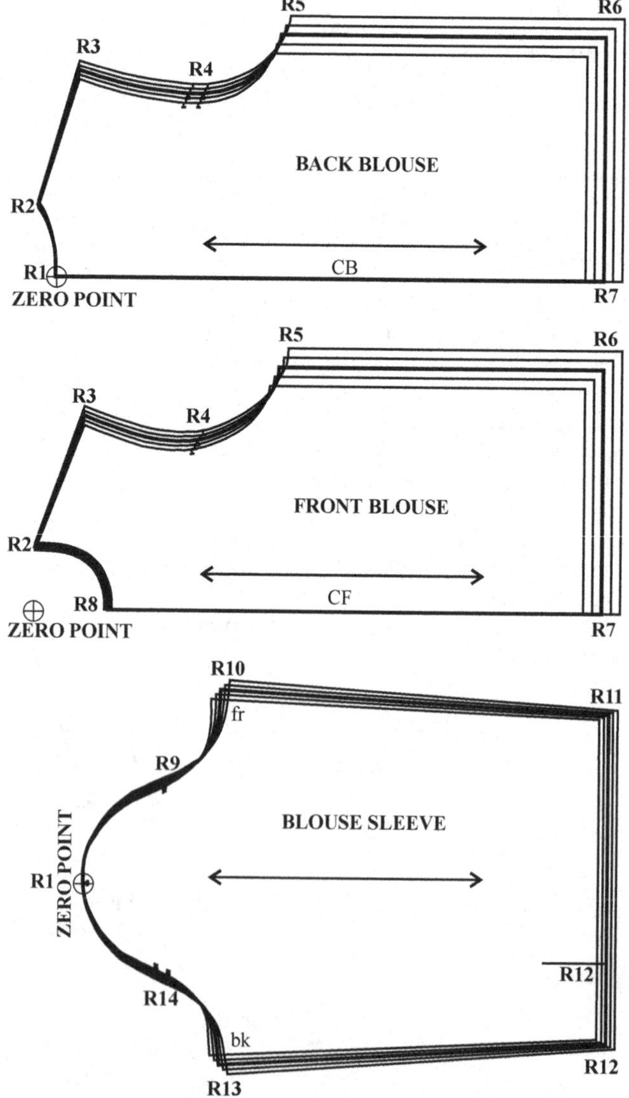

Figure 2.19 Grade point movement and grade rules for a blouse increasing size 12 to 14 for a 4 cm grade

Table 2.13 Grade rule table for a blouse with a 4.0 cm girth grade for sizes 8 to 16

RULE TABLE:	BLOUSE 4 cm	RULE METHOD:	Incremental Small to Large
SIZE RANGE:	8–16	SIZE NAMES:	Numeric
SAMPLE SIZE:	12	NOTION:	Metric

RULE NUMBER:	1		2		3		4		5	
COMMENT:	zero		neck width		shoulder width		Xback Xfront		bust/waist width	
POINT ATTRIBUTE:	N		N		N		N		N	
BREAKS	X	Y	X	Y	X	Y	X	Y	X	Y
8 to 16	0.0	0.0	0.0	0.2	0.0	0.5	0.25	0.5	0.5	1.0
RULE NUMBER:	6		7		8		9		10	
COMMENT:	hip/hem width		CF CB length		CF neck		fr slv crown width		fr upper arm width	
POINT ATTRIBUTE:	N		N		N		N		N	
BREAKS	X	Y	X	Y	X	Y	X	Y	X	Y
8 to 16	1.0	1.0	1.0	0.0	0.2	0.0	0.25	0.25	0.5	0.5
RULE NUMBER:	11		12		13		14			
COMMENT:	fr wrist width		bk wrist width		bk upper arm width		bk slv crown width			
POINT ATTRIBUTE:	N		N		N		N			
BREAKS	X	Y	X	Y	X	Y	X	Y		
8 to 16	0.5	0.25	0.5	−0.25	0.5	−0.5	0.25	−0.25		

NB POINT ATTRIBUTE N means no smoothing at the grade point.

PATTERN PREPARATION FOR DIGITISING

Accurate preparatory work and verification is important for successful grading and digitising. The main procedures are examples using a Gerber Accumark system. However, the principles apply to other CAD systems, and terminology will vary between system type used.

- Setting up parameter tables
- Checking the master pattern
- Building grade rule tables
- Digitising the pattern pieces
- Retrieving and displaying pieces
- Verifying pattern pieces
- Plotting pattern pieces

Setting up parameter tables

Before any data can be input into the system certain parameters have to be set up that are generally in the form of tables. Each system will have its own format and types of commands for which the operator has to be trained. The following information is generally required and input into the computer:

- Creating a storage area or a style
- Setting up a user environment
- Creating parameter tables
- Determining and entering grade rule tables

STORAGE AREA
Some systems require a storage area that is a user-defined workspace. It groups related work such as patterns and markers. The storage area can be exclusive to an individual or defines a certain group of garments generally. Some systems group the pieces into styles while others require the operative's initials.

USER ENVIRONMENT
Certain information is required to be input into the system before it can be operated. The user environment is often in the form of tables that have to be completed. Each system will have its own requirements. Some typical information is:

- Notation: measurements can be recorded in metric or imperial
- Decimal precision places: specifies the number of decimal points in a measurement
- Seam allowance: the amount added
- Grading method: (relates to the type of grade rule tables, described previously).

PARAMETER TABLES
Certain information has to be put into the computer before patterns can be digitised, created or plotted or markers made. The essential ones are:

- Notch parameter table
- Drill hole parameter table
- Pattern annotation table
- Pattern plotting parameter table

Notch parameter table
A **notch** is a slit or V-shaped cut on the perimeter of a pattern or cut garment part to identify adjoining seams, balance marks, CB, CF, seam or hem allowances, stitching lines of darts, pleasts, etc. A variety of different types can be quoted, the most usual ones are:

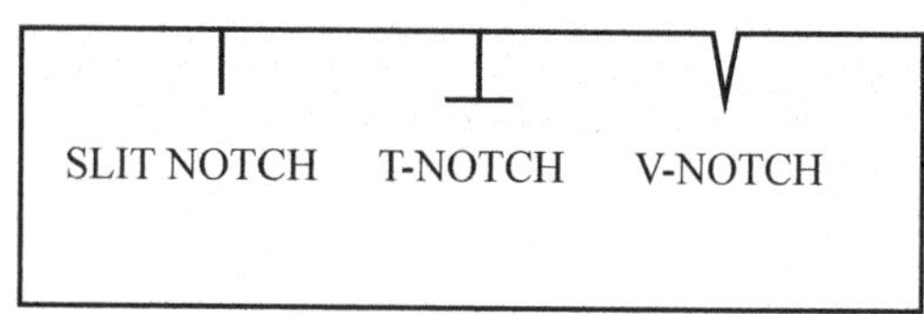

Notches should not penetrate the stitching line but should finish approximately 3 mm to 5 mm away. These notches are generally numbered and placed in a table for selecting when digitising patterns or for adding notches when creating a new pattern. Various notch lengths can be quoted for different seam widths. The width of a V notch can also be varied. Some systems offer a double notch option that can be useful for identifying the back and front of garments.

Drill hole parameter table
A **drill hole** marks an internal position within the cut fabric, e.g. pocket position, the apex of a dart. The holes must be covered by stitching, therefore placed 0.5 cm or 1 cm in from the stitching line. They also have to be concealed and not visible on the outer face side of the fabric. Most systems have available various drill hole symbols such as:

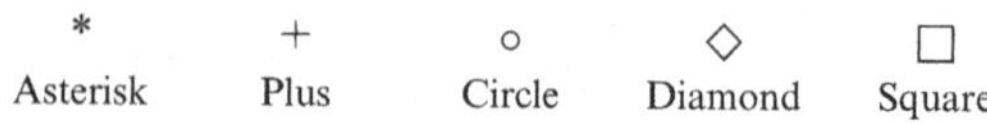

Pattern annotation table
Annotation refers to the information required on plotted pattern pieces or markers. The grain line is generally automatically marked on plotted pattern pieces but not markers. Each computer system has its own method for identifying individual pattern pieces, for example:

- **Name** or code (either numeric or words)
- **Size** (either numeric or words)
- **Category** or code (either numeric or words) e.g. back, front, sleeve
- **Description**, whatever may be relevant, e.g. lining, type of garment
- **Grade rule table name**

Pattern plotting parameter table

This table determines how a pattern piece will be plotted. It may have a different title and the information required will vary. The most useful items to change are:

- **Rotation**: This can be set at 90% from the original digitised position so that the full width of the paper can be used for economy.
- **Scale X: 100% scale Y: 100%**: This scale is for plotting full-scale patterns. If a reduced scale is required the percentage can be altered, e.g. 50% for half scale, 20% for fifth scale.

Checking the master pattern

Before grading and digitising, the master pattern has to be carefully checked. First, a sample garment made from the master pattern has to be examined for correct style, size, fit and balance. The measurements should be according to the size chart specifications. It should be assembled by the correct production methods. It is advisable to keep this sample garment beside you while grading. Patterns cut in card or paper can be digitised into the computer. Alternatively the pattern can be drawn more accurately on to pattern tracing paper (similar to marker paper). When drawing it is advisable to rule all straight lines and draw curves smoothly with a French curve or similar tool. The intersection at the corners must be completed. It is helpful if the corner lines crossover. Only half a pattern needs to be digitised as the computer will produce the other side as a mirror image or a pair (Figure 2.20).

In preparing a master pattern the following points should be checked:

- The *grain line* is accurately drawn representing the warp threads. Pattern pieces which are to be cut on the bias will have a grading reference line which differs from the warp threads.
- The *stitching line* is measured to ensure that all the major girth and length measurements conform to the size chart.
- Adjoining *seams* have to match in length and width.
- The *perimeter edge* is accurately cut or drawn with the corners of adjoining seams at the correct angle, the straight lines are ruled truly straight, and the curved edges smooth.
- The positions of the *notches* and *drill holes* have to be carefully checked and kept to the minimum.
- All the correct *information* has to be written on each pattern piece, generally identifying the style number or name, size, a category, description of the piece, and rule table name according to the requirements of the computer program.
- The pattern is marked with the *relevant codes* for

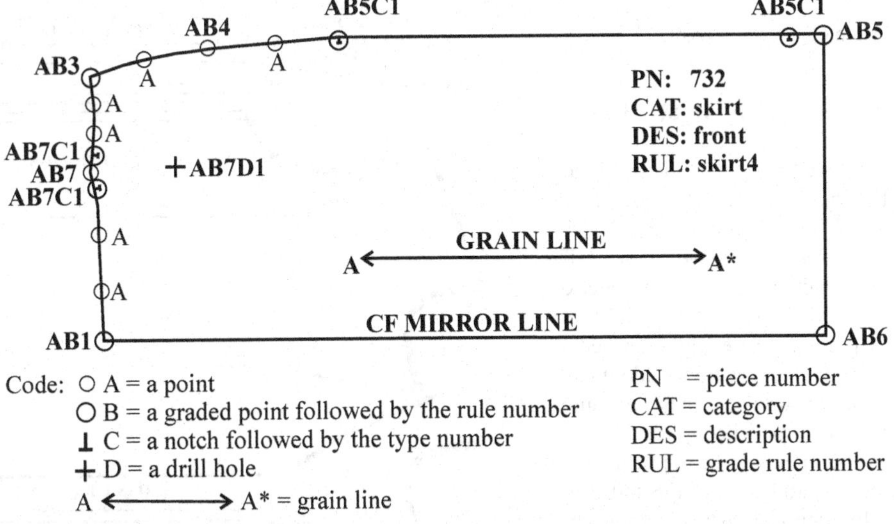

Figure 2.20 Pattern prepared for digitising (**NB** This example is not related to any specific computer system; each system has its own method of coding.)

the perimeter cardinal points, grade rules, intermediate curve points, notches and the internal drill holes and internal lines (the type of code required will vary with each computer system).

- The *matching points* for the fabric design have to be identified by a code where required. (See Placement strategies for fabric type and matching in Part 5).
- The *numbers of pattern pieces* have to be checked and grouped according to the type of fabric.

INTERMEDIATE CURVE POINTS

The number of intermediate curve points depends on the depth of the curve. Deep curves require more points than shallow curves. As a guide a ruler can be placed against the curve to find the deepest point, which is then marked by A (Figure 2.21). The ruler is then moved to the next section to identify another deepest point. This process is repeated for the whole curve. Avoid excessive use of points on curves as they slow the computer or could distort the contour. Shallow curves require fewer intermediate points.

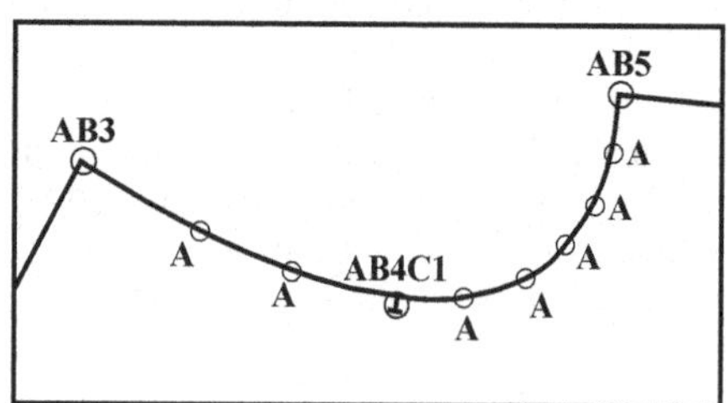

Figure 2.21 Intermediate curve points

MATCHING OF IDENTICAL SHAPED PATTERNS

Where adjoining patterns have identical contours it is advisable to draw them both in the same direction. This facilitates them having common grade rules and curve points. The contour of these matching seams can be checked and made identical when viewed on a light box. Common grade points, notches and intermediate curve points can then be positioned identically. This enables the contour of these adjoining seams to remain the same when graded.

To maintain the matching of the neck shape of both this bodice and facing, the pattern is folded. Then the identical curve is drawn and the same intermediate points marked (Figure 2.22).

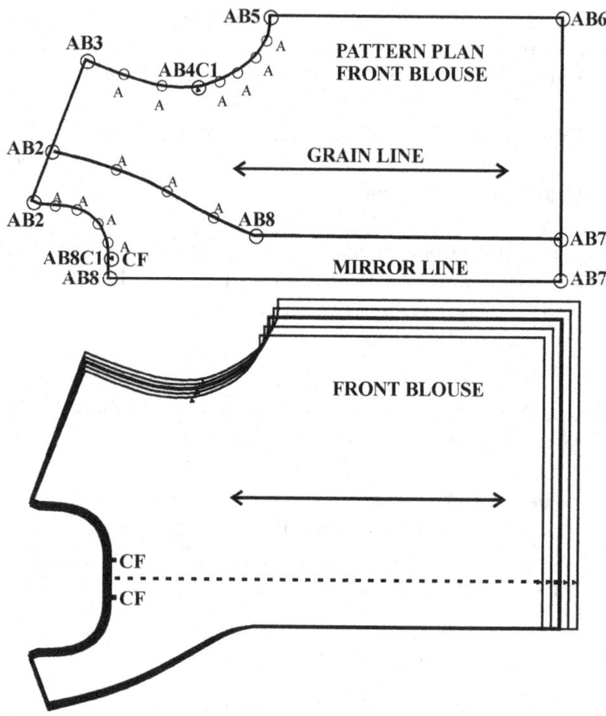

Figure 2.22

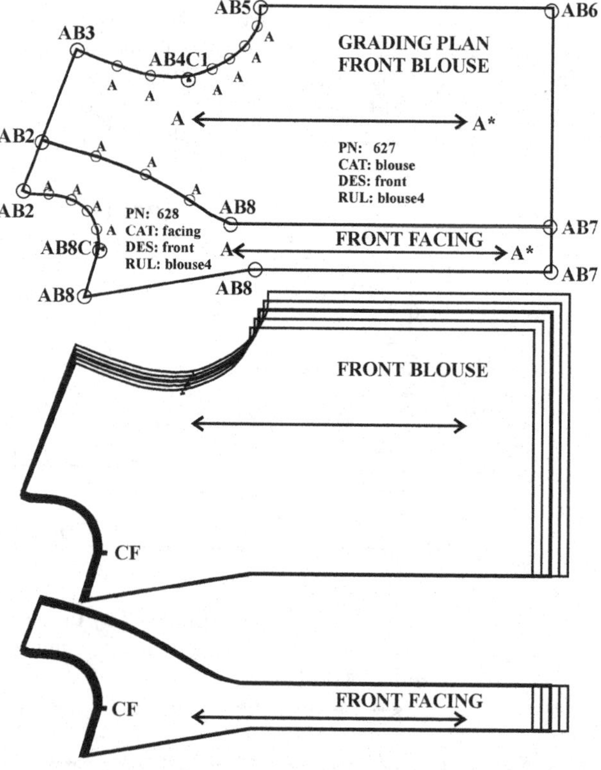

Figure 2.23

Where a separate facing is required it can be superimposed and drawn on to the bodice. The bodice and facing can then be given the same grade rules and curve points. Then both pieces can be digitised separately from the same tracing (Figure 2.23).

Building grade rule tables

This procedure has been described in detail at the beginning of this Part.

Computer digitising

DIGITISER

The **digitiser** consists of three component parts:

- The **table top** to which the pattern pieces are secured by masking tape. The grain reference line should be approximately parallel with the lower edge. This is because under the surface is a grid of wires that represents the X and Y axes. The table top is adjustable in both height and angle.
- The **cursor** is used to trace the pattern pieces and input information into the system. It is activated by pushing the relevant keys. The position of the cross hairs of the cursor registers the X and Y coordinates on the table top wire grid. The pattern information and outline are converted into a numerical language that is understood by the computer.
- The **digitising menu** is used in combination with the cursor to input pattern information and select commands.

DIGITISING PROCESS

- The pattern piece is placed on the digitising table with the grain reference line horizontal, that is approximately parallel with the lower edge of the table, and secured by masking tape (Figure 2.24).
- To start digitising the pattern piece name or number and relevant information are generally input by the cursor cross hairs being placed over the relevant boxes on the digitise menu and a key being pressed on the cursor.
- The grain line is digitised.
- The perimeter information is digitised by starting at the left-hand corner and working clockwise. This is required by most computer systems. The piece is then closed or mirrored by selecting the command from the menu. The systems are generally programmed to connect last and first points with a straight line.
- The internal drill holes are recorded. This process is then completed by selecting the 'end input' box in the menu. The command and sequence of digitising will vary with different computer programs, but are fairly similar.

DIGITISING A GRADED NEST OF PATTERNS

Digitising a nest of graded patterns is useful for a complex-shaped pattern piece where the grade is uneven between the sizes or the graded pattern has been modified due to fabric reaction. These nests of graded patterns are generally graded by hand. Some pattern pieces of a draped style may require soft folds positioned on the bias grain; therefore each size may require modelling on a dress workroom stand. The modelled section of each size can be drawn as a nest of patterns for digitising. The remainder of the pattern pieces can be digitised as previously explained. The method of digitising a nest of grades will vary according to the computer program.

DIGITISING A LARGE PATTERN PIECE

Some pattern pieces are too large to fit on to the digitising table. This means they have to be split. It is advisable to split them into as few sections as possible horizontally and vertically. The coding required will vary according to the specific computer program, but generally each section will have to be numbered. If the computer program does not have this facility the sections can be digitised as separate pieces and then merged together in the pattern design system.

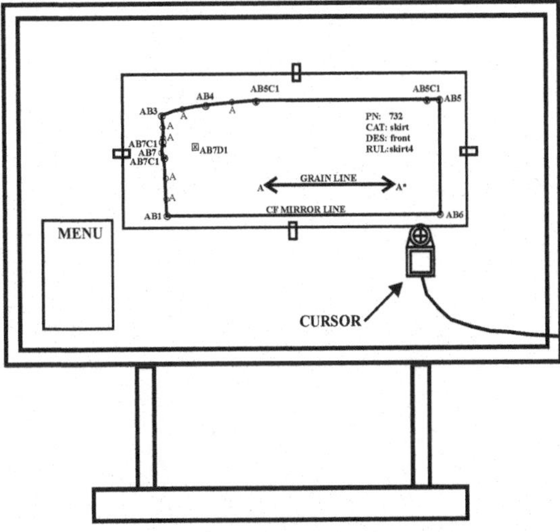

Figure 2.24 The digitising table

Verifying the pattern grading

To verify the graded pattern after digitising, a nest of grades can be displayed on the visual display unit. The grade rules can be checked and edited if revision is required. The pattern can be transferred to the pattern design system (PDS) so that the adjoining graded seams can be measured for each size to ensure accuracy. (The method of measuring varies according to the type of computer program.)

CHECKING NESTS OF FULL SCALE GRADED PATTERNS

The plotted full scale nest of graded patterns can be viewed and checked on a light box by superimposing adjoining seams to ensure that the seams and notches match and that all the grading increments are correct (Figure 2.25).

Methods of grading vary according to different CAD systems that are available to the clothing industry. It is difficult to present the procedures employed with every CAD system available.

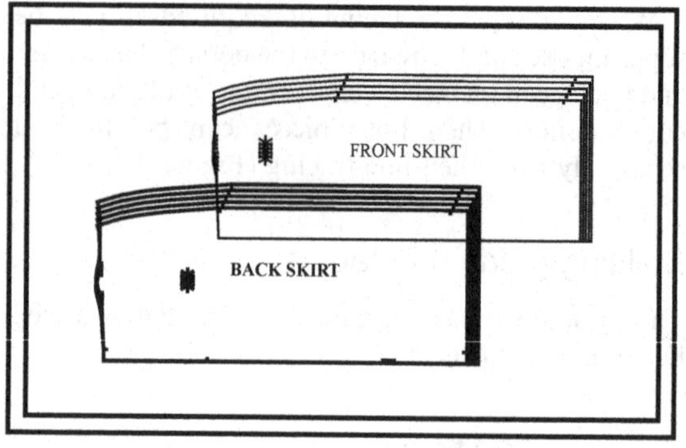

Figure 2.25 Checking a nest of graded front and back skirts using a light box

Part 3
Pattern designing and grading

In Part 3 various methods are explained for designing patterns for different styles. This is because one method may be more applicable to the computer system being used. This Part covers:

- Pattern design procedures
- Skirt styling including gathering, pleats, flared and gored skirts, circular skirts, details of yokes, pockets and waist band
- Bodice styling by manipulating darts into various positions of gathers, yokes, draped folds and panels
- Modifying for a sleeveless armhole and a cut-away neckline
- Collar styling including flat, semi-stand, high-stand and grown-on collars
- Sleeve styling including set-in sleeves with a cuff, puff sleeves, raglan and kimono
- Details for completing the pattern by adding seams, hems and facings.

To design a pattern successfully several factors have to be considered simultaneously:

- The design features have to be related to the body shape, size and proportion
- The function of the garment, e.g. dress, coat for formal, work or leisure wear
- The behaviour and design of the fabric
- The economical use and cost of the fabric and trimmings
- The method of garment assembly.

Pattern designing discussed in this Part is mainly by the adaptation and manipulation of block patterns. This can be undertaken using a computer, or manually produced patterns can be digitised. For complex styles where the drape of the fabric is involved it is better to model those sections manually. Whichever method is used the grading can be incorporated by either using graded blocks stored in the computer, or adding grade rules to the digitised pattern. When creating the pattern it is preferable not to include the seams, hems or facings, as it is easier to measure and match the adjoining seam stitching lines without the seams. It is therefore more accurate to add them after the construction of the pattern has been completed. However, commercial use within industry often includes seams and hems on the block and production patterns. Seam types may need modification when creating new patterns.

PATTERN DESIGN PROCEDURES

When designing a pattern certain steps have to be followed. The procedure will vary according to the experience of the pattern technologist. Suggested below is a typical plan of action.

(1) A clear **sketch of design details** is required with both front and back views of the garment (see Figure 3.1). This should illustrate and state clearly the designer's intentions.

(2) (This step is optional as it is for those who find it difficult to visualise the garment three-dimensionally). A sample of the block pattern can be placed on a workroom stand so that the style features can be taped in the appropriate position.

(3) The **pattern is planned** by selecting a block pattern which is nearest in silhouette to the sketched style. The block is copied and adapted further if required. The style lines are then superimposed on to the **pattern plan** (see Figure 3.2). These style lines can correspond to the taped lines planned on the workroom stand.

(4) Each pattern piece is traced off the plan and the matching of adjoining seams and notches is checked. The seams, hems and facing are then added for the **final pattern**.

(5) A **lay plan** of the pattern pieces is planned for cutting a sample garment. This also gives an indication of the amount of fabric required for costing purposes.

(6) A **sample garment** is then made and tested for style, size, fit and method of assembly. If necessary the pattern is corrected. Major alterations may require another sample garment to be made.

(7) The correct pattern can now become the **production pattern** and the master pattern for grading. Where a graded block pattern is used within the

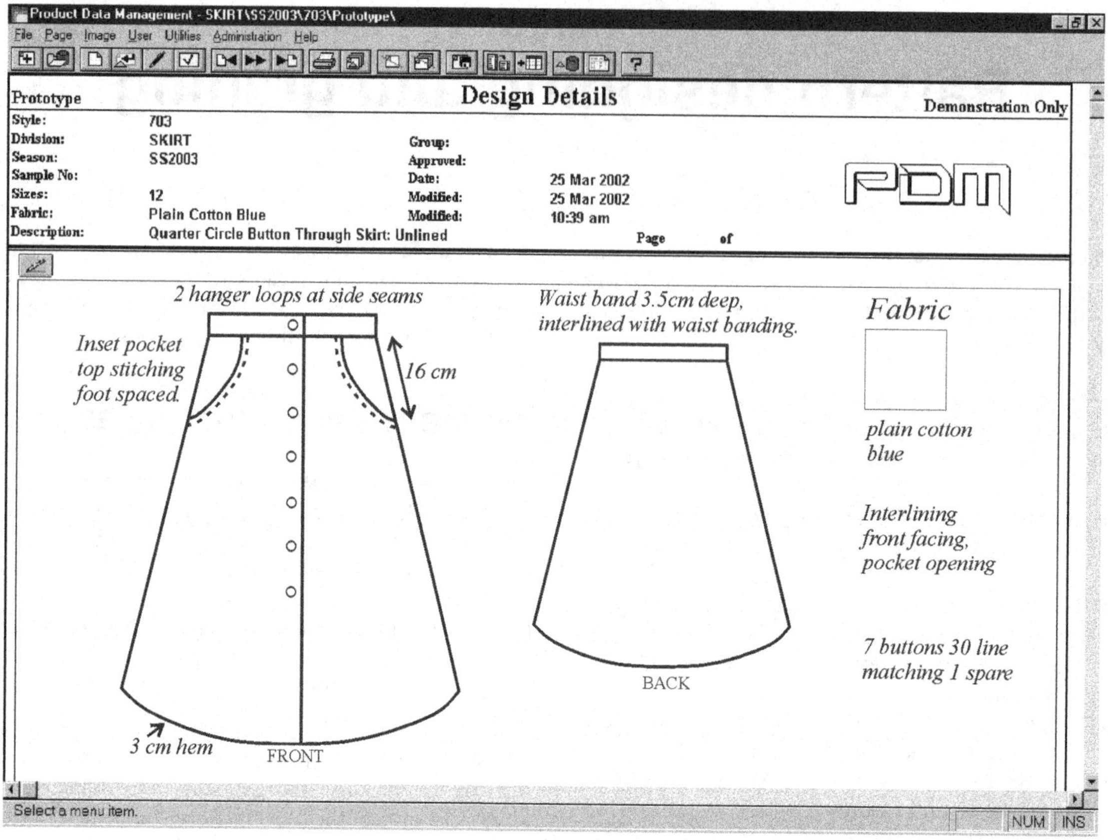

Product Data Management - SKIRT\SS2003\703\Prototype\
File Page Image User Utilities Administration Help
Prototype
Design Details
Demonstration Only
Style: 703
Division: SKIRT
Season: SS2003
Sample No:
Sizes: 12
Fabric: Plain Cotton Blue
Description: Quarter Circle Button Through Skirt: Unlined
Group:
Approved:
Date: 25 Mar 2002
Modified: 25 Mar 2002
Modified: 10:39 am
Page of
PDM
2 hanger loops at side seams
Inset pocket top stitching foot spaced.
16 cm
3 cm hem
FRONT
Waist band 3.5cm deep, interlined with waist banding.
BACK
Fabric
plain cotton blue
Interlining front facing, pocket opening
7 buttons 30 line matching 1 spare
Select a menu item.
NUM INS

Figure 3.1 Design details

pattern plan
front skirt
style 703
CF

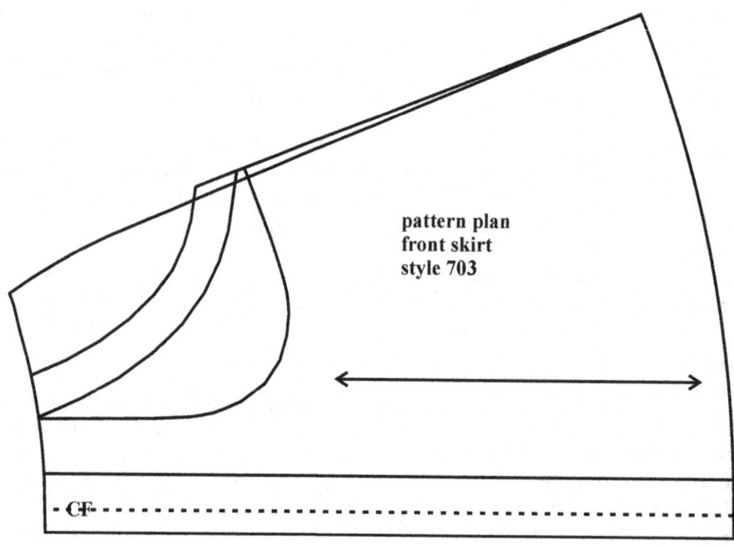

Figure 3.2 Pattern plan

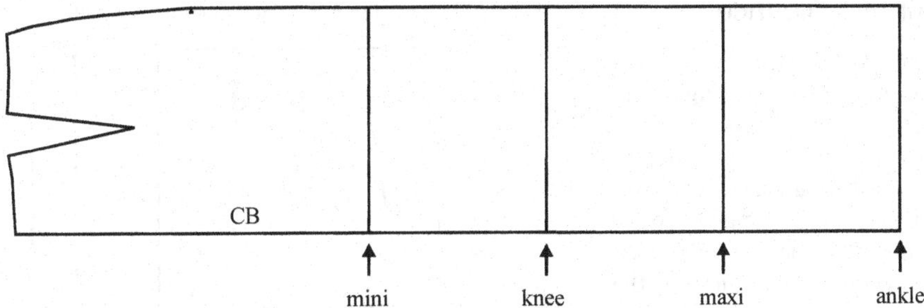

Figure 3.3 Various skirt lengths

computer pattern design system, the grade rules will automatically be transferred to the new pattern. Grades may have to be revised on some pattern pieces or new ones added by using the grade rule table.

By designing the pattern within the computer system the patterns for developing the style and the final graded patterns for production can be the same, whereas often with manual patterns there is the designer's pattern, and the production pattern that requires grading.

SKIRT STYLING

The first consideration is to select the correct basic block for the style. A copy of the selected block can be altered in length according to current fashion (Figure 3.3).

Flaring, gathering or pleating can add extra fullness. A comparison of skirt silhouettes and the corresponding patterns of a half front skirt are illustrated in Figure 3.4. The type and positioning of the opening, allowing the wearer to get in and out of the garment easily, are also important.

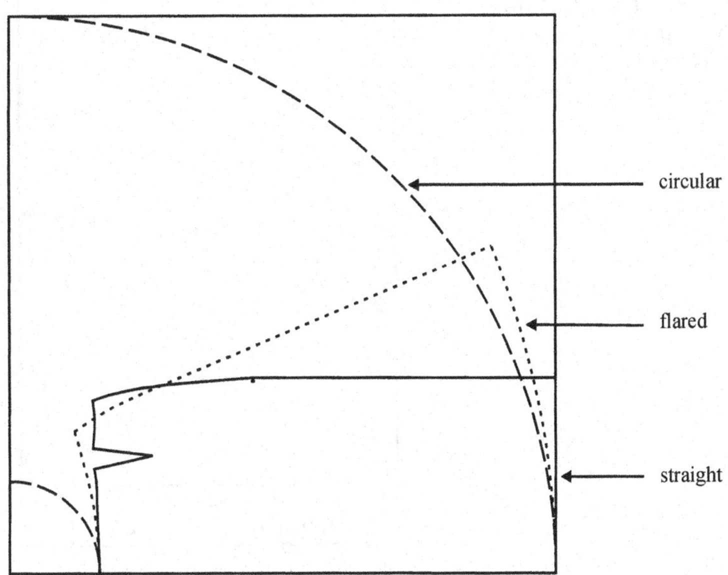

Figure 3.4 Comparisons of hem widths and patterns for half the front skirt

Gathered skirt with flounce

The simplest skirt to construct is a gathered rectangle of fabric. Constructing large rectangular pattern pieces by computer is far more accurate than cutting card patterns by hand. **Gathering** along the weft grain or true bias of the fabric gives softer folds. The amount of gathers is influenced by the weight, structure and finish of the fabric. The following is suggested:

- Sheer fabric: 3 times the finished size
- Medium weight: 2 times the finished size
- Heavier weight: 1.5 times the finished size

The fabric width also has to be considered to maximise fabric utilisation and garment cost. For example, for fabric 140.0 cm wide this skirt could have one width for the skirt and two widths for the flounce. The distribution of the gathers has to be controlled by a series of notches when using a basic lock stitch machine. The example pattern for a quarter of the skirt indicates the matching of the notches **A**, **B**, **C**, **D** and **E** that distribute the flounce gathers on to the skirt. However, there is machinery available which will gather the lower ply of fabric on to a flat top ply. The ratio for the gathers has to be pre-set according to the exact amount allowed on the fabric.

Gathered skirt and flounce cut at 2 times the finished size

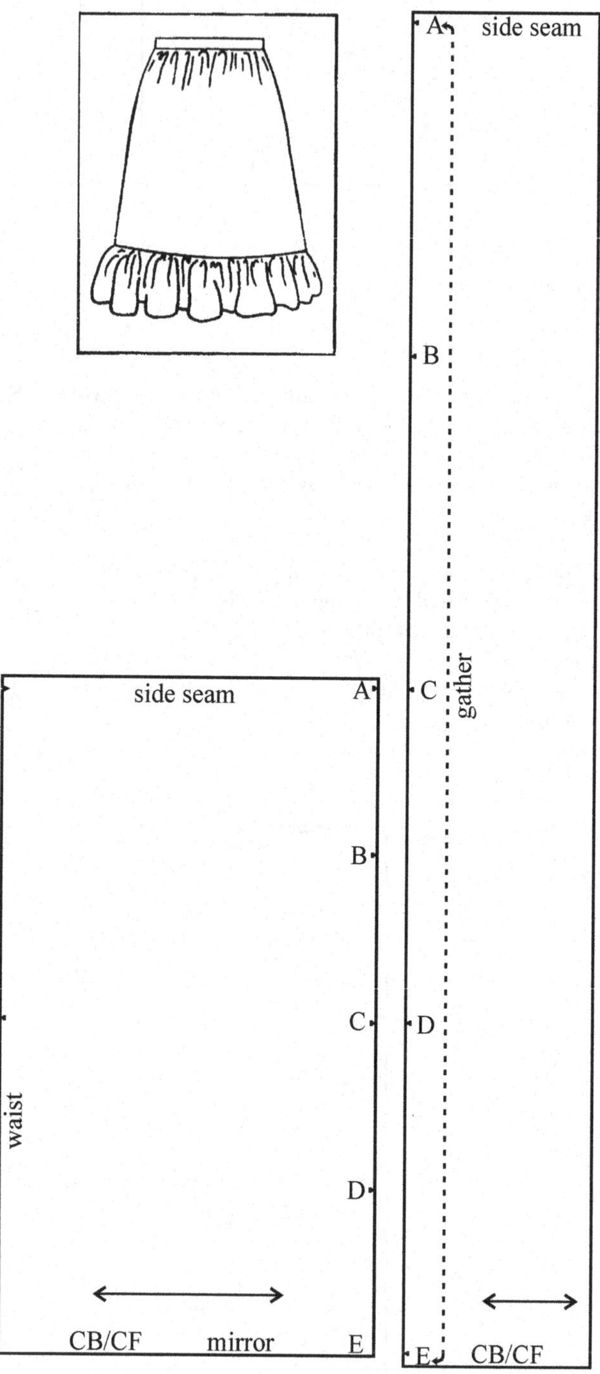

Pleats

Pleats add fullness to a skirt hem as the flattened folds double back upon themselves, forming three layers of fabric. The top layer is visible and part of the upper surface of the garment. The two lower layers are known as the **underlay**, which gives the extra fullness. The folds of the pleats are generally sharply pressed, but if they are not they are known as unpressed pleats. All these types of pleats can be created using a computer. First, the position of the pleat's top fold line has to be planned, then the depth of the underlay and the direction in which this will lie.

There are various configurations of pleats. A single pleat is known as a **knife pleat**. A **box pleat** is where two knife pleats are pressed in opposite directions. An **inverted box pleat** is the reverse side of a box pleat where two knife pleats are pressed towards each other.

These pleats can either have the folds lying parallel, or can be shaped by widening the visible layer at the hem and reducing the underlay. A different form of pleats is **sunray pleating** where opposing folds radiate from one point and form a section of a circle.

Sunray pleated quarter circle stitched into a panel

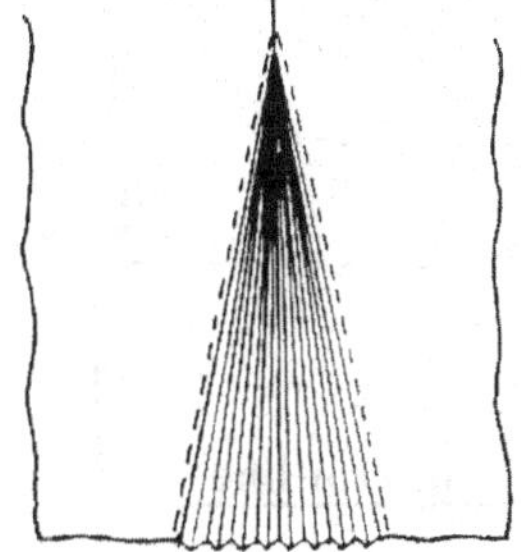

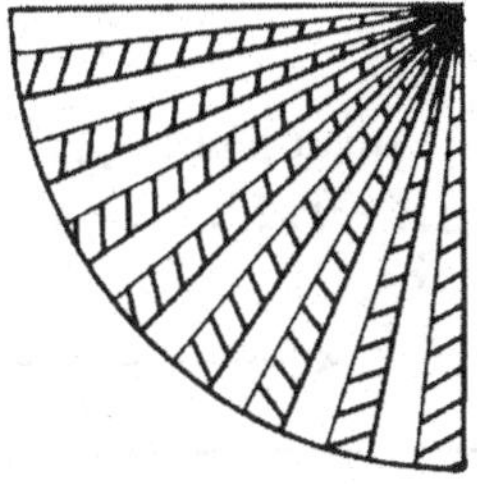

Knife pleats

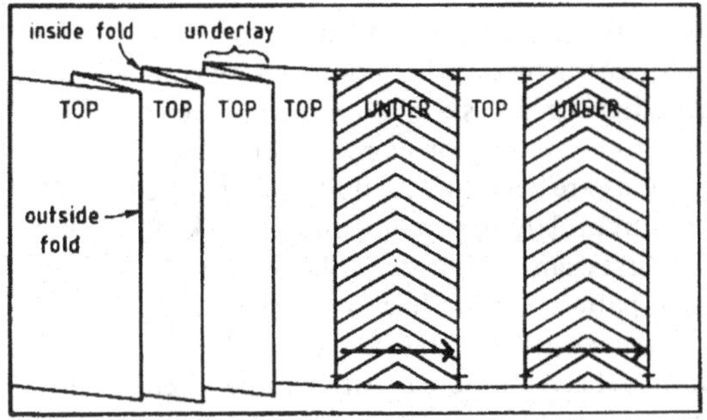

Box pleats

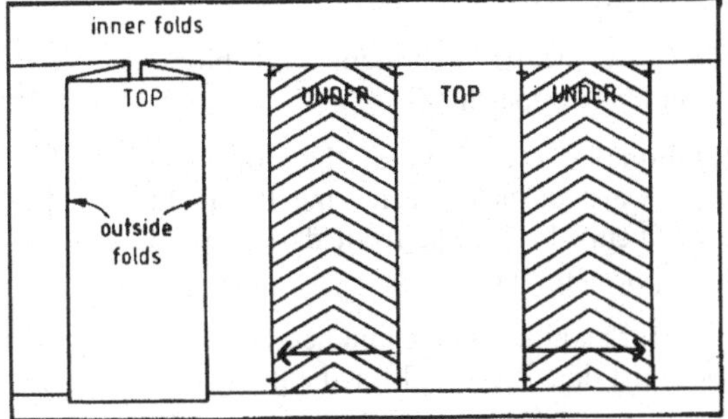

Inverted box pleats

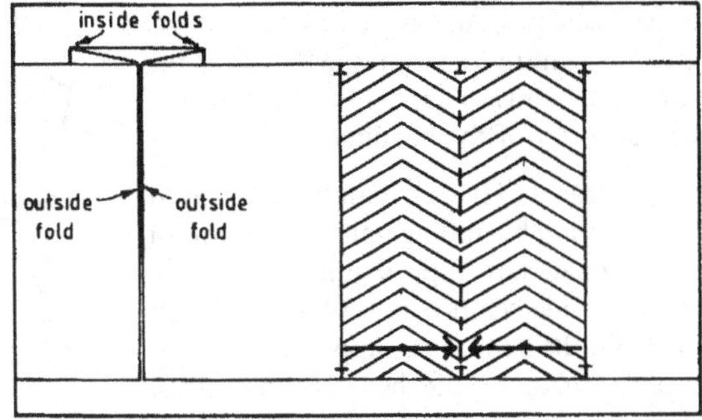

Shaped knife pleats

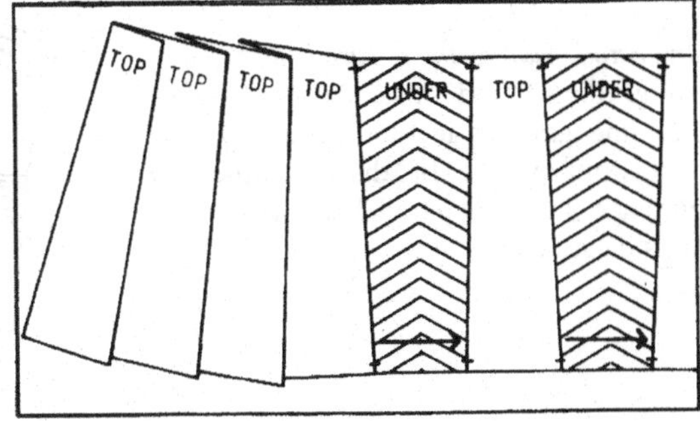

Flared and gored skirts

Extra fullness can be added to the hem width either by dart manipulation or constructing a gored skirt. Both methods can produce a flared or gored skirt. These methods are generally used for a moderate amount of flare to produce an A-line skirt. The first example illustrated below has the same pattern for the back and front skirt. The second has a separate back and front skirt that gives a more accurate fit between the waist and hips.

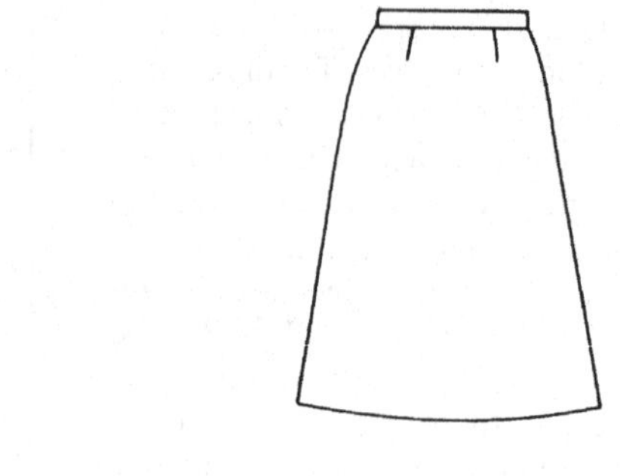

FLARED SKIRT CONSTRUCTION BY DART MANIPULATION

(1) Copy the front straight skirt block. Include the upper hip line.

(2) Equalise the width of the front and back waist darts by increasing the front dart (1 cm). Lower the front dart apex to the upper hip line.

(3) Mark a point on the mid hem to where the waist dart is pivoted for the flare.

(4) Select a line that will hold the pattern in the correct position according to the grain line (in this case the CF/CB).

(5) Pivot a proportion of the waist darts into the hem (in this example 8.0 cm at the hem). Then draw a curved line to complete the perimeter of the hemline.

(6) (Optional) The hem can be widened at the side seam by adding half the amount of the original flare (approximately 4.0 cm). The new side seam connects in a straight line from the hem to the upper hip level. A smooth curve to the hem is maintained when the corners are at right angles.

(7) Mirror the CF/CB line to complete the pattern.

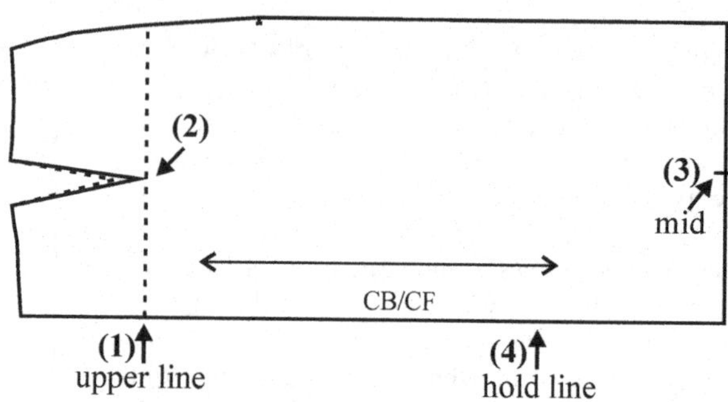

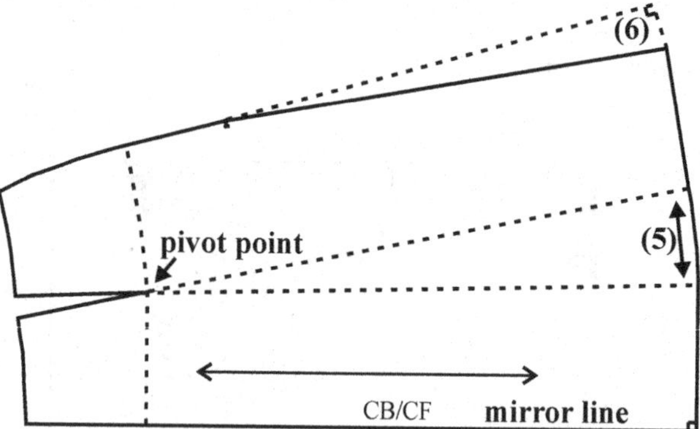

Suggested grading for a flared skirt

For grading flared skirts the multi-reference line grading method would be the most appropriate. In this example the reference lines are parallel to the CF/CB and side seams.

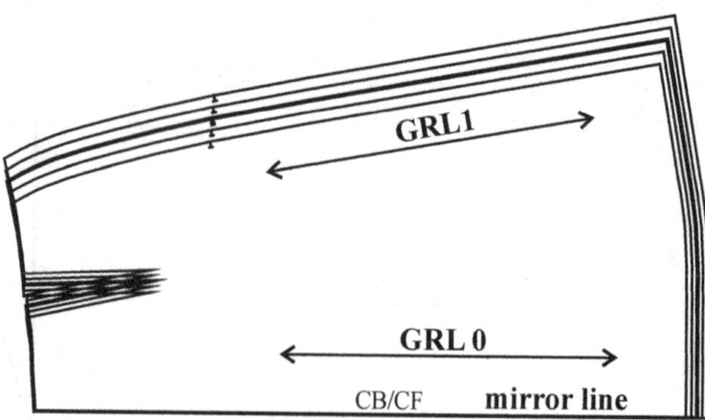

EIGHT GORED SKIRT CONSTRUCTION

In this example the skirt is divided into 8 equal width panels. The same method can be used for any number of gores that is required. The same fit of the back and front skirt is maintained between the waist and upper hip as on the original straight skirt block. The gores can be merged together to form a simple flared skirt.

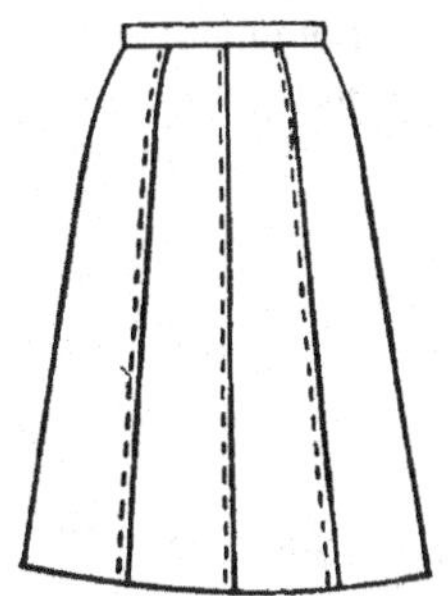

(1) Copy the straight skirt block; include the upper hip level. Position a line midway between the CB, CF and side seams.

(2) Reposition the back and front waist darts so that their apexes meet the midway line. Extend the front dart apex to the upper hip level. The back dart remains the original length.

(3) Split the back and front skirts at the midway line.

(4) Add the required amount of flare to the split sides of each panel at the hem (this example 4.0 cm). Connect with a straight line between the hem and original dart apex and at the side seams to the upper hip. A smooth curve to the hem is maintained when the corners are at right angles.

(5) Draw the new centre front seam in a straight line from the hem through the upper hip level to the waist. Measure the distance between the original block CF and the new seam at **A**. Add this amount to the other side of the gore at **B**. (In this example the difference is 0.75 cm.) This measurement at the waist will vary according to the amount of flare that has been added. The modification is to maintain a straight centre seam.

(6) Repeat the instructions of (5) for the new CB seam.

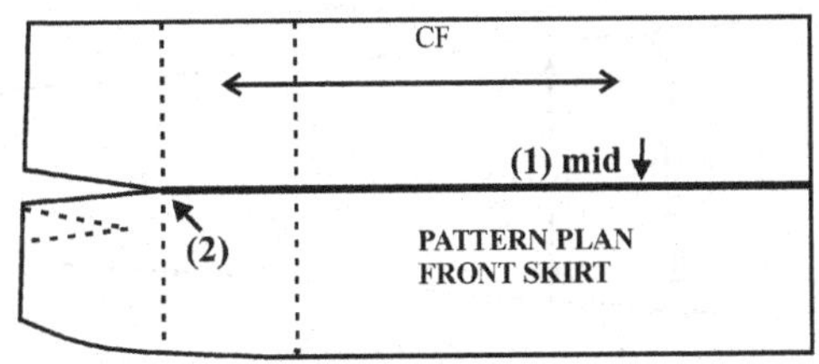

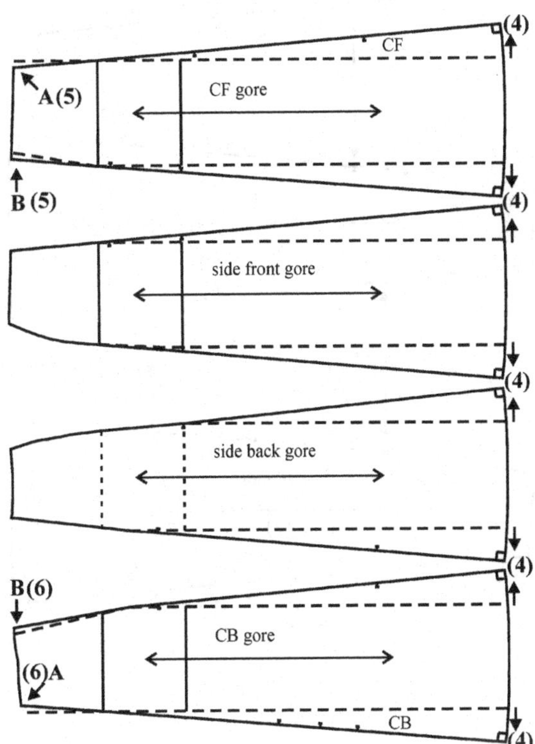

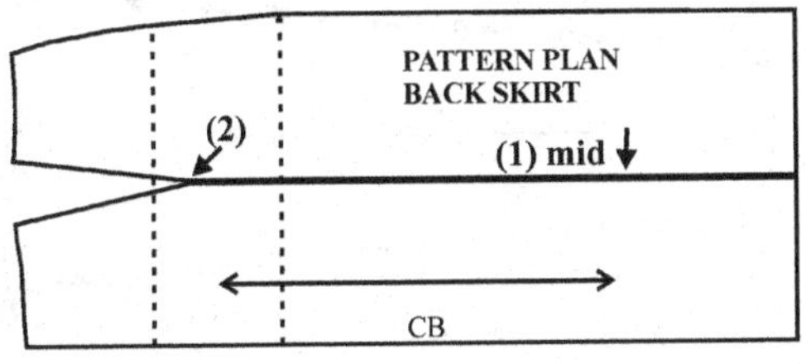

(7) Position a central grain line on each gore.

(8) Notch the adjoining seams in a series that differentiates between the gores.

(8) (8) CF
CF gore
(7)
(8) (8)
side front gore
(7)
(8)
side back gore
(7)
(8) (8)
CB gore
(7)
CB
(8)

Suggested grading for an 8-gored skirt

Each gore can be graded the same amount either side of the central grain line. In this example the skirt has a full 4 cm grade in width. A grading increment can be calculated as $\frac{1}{16}$ of the full waist, hip and hem girth grade. This is ± 0.25 cm in width either side of each gore per size. The length at the hem has an increment of 0.5 that can be optional.

Care has to be taken about the adjoining notches when grading. In this example all the matching notches have been fixed by a code to the grade points at the junction of the seams and hem (see the section on grading notches in Part 2).

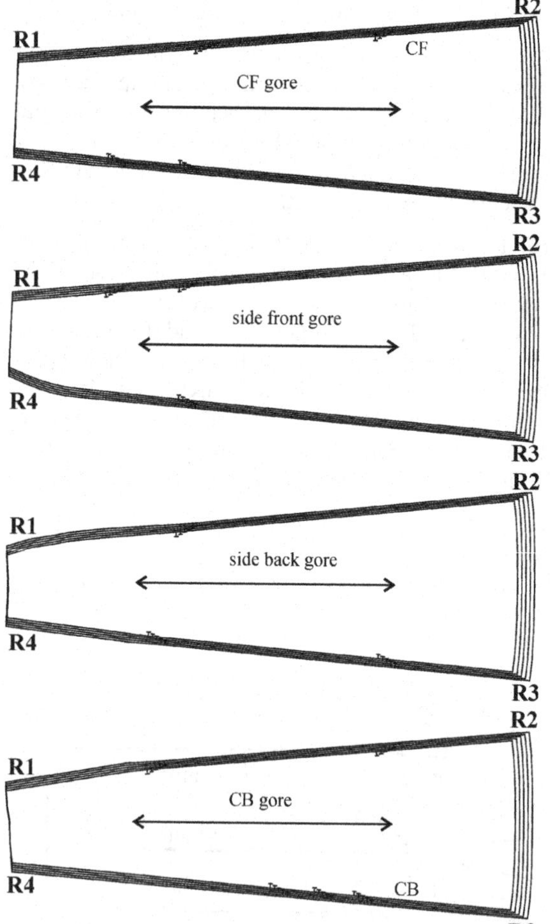

Construction of circular skirts

The construction of this group of flared skirts is based on a segment of a circle. The silhouette and width of the hem varies according to whether the skirt is a full, half or quarter circle. These skirts do not need any dart suppression for shaping from the hips to the waist, therefore the back and front of the skirts can be the same. Only a quarter of each skirt needs to be constructed; for this a waist and length measurement is required, except for the quarter circle skirt that also requires the upper hip measurement.

MEASUREMENTS REQUIRED FOR SIZE 12

	Body	*Ease*	*Skirt*
Waist	70	+4	74
Length	60	—	60

CALCULATION FOR CIRCULAR SKIRTS

To construct these skirts the radius of the waist is required. This is calculated with the formula:

$$\text{Radius} = \frac{\text{circumference}}{2\pi}$$

$$\text{or} \quad R = \frac{\text{waist plus ease allowance}}{2 \times 3.14}$$

FABRIC CONSIDERATIONS

When circular skirts are cut in woven fabric they generally mould around the waist to hip area, as much of the fabric is on the bias. Generally woven fabric does not stretch in the warp, stretches a little in the weft and a considerable amount on the bias. Certain knitted and stretch fabrics also do not require darts for shaping, as their elasticity is sufficient for moulding.

These skirts generally require large pattern pieces. Consideration has to be given to the most efficient placement of the patterns to use the minimum of fabric when cutting. This will influence the skirt style and the positioning of the seams and grain lines.

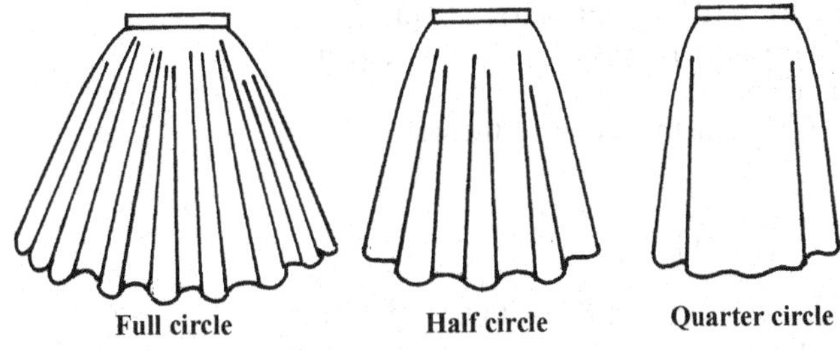

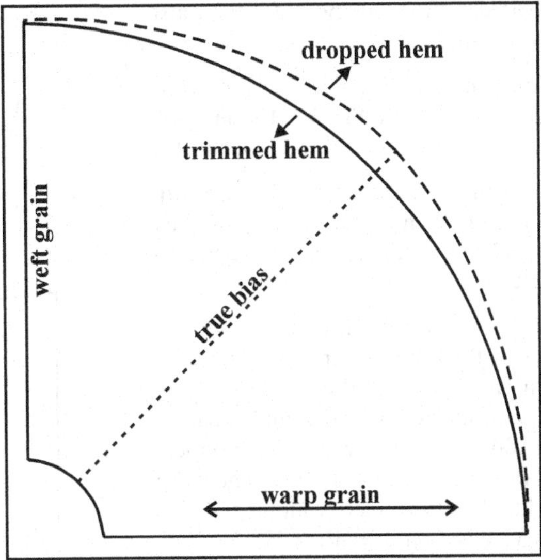

The pattern may need adjustment for fabric reaction. Most of the circular skirt is on the unstable bias grain, which can narrow in width and stretch in length. The cut sample skirt should hang for a minimum of 24 hours, before the surplus fabric is trimmed from the hem and the pattern shortened.

FULL CIRCULAR SKIRT SIZE 12

This pattern is constructed within a quarter of a circle.

Calculate the radius of the waist:

$$R = \frac{74}{2 \times 3.14} = 11.8$$

(1) Construct a square with the sides equal to the waist radius plus the skirt length, **11.8 + 60 = 71.8 cm**. The lower left corner can be used as the centre point for the waist and hem radii.

(2) Construct a full circle for the **waist circumference with a radius of 11.8 cm**. (The diagram has been reduced to a quarter.)

(3) Construct a full circle for the **hem with a radius of 71.8 cm** (waist radius plus skirt length). (The diagram has been reduced to a quarter.)

(4) Trace off the pattern that represents a quarter of the full circular skirt. The final pattern can be cut in three sections, one half circle and two quarters, and then the seams added. This is for more economical use of fabric when lay planning. The warp grain line has been placed along the side seam. A notch identifies the CF.

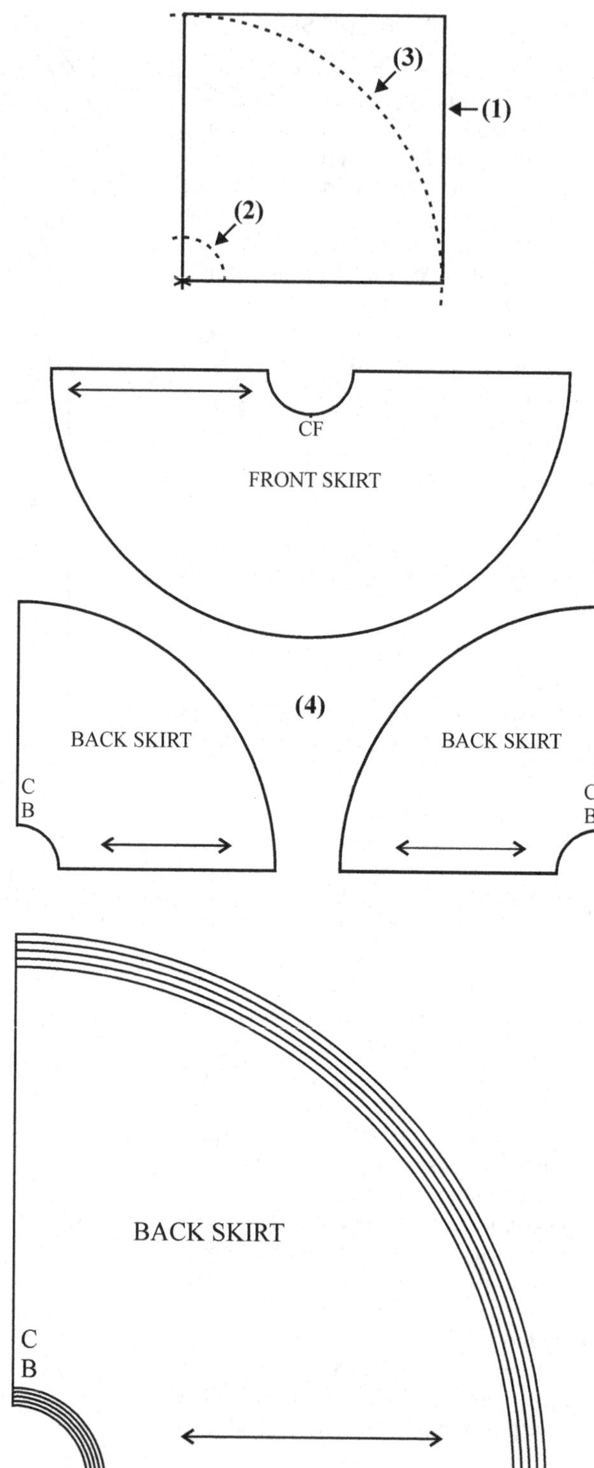

Suggested grading

The grading increment can be calculated as the difference between the radii of two waist sizes, e.g.:

size 14 waist radius = 12.4 cm
size 12 waist radius = 11.8 cm
difference = 0.6 cm

When this increment is used at the beginning and end of a curved waistline the graded curve will be parallel to the original size. (In this example the skirt length grade is 0.5 cm.)

HALF CIRCULAR SKIRT SIZE 12

This skirt is constructed within a quarter circle. The waist circumference for a half circular skirt is twice the skirt waist plus ease. Calculate the radius for the skirt waist:

$$R = \frac{2 \times 74}{2 \times 3.14} = 23.6\,\text{cm}$$

(1) Construct a square with the sides equal to the waist radius plus the skirt length, **23.6 + 60 = 83.6 cm**. The lower left corner can be used as the centre point for the waist and hem radii.

(2) Construct a full circle for the **waist circumference with a radius of 23.6 cm**. (The diagram has been reduced to a quarter.)

(3) Construct a full circle for the **hem with a radius of 83.6 cm** (waist radius plus skirt length). (The diagram has been reduced to a quarter.)

(4) Trace off the pattern that represents a quarter of the half circular skirt. The final pattern can be cut in two sections that each represent two quarter circles, and seams added. This is for a more economical use of fabric when lay planning. The warp grain line has been placed along a side seam. The CB and CF are identified with notches.

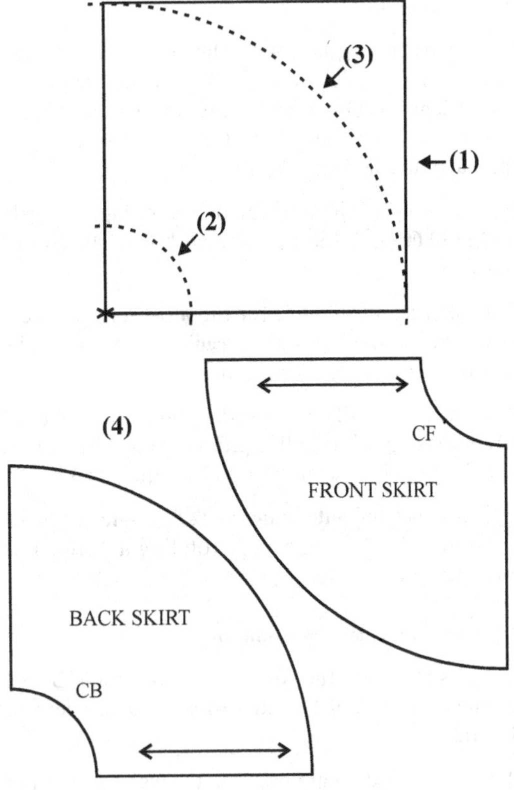

Suggested grading

The same method can be used as explained for the full circular skirt.

QUARTER CIRCULAR SKIRT BLOCK SIZE 12

This skirt is constructed within a sixteenth of a circle.

Measurement required

	Body	Ease	Skirt
(a) Waist	70	+4	74
(b) Upper hip	90	+4	94
(c) Waist to upper hip	10	—	10
(d) Upper hip to knee	50	—	50

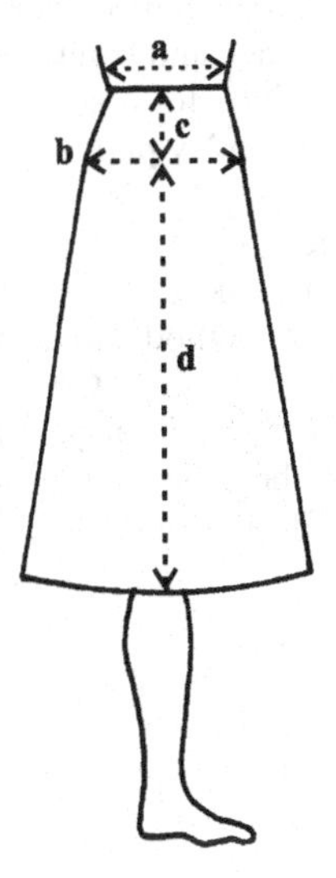

This skirt fits more closely between the waist and upper hip and therefore needs shaping at the side seam. To construct this skirt the basic circumference of the circle is equal to four times the skirt upper hip measurements:

$$R = \frac{4 \times 94}{2 \times 3.14} = 60\,\text{cm}$$

Construction lines

(1) Construct a square with the sides equal to the upper hip radius plus the upper hip to knee length (or desired length), **60 + 50 = 110 cm**. The lower left corner can be used as the centre point for the upper hip, waist and hem radii.

(2) Construct a full circle for the **upper hip level with a radius of 60 cm**. (The diagram has been reduced to a quarter.)

(3) Construct a full circle for the **waist level** at 10 cm above the upper hip **with a radius of 50 cm**. (The diagram has been reduced to a quarter.)

(4) Construct a full circle for the **hem with a radius of 110 cm**, or upper hip radius plus desired skirt length. (The diagram has been reduced to a quarter.)

(5) Construct the **side seam at 22.5° angle** from the lower side of the rectangle (by rotating a drawn line from the centre of the circle).

Waist line and side seam shaping

(6) Measure along the waist level from the CB and CF line a quarter of the skirt waist measurement of **18.5 cm**.

(7) Curve the side seam between the waist and upper hip. Some figure types require the side seam to extend above the waist approximately 1 to 2 cm. The waist line is then curved down towards the CB/CF.

(8) Trace off the pattern that represents a quarter of the skirt. This pattern can be mirrored at the CF/CB line to give a total back and front skirt. Alternatively it can be divided into four gores that give better fabric utilisation when lay planning.

Suggested grading

The quarter circular skirt can be graded using the multi-reference line method similar to that used for flared skirts with a mirrored CF and CB line. The second example is for a four-gored skirt with the reference lines at the central grain line and parallel lines to the CF/CB seam and side seam.

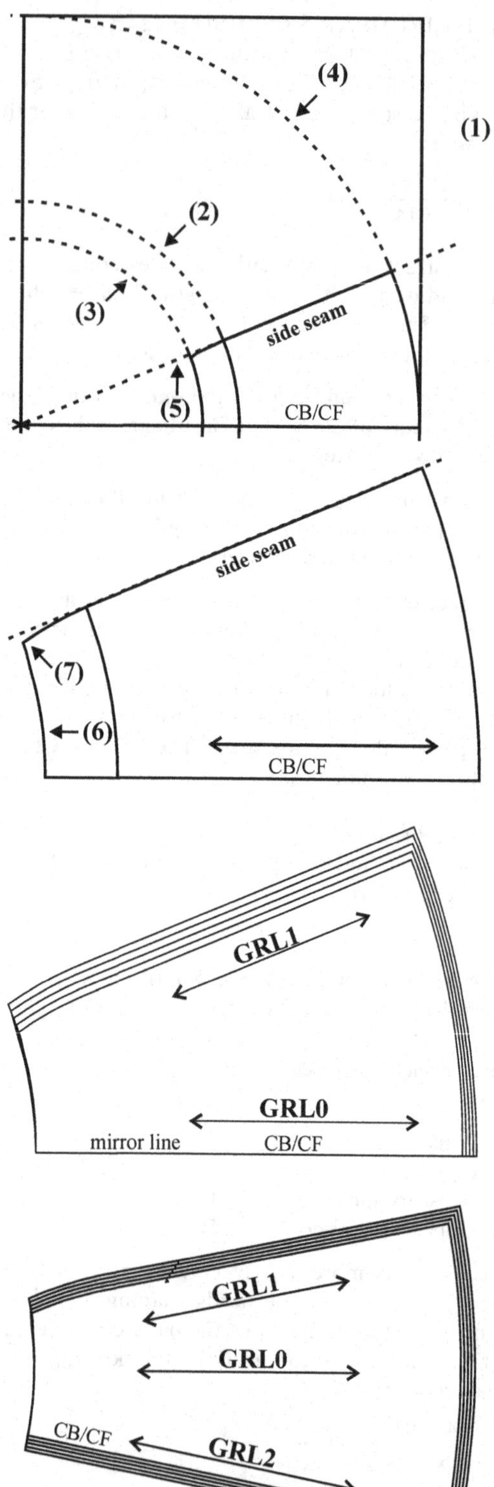

Skirt style with yoke and pleats

This style of a front skirt has a shaped yoke and a series of knife pleats. All the pleats are pressed in the same direction. This means that the right and left sides of the pleated section are different so that the full front skirt has to be constructed because it cannot be mirrored. The yoke on the other hand can be mirrored at the centre front, as the right and left sides are the same.

For permanent pleating a rectangle of fabric is sent to be pleated that is sufficient for the total front skirt and has been hemmed. A pattern can be traced from the skirt plan that includes all the seams but no hem. It is used as a template for re-cutting the pleated front skirt.

(1) Plan the style lines (pleat width 3.5 cm). The apex of the waist dart is lowered to the yoke line.

(2) Trace off the front and side sections of the yoke and merge to close the dart. The CF becomes a mirror line (retain the original grain line of the side yoke for the grade reference line).

(3) For permanent pleating construct a template for re-cutting the pleated section by tracing off the skirt section and mirror at the CF. (All the 9 lines for the pleat positions can be included.) Add all the seams but not the hem.

(4) For hand pleating insert pleats with underlay approximately twice the top pleat width and all folding to the left. This pattern has to be laid face-side up as the right and left side underlays differ.

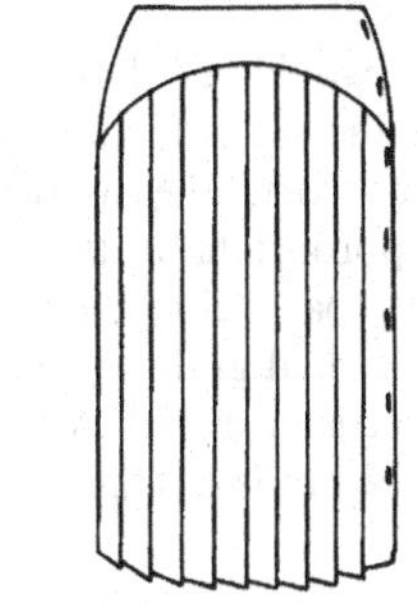

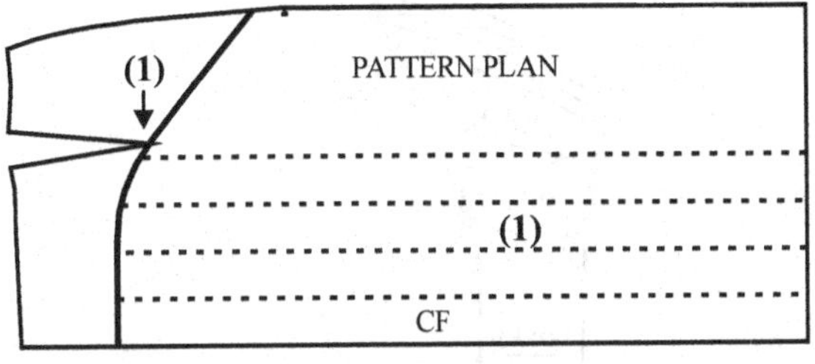

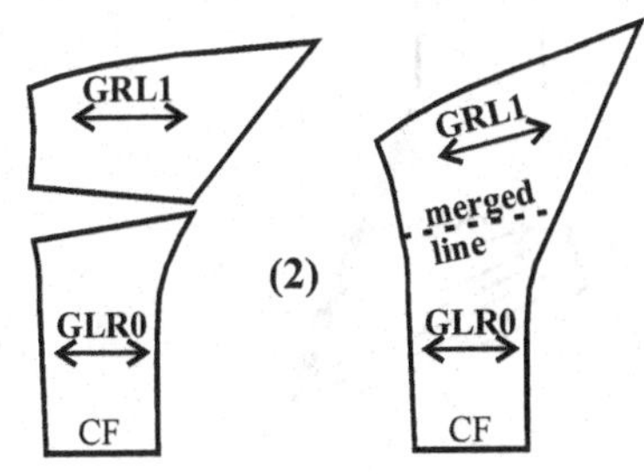

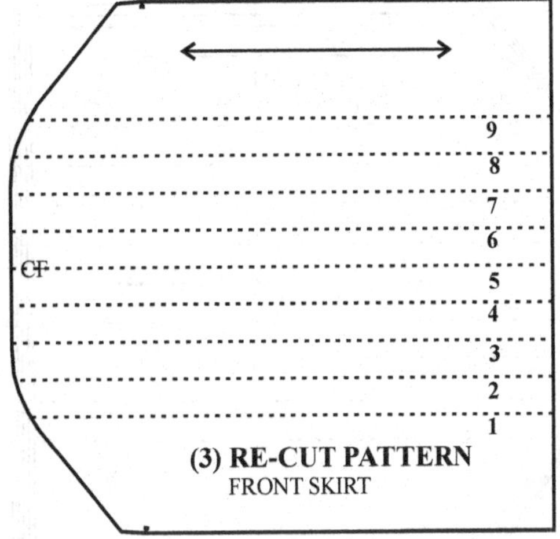

Suggested grading

To grade this style it would be advisable to leave the width of the pleats the same for all sizes and only grade the sides of the skirt. The multi-reference line grading method is appropriate for the yoke. The first grade reference line is parallel to the CF; the second was parallel to the original grain line before the dart was closed. This piece is mirrored. A 're-cut' pattern would also have to be graded. This example has the CF line mirrored.

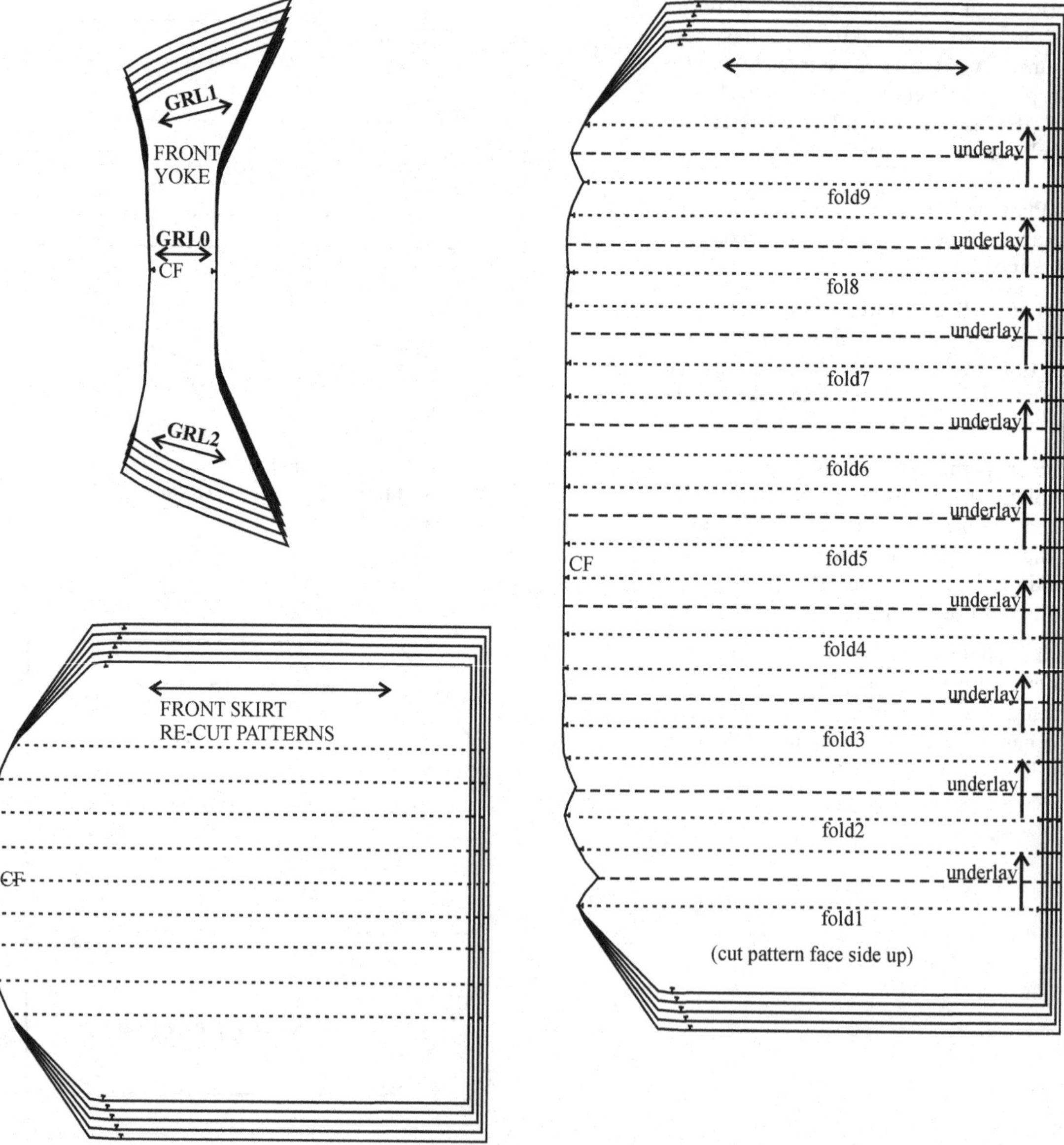

POCKETS

Pockets are both functional and decorative. They can be either applied or inserted into a garment. The pocket should be positioned for easy access by the hands. There are basically three groups of pockets: patch, piped or jetted, and inset.

Patch pocket

A **patch pocket** is applied to the surface of the garment.

(1) Copy the front skirt and make a pattern plan of the patch pocket shape and position.

(2) Trace off the pocket and facing from the pattern plan.

(3) Position a drill hole and notches on the front skirt pattern to indicate the position where the pocket will be attached. (The drill holes have to be placed approximately 0.5 cm away from the stitch line at the pocket corner so that they are concealed when the pocket is stitched.)

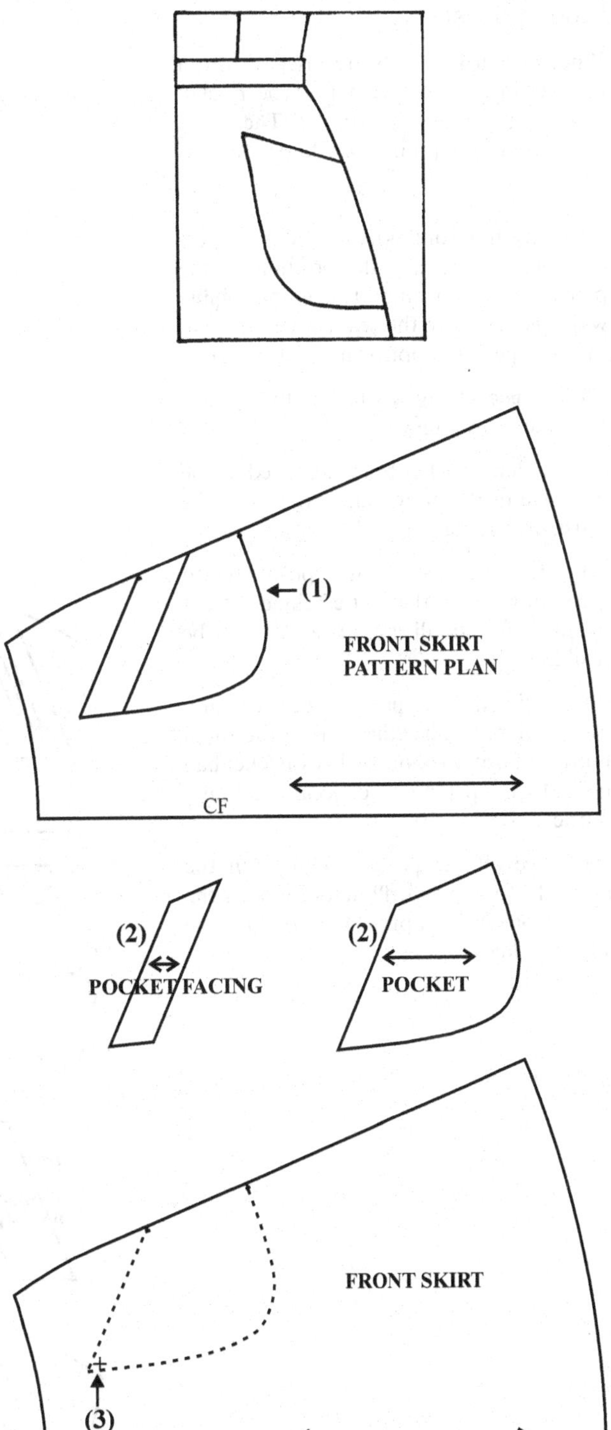

Piped pocket

Piped or jetted pockets are inserted into a cut within the garment part. The pocket bag hangs inside the garment. The edges are bound by a piping also known as jetting.

(1) Copy the front skirt and make a pattern plan to indicate the position of the piped pocket opening. The visible **piping** width is generally the seam width around the cut opening (approximately 0.7 cm).

(2) The **pocket bag** is attached to the edge of the lower piping.

(3) The **back pocket bag** is attached at the top edge of the piping and extends to the waist, stabilising the pocket bag.

(4) A reinforcing **stay** can be positioned to the wrong side of the fabric behind the cut opening, for stabilising. (The stay can be fusible.)

(5) (Optional if the pocket bag is cut in a lining fabric.) A **backing**, cut in the main fabric, is positioned on the back pocket bag behind the opening to conceal the lining fabric.

(6) Trace off the pattern pieces. On the front skirt position **drill holes** to mark the pocket opening (approx. 0.5 cm from the stitching line).

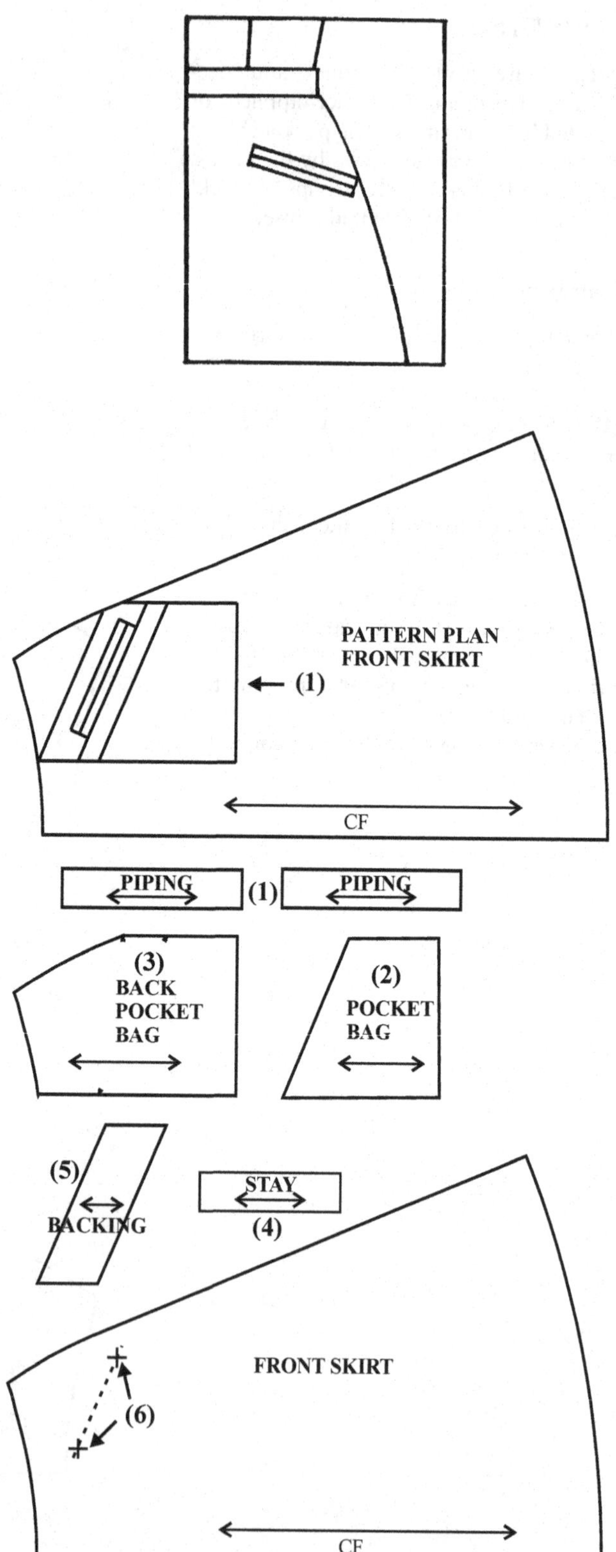

Inset pocket

Inset pockets are inserted into a seam or style feature. In this example of a front skirt the pocket is inserted into a curved seam. The pocket opening is faced by the pocket bag. The back pocket bag is an extension of the side waist to hip section.

(1) Copy the front skirt and make a pattern plan for the position of the pocket opening.

(2) Increase the width of the skirt at the side seam (approx. 1 cm) at the lower end of the opening tapering to nothing at the hem. This allows the faced curve to lie smoothly over the hip section.

(3) The lower edge of the opening is faced by the **pocket bag**. The top of the pocket bag is stabilised by stitching it into the waist seam.

(4) The faced curve can be **interlined**.

(5) The **back pocket bag** has the same side seam as the original skirt.

(6) Trace off the pattern pieces. The opening position has to be notched on both the front hip section and the back skirt side seam.

Grading of pockets

It is advisable to avoid grading pockets as their shape can become distorted. If necessary it is better to reposition them by grading the surrounding area.

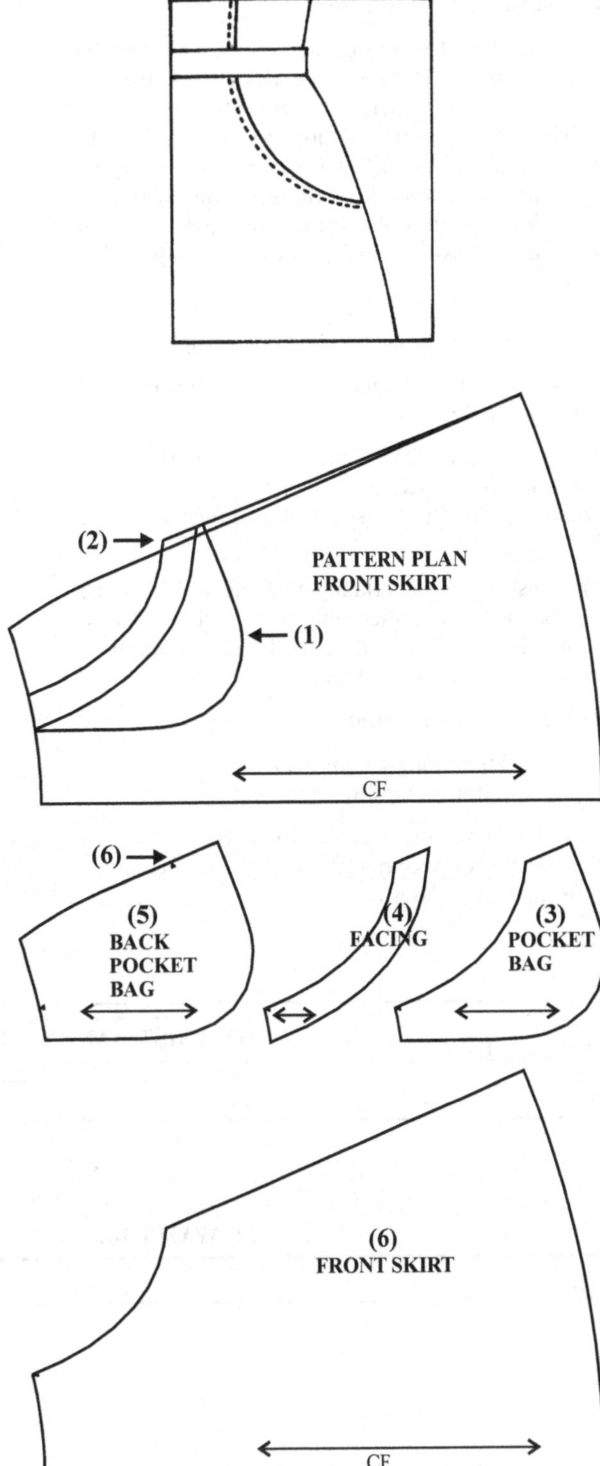

Waist band and placket

The waistline of separate skirts can be finished by several methods. The most general is a waistband and a placket opening fastened by a zip. The waistband is stabilised by interlining or slotted waist banding. In this example a button fastens the waistband. A **zip guard** protects the back of the zip. The skirt waist generally eases on to the waistband (approx. 2.0 cm) to enable the skirt to mould around the upper hip area.

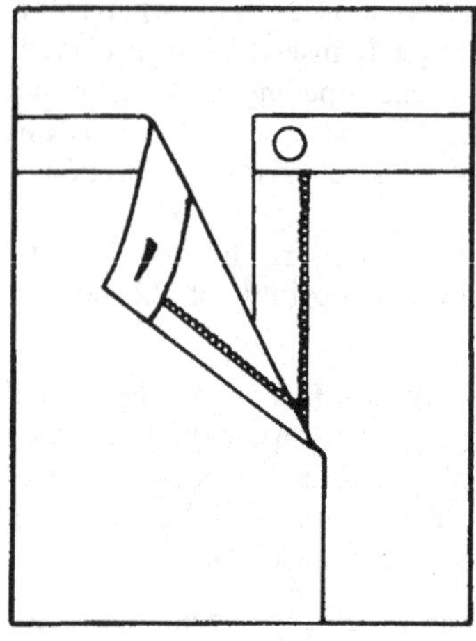

(1) For the **waist band** construct a rectangle:

length = waist + 2.0 cm ease + 3.0 cm button stand
width = band width + facing.

(1 cm seam allowance can be added to the calculation or to the pattern later.)
Notch the CB, side seams and CF and the facing fold line.

(2) Construct the interlining half the width of the waistband. Fusible interlining only requires 0.4 cm seam allowances to reduce bulk. (No pattern is needed for slotted waist banding.)

(3) Construct a **zip guard**:

length = 5.0 cm longer than the zip
width = button stand width (3.0 cm).

Curve the lower corner of the zip guard.
(1 cm seam allowance can be added to the calculation or to the pattern later.)

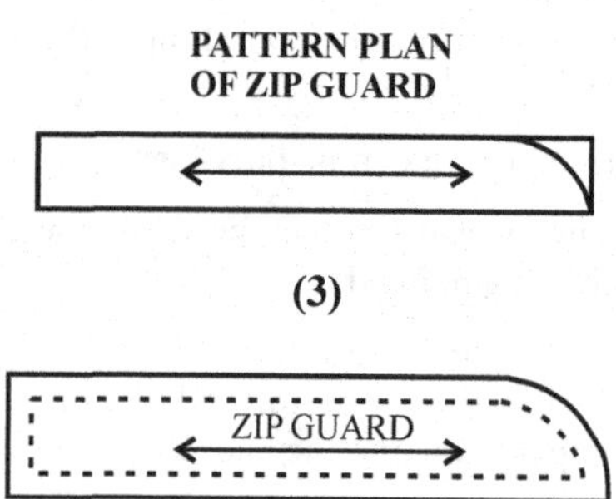

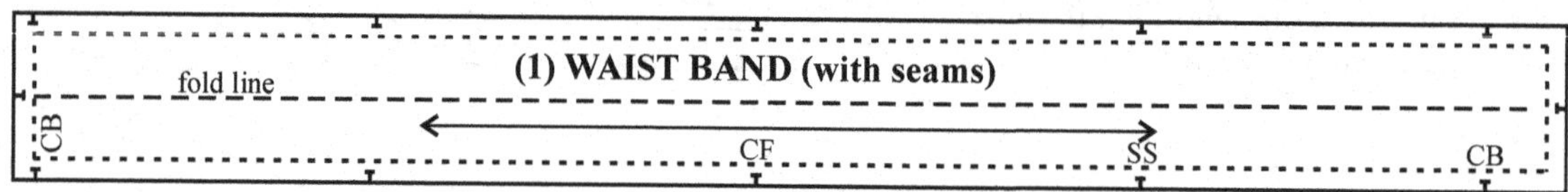

(2) WAIST BAND INTERLINING

CB SS CF SS CB

BODICE STYLING

The style features of a bodice often incorporate suppression in various decorative forms, such as darts, gathers, flare, panels or drapery. To achieve these style features the principles and method of dart manipulation have to be understood. Suppression radiates from the prominence of the body, as illustrated in Figure 3.3. The front darts can be combined into one dart as they radiate from the same prominence of the bust, whereas at the back the shoulder blade prominence is a long prominence requiring two separate darts that cannot be combined (Figure 3.5).

Technically the front darts can be positioned from any perimeter edge of the pattern, but there are some positions which are considered more complimentary to the wearer. Typical front dart positions are illustrated in Figure 3.3.

The *shoulder dart 1* and *waist dart 5* are positioned when constructing the bodice block to maintain the bust level parallel to the ground. *Shoulder dart 1* and *armhole dart 2* are generally incorporated into panel seams with *waist dart 5*. Darts themselves are not always considered decorative and so shoulder darts are often transferred to *underarm 3* or *side seam 4* positions where they are partly concealed by the arm. It is more complementary to the figure shape for side darts to slant up towards the bust prominence. The *waist dart 5* is generally used for styles fitting at the waist. The *centre front dart 6* and *neck dart 7* are more usually incorporated into another style feature. For some jackets the *neck dart 7* can be positioned under a wide lapel or collar.

For a smooth contour the apex of all bust darts should finish 2.0 cm to 5 cm from the point of the bust prominence. This depends on the size of the bust (Figure 3.4). The dart apex for fuller busts finishes further from the bust prominence. The bust prominence (BP) is the pivot point when manipulating the darts into another position.

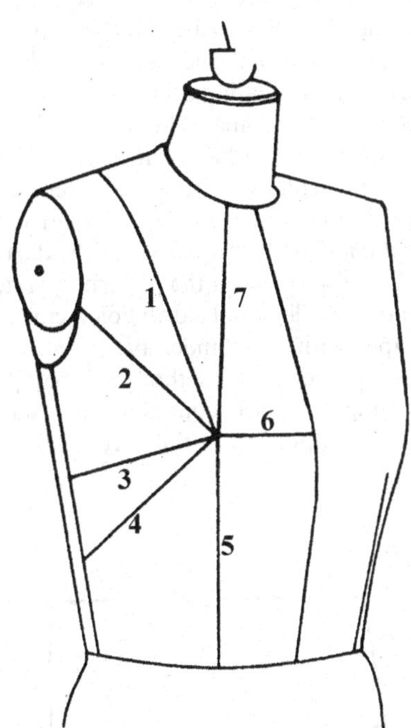

Figure 3.3 Front bodice dart positions

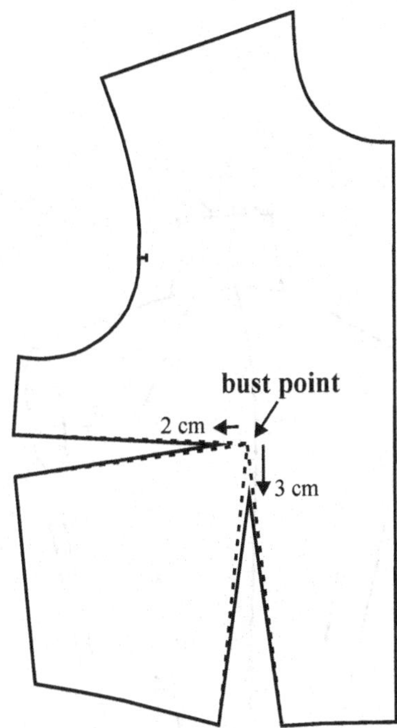

Figure 3.4 Completed front bodice darts

The two darts on the back bodice cannot be combined as they do not meet at the same prominence (Figure 3.5). The *shoulder dart 2* can be pivoted as a *neck dart 1*. This is useful for cut away necklines and sleeveless styles with narrow shoulders. When the dart is pivoted into the *armhole 3* it is generally incorporated into a yoke seam. In some patterns for a wearer with a flat back or soft moulding fabric the *shoulder dart 2* suppression can be eased on to the front shoulder seam. To control this, notches have to be placed so that the fullness is directed over the shoulder blade (Figure 3.6) and does not interfere with the neck or armhole. The *waist dart 4* is used for garments fitted at the waist. It can be incorporated into a panelled seam with *shoulder dart 2*, providing the seam runs over the shoulder blade prominence.

Dart manipulating is easy and quick when using most computer pattern design systems. There are several points which have to be remembered.

First, the computer has to recognise the dart generally as a perimeter line and not folded. If it is folded it will have to become unfolded and the drill hole at the apex removed. The terms related to the dart (Figure 3.7) are as follows:

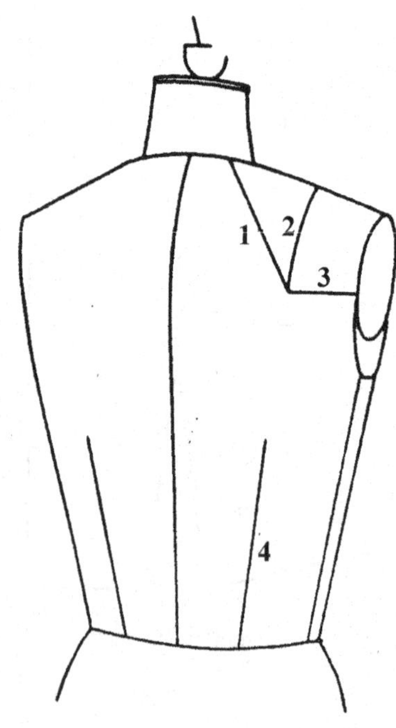

1 neck
2 shoulder
3 armhole
4 waist

Figure 3.5 Back bodice dart positions

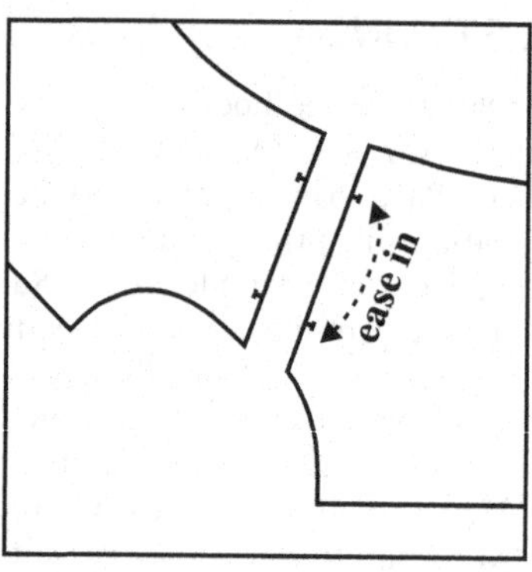

Figure 3.6 Suppression of the back shoulder seam 'easing in' on the front shoulder seam

The *dart apex 1* is the pointed end of the dart. The *dart legs 2* have to be of equal length for stitching together. The *pivot point 3* is not necessarily at the apex, but the apex has to be extended to the *pivot point* for manipulation. The *hold line 4* has to be selected when manipulating darts to keep the pattern on the correct grain line. The *dart underlay 6* is concealed after the dart has been stitched. The perimeter edge is calculated by the computer according to the direction in which the dart *underlay* is folded. The *central line 5* is the inner fold of the dart *underlay*. The dart is marked on the pattern by notches at the end of the dart legs and a drill hole 1 cm to 0.5 cm from the apex within the underlay.

There are three basic methods of manipulating darts: pivoting one dart, combining two darts and distributing a proportion of the dart. These are illustrated below.

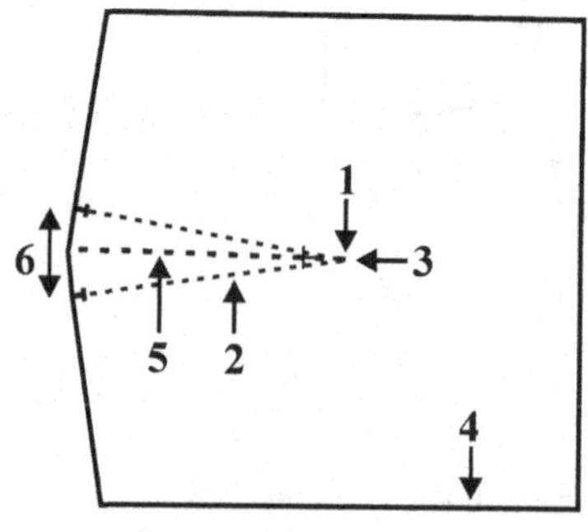

1 apex
2 dart leg
3 pivot point
4 hold line
5 central line
6 dart underlay

Figure 3.7 Sections of a dart

Pivoting a dart

This is an example of pivoting the front bodice *shoulder dart 1* (Figure 3.3) to the *underarm dart 3.*

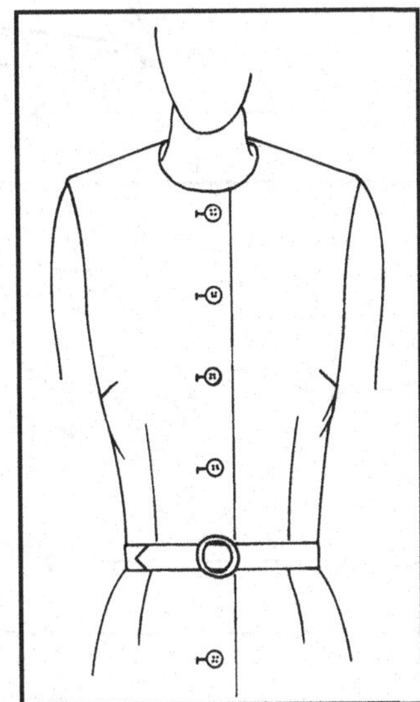

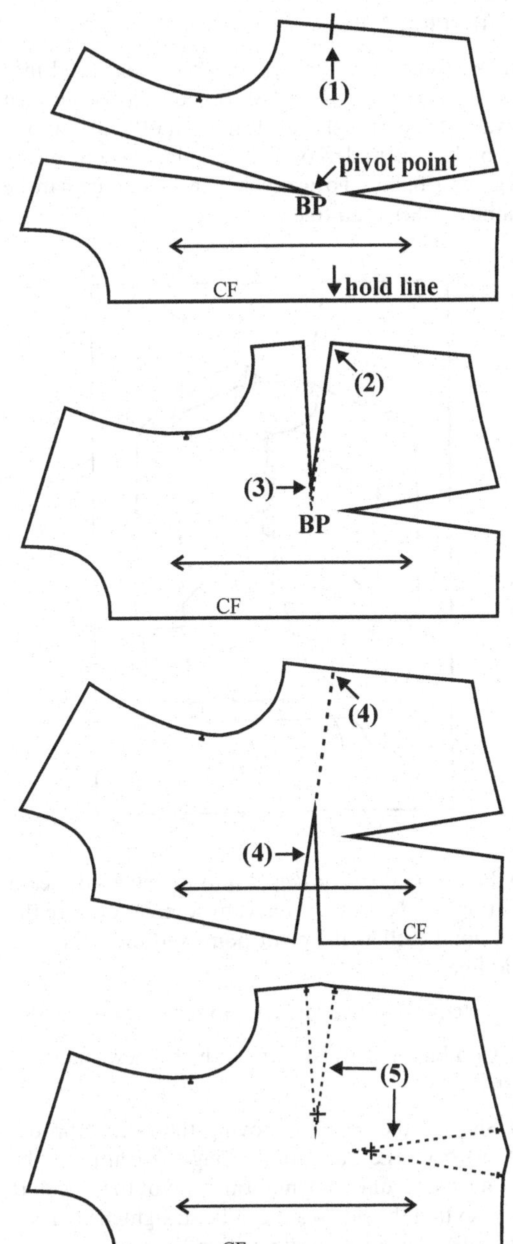

(1) Mark the new underarm dart position. Select the apex of the shoulder dart at the BP as the pivot point and the CF as the hold line.

(2) Pivot the *shoulder dart 1* into the new underarm position by selecting the relevant commands.

(3) Shorten the dart by moving the apex (approx. 2.0 cm to 5.0 cm) along the central line of the dart. The dart legs have to be of equal length.

(4) If the seam at the new opening has to be a straight line the new dart will have to be temporarily pivoted and the line corrected. The dart is then pivoted back to the underarm position.

(5) The **dart underlay** can now be completed. The computer program calculates the perimeter edge of the folded dart according to the direction in which the dart will lie after stitching. In this example the inner fold of the underarm dart lies towards the waist, the inner fold of the waist dart lies towards the CF.

Suggested grading

The same grade rules can be applied to this pattern as to the original block. The amount that the shoulder and armhole have moved is so slight that the alternative grade reference line grading method is not applicable.

Combining darts

The shoulder and waist darts can be combined into one large diagonal dart. As this dart is positioned towards the bias of the fabric it will mould better over the body. The underlay of such a dart becomes very large and bulky. To overcome this the dart can be seamed rather than folded.

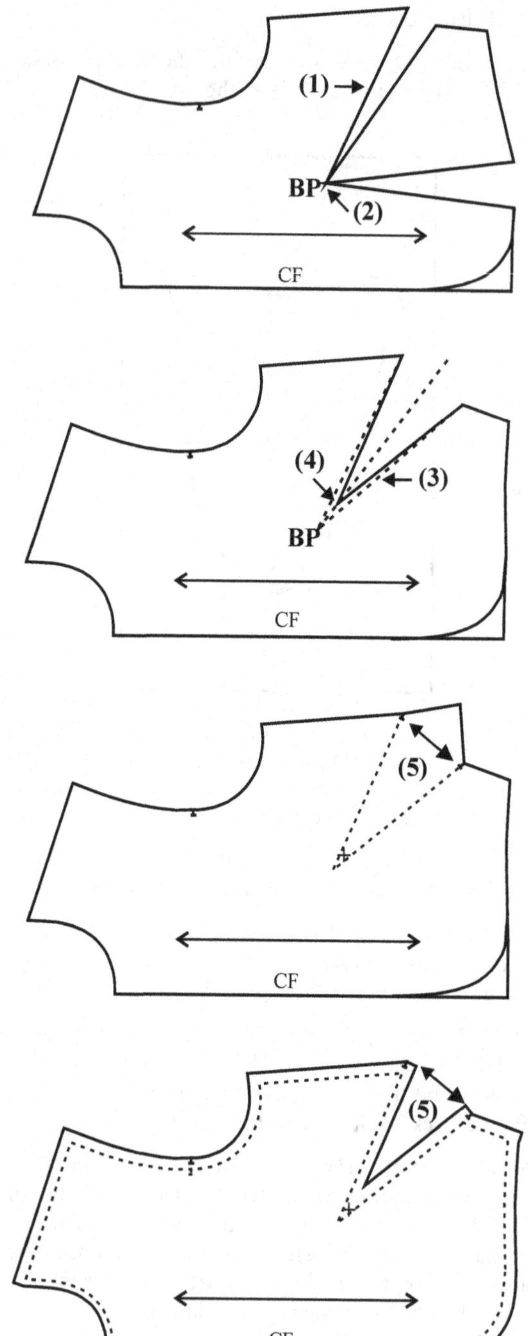

(1) Pivot the *shoulder dart 1* into the new side seam position by selecting the relevant commands. Use the bust point (BP) as the pivot point and the CF as the hold line.

(2) Extend the waist dart apex to the bust point (BP).

(3) Combine the waist dart with the new diagonal dart.

(4) Shorten the dart by moving the apex (approx. 3.0 cm) from the bust point along the centre of the new dart. (If required, temporarily pivot the new dart away so that the side seam can be straightened. Then return the dart to its new position.)

(5) If the new dart is very large it may be more appropriate to seam the dart.

Suggested grading
The original block grade rules can be applied, except to the diagonal dart when they may require editing (see 'Computer grading techniques', p. 66).

Combining darts for gathering

Suppression can be in the form of gathers, for example at the waist, achieved by combining the shoulder and waist darts. This changes the direction of the side bodice and means that a second grade reference line (GRL1) is required for grading, parallel to the original CF grade reference line (GRL0)

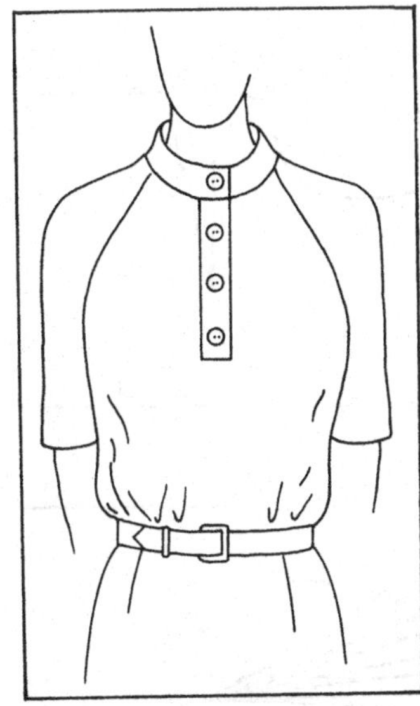

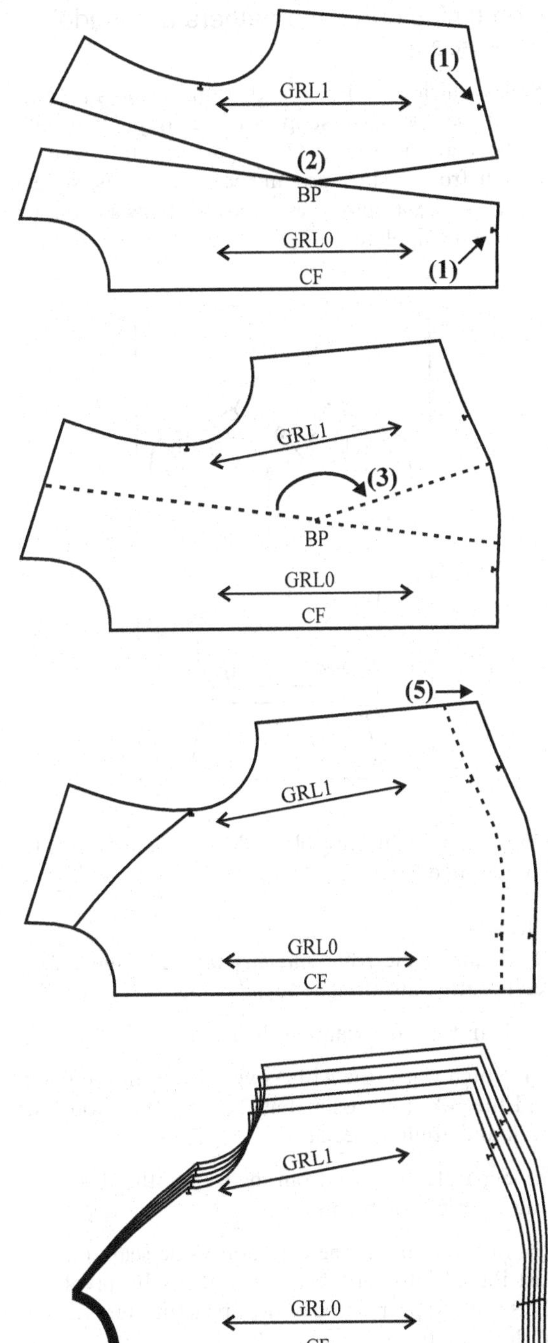

(1) Plan the position of the area to be gathered. This has to position the gathers under the bust prominence.

(2) Extend the *waist dart 5* (Figure 3.3) apex to the bust point (BP).

(3) Combine the *shoulder dart 1* with the *waist dart 5*. Use the bust point (BP) as the pivot point and the CF as the hold line.

(4) Re-curve the waist seam, eliminating the dart.

(5) To increase the blousing effect the new waist seam can be lowered (approx. 3.0 cm to 5.0 cm).

Suggested grading
For grading a small size range the multi-grade reference line grading method could be used. Using two grading reference lines positioned from the original bodice block before the darts were combined.

Combining darts into gathers at a saddle yoke seam

Styles which have horizontal seams such as a yoke, can have the dart suppression manipulated into gathers at the seam. These gathers still have to radiate from the bust prominence. This blouse style is based on the straight dress block as it has a shoulder dart to manipulate.

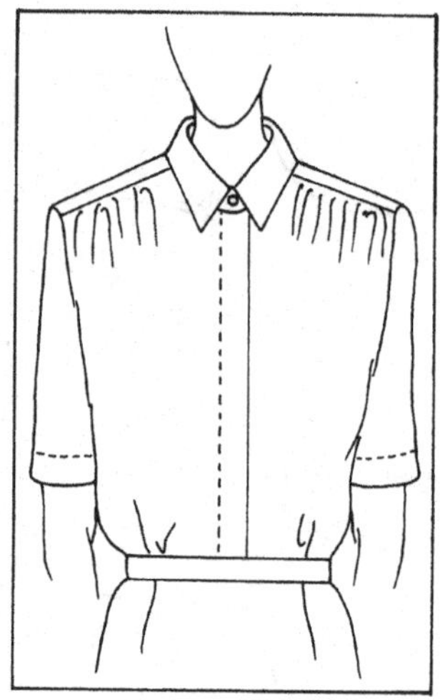

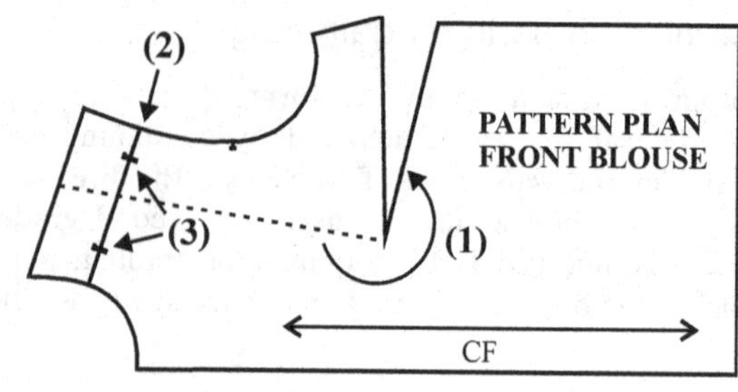

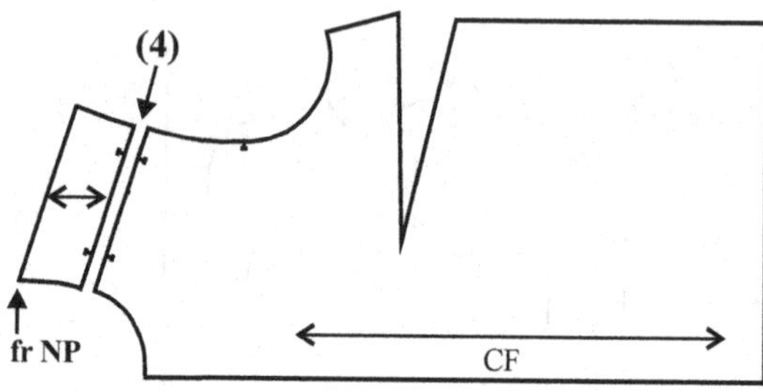

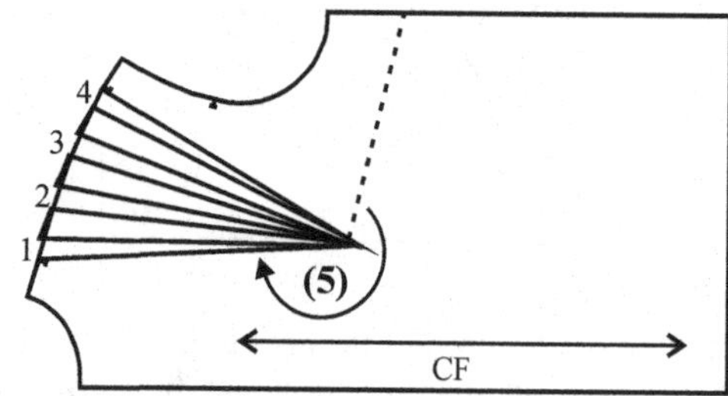

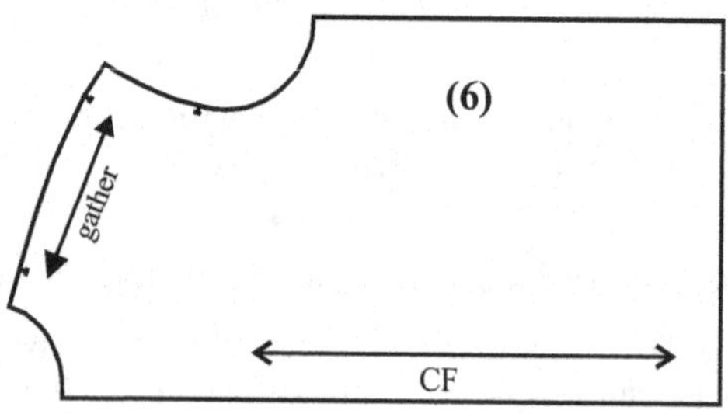

Copy the straight dress block. Move the hem level to the required length.

FRONT

(1) Transfer the *front shoulder dart 1* (Figure 3.3) temporarily into the side seam.

(2) Plan the yoke seam on the front.

(3) Notch where the gathers will finish on the front and the yoke (approximately 2 cm or 3 cm from the neck and armhole seam.

(4) Separate the yoke from the front either by splitting the block or tracing.

(5) In this example the temporary side seam dart is distributed into four small even darts by pivoting. This is in the proportion of the first a quarter, second a third of the remainder, the third a half of the remainder and finally the remainder. A smooth curved seam line is drawn across the ends of the four small darts. (Some computer programs may have the facility to distribute the temporary side dart evenly between the shoulder notches.)

(6) Trace the new front pattern.

BACK
Trace the back blouse. The CB becomes a mirror line.

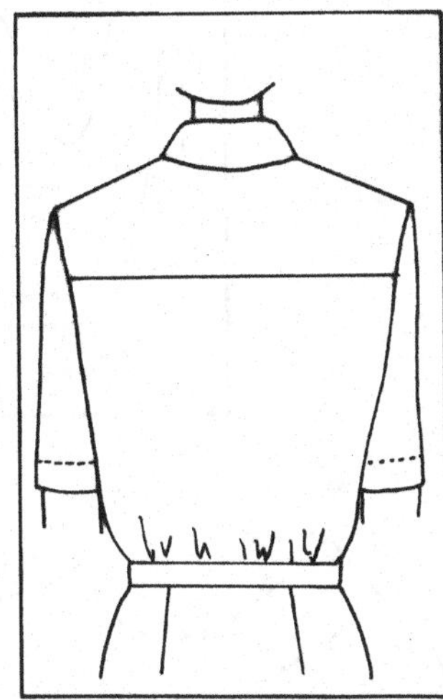

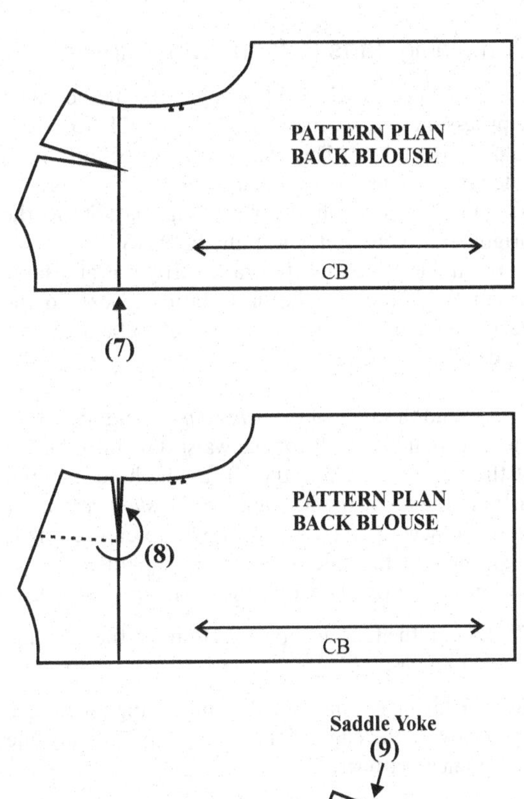

(7) Plan the back yoke seam.

(8) Pivot the *back shoulder dart 2* (Figure 3.5) into the armhole at the seam position. The left leg of the dart becomes the yoke seam, the right leg the back bodice seam. The yoke seam appears curved on the flat pattern but when positioned on the three-dimensional body it will appear horizontal and parallel to the ground.

(9) Trace the back yoke. Flip over and pivot the front yoke so that it can be merged with the back yoke along the shoulder line, matching at the neck points. The CB becomes a mirror line for the yoke and the direction of the grain line.

(10) Trace the back blouse. The CB becomes a mirror line.

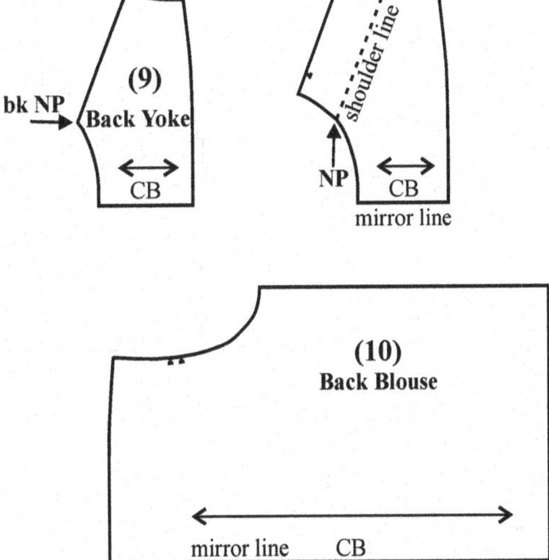

Suggested grading
The back and front bodices will be graded as the original block. As the front yoke is small the total yoke can grade as the original back bodice block.

Combining darts to form draped folds

For a draped cowl neckline style the bodice dart suppression can be transferred to the CF to create folds in the fabric. This changes the direction of the side bodice. This means that a second grade reference line (GRL1) is required for grading, parallel to the original CF grade reference line (GRL0).

To enable these folds to drape softly it is advisable to cut the bodice on the bias grain and extend the neckline facing. This grain line is not to be used as a grade reference line.

(1) Extend the apex of the *waist dart 5* (Figure 3.3) to the bust point (BP). Pivot the waist dart into the CF at the bust level. Using the CF as the hold line and (BP) as the pivot point, combine the *shoulder dart 1* with the new dart. (A second grade reference line is positioned at the side of the bodice, parallel to the original CF (GRL1) before the darts are pivoted.)

(2) Extend the CF line upwards to meet the neckline at right angles.

(3) The facing is constructed by mirroring the bodice and drawing the edge of the facing. Split at this line for the new pattern.

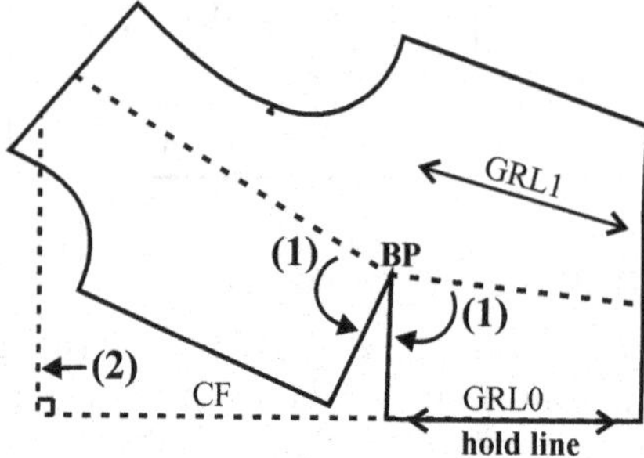

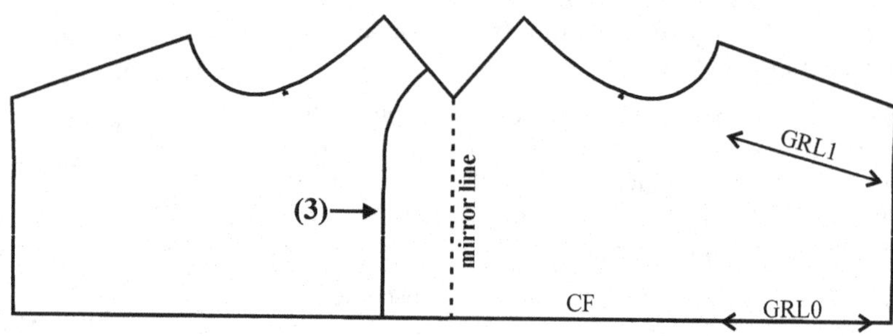

(4) Draw a new grain line at 45° to the CF. This is different from the grading reference lines.

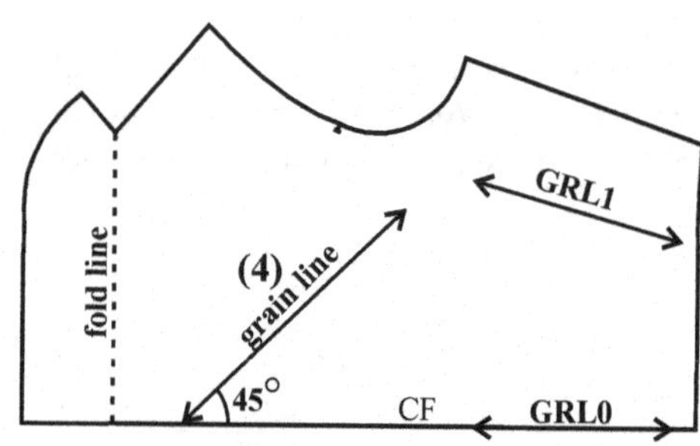

Suggested grading
For grading a small size range, the alternate grade reference line grading method could be used. This example illustrates the mirrored pattern using the CF as the mirror line. The GRL1 and all the grade rules have also been mirrored. The computer is programmed to rotate the pattern so that the grain line is on the X-axis when lay planning.

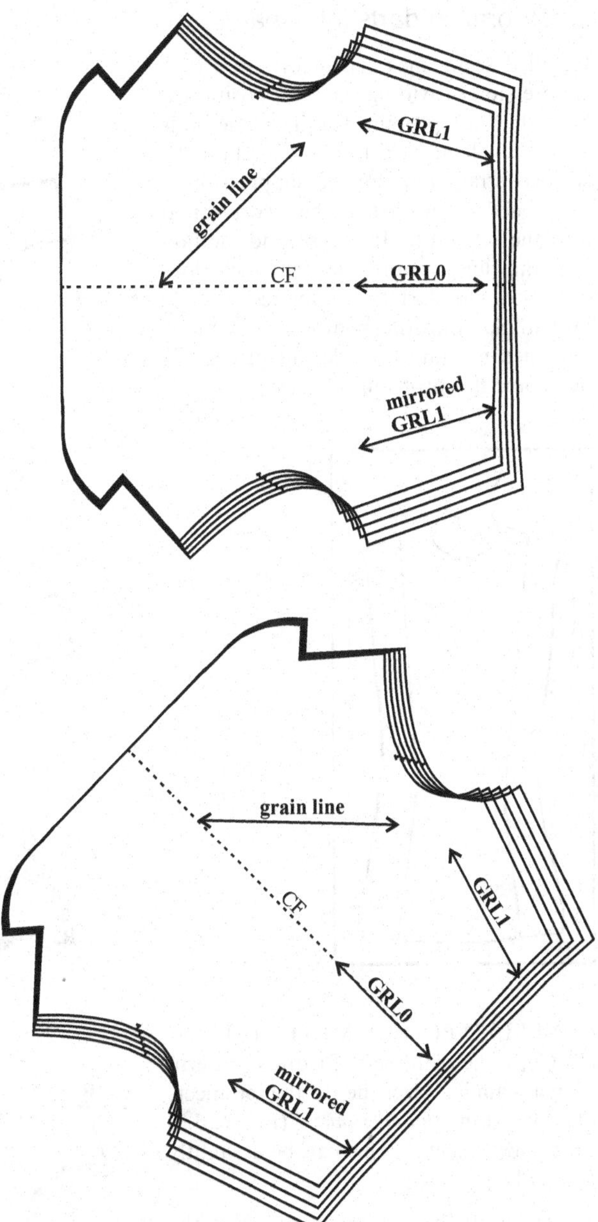

Incorporating darts into seams

Panelled seam lines can incorporate dart suppression providing they are positioned over the bust prominence. Two methods will be explained. In the first method the seam is straight where the suppression is positioned so the dart can be incorporated into the seam line. In the second method the seam line is more curved, which does not allow the dart to be pivoted directly into it. Splitting the bodice into sections and then merging the relevant pieces can overcome this problem.

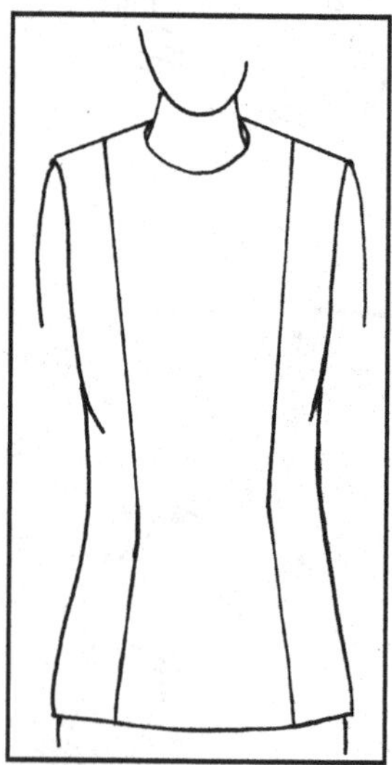

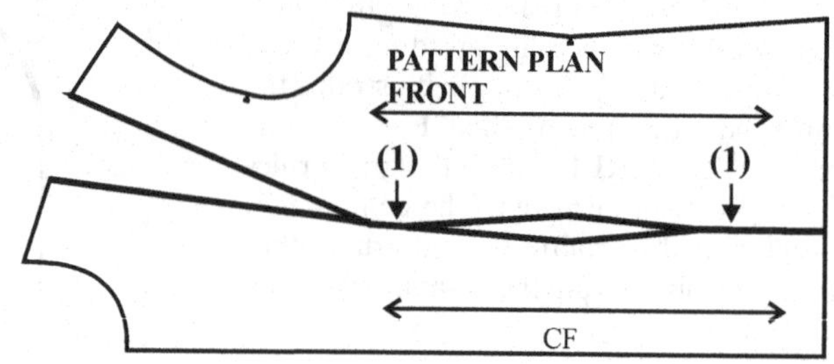

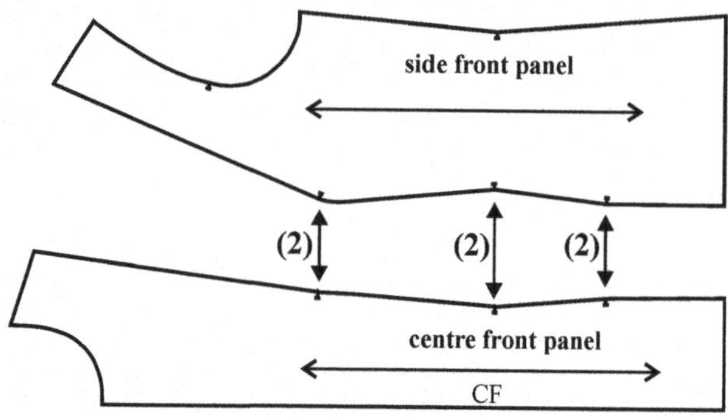

PANELLED BODICE METHOD 1

(1) Copy the front semi-fit dress pattern. Plan a seam line over the BP that connects the apex of the *shoulder dart 1* (Figure 3.3) to the *waist dart 5*. This can be a slightly curved line.

(2) Trace off the pattern pieces from the pattern plan. Position notches at the BP and waist to match the new seam line when stitching.

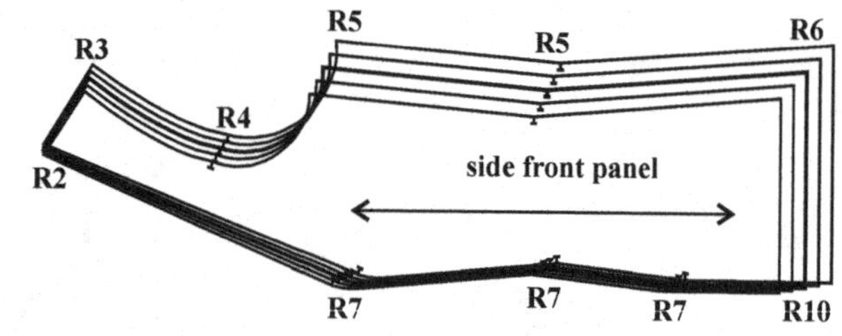

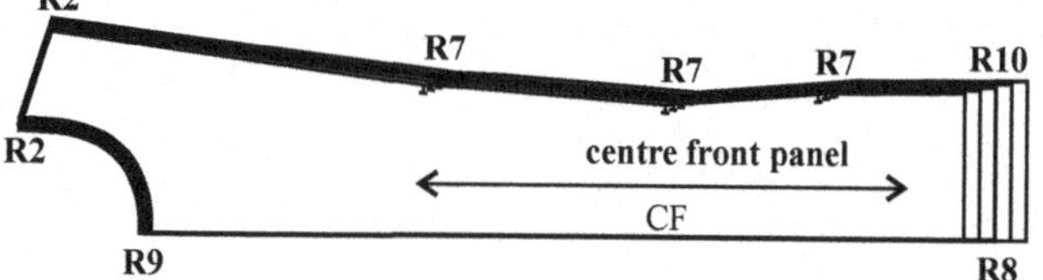

Suggested grading

Some computer programs allow the panel patterns to be graded proportionally, as for the original block. This is because the panels were constructed by splitting the semi-fit dress block. The matching point at the panel seam, where the block was split, will have the same grade rule numbers.

PANELLED BODICE METHOD 2

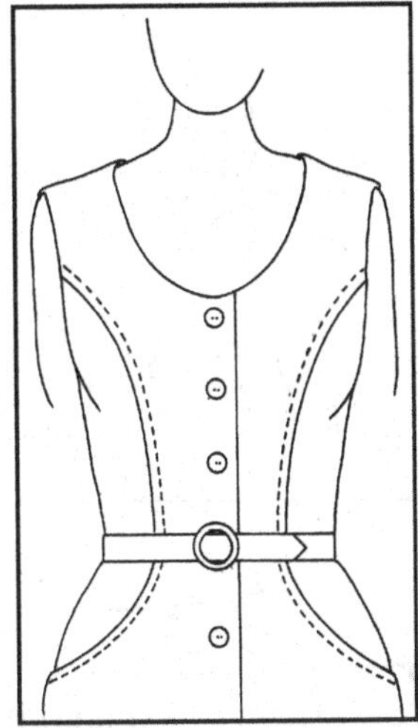

(1) Copy the front bodice block. Copy the grain line on to planned side panel.

(2) Plan the seam line, which will curve from the mid armhole to the bust prominence.

(3) For a closer fit under the bust the seam can be curved in approximately 1 cm at 8 cm from the waist.

(4) Copy the plan, then trace or split it into three sections.

(5) Merge the armhole section to the centre front section at the original shoulder dart position. The side section is the new side panel.

(6) Position notches on the new seam line at the bust level and at 8 cm above the waist.

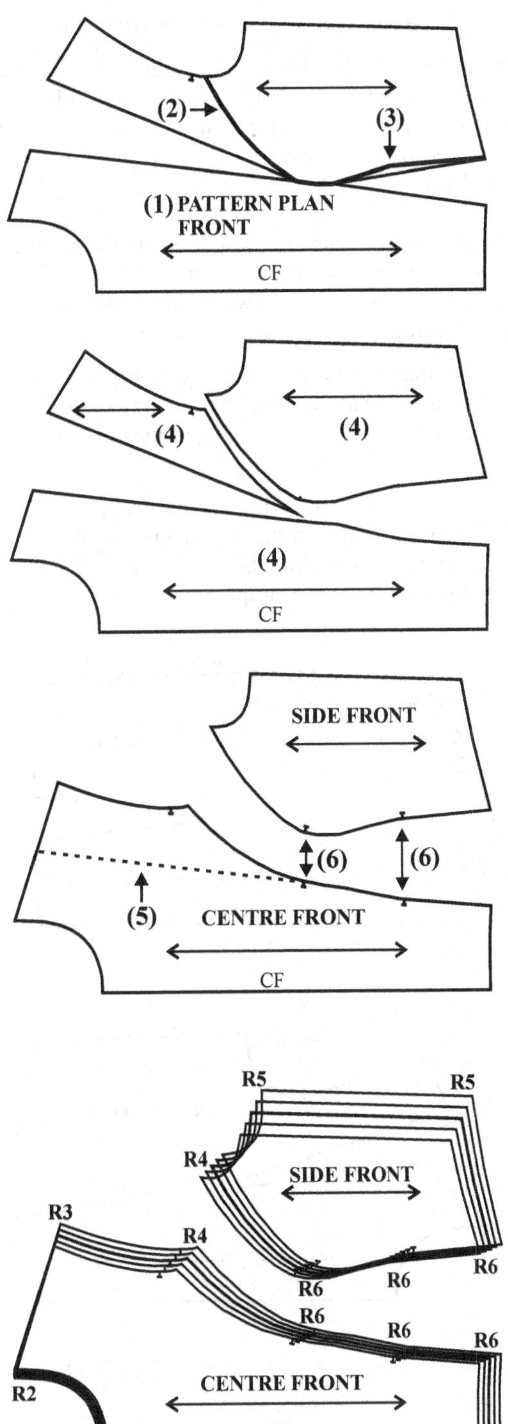

Suggested grading

Some computer programs allow the panel patterns to be graded proportionally, as for the original block. This is because the panels were constructed by splitting the bodice block. Consequently the matching point at the panel seam, where the block was split, will have identical grade rule numbers. If this facility is not available the adjoining panel seam can be edited and the same grade rule applied.

Modification for a sleeveless armhole

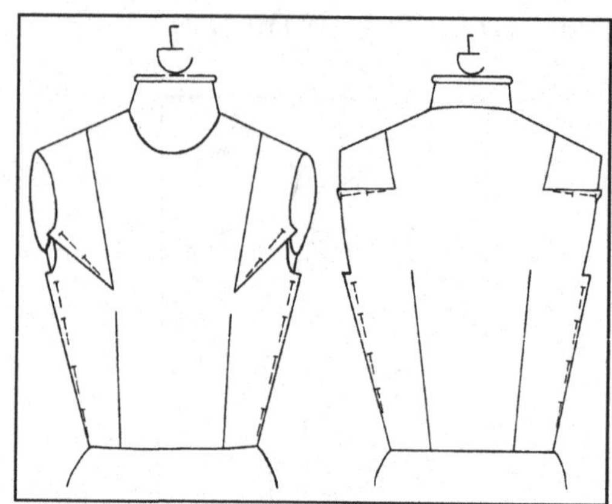

A sleeveless bodice requires a closer fitting armhole that does not gape. Therefore the ease allowance related to a sleeve and arm movement can be removed from a bodice block constructed for a 'set-in' sleeve.

(1) Reduce the ease allowance from the back and front side seam (approx. 1.0 cm) at the underarm tapering to nothing at the waist. Raise the armhole at the underarm seam (approx. 1.0 cm to 3.0 cm).

(2) On the back bodice insert a dart (approx. 1.0 cm to 2.0 cm wide) at the armhole, tapering to nothing at apex of the shoulder dart. Combine this new dart with the shoulder dart.

(3) On the front bodice insert a dart (approx. 1.0 cm to 2.0 cm wide) at the armhole, tapering to nothing at BP. The new dart can then be combined with an existing dart. In this example it is the shoulder dart.

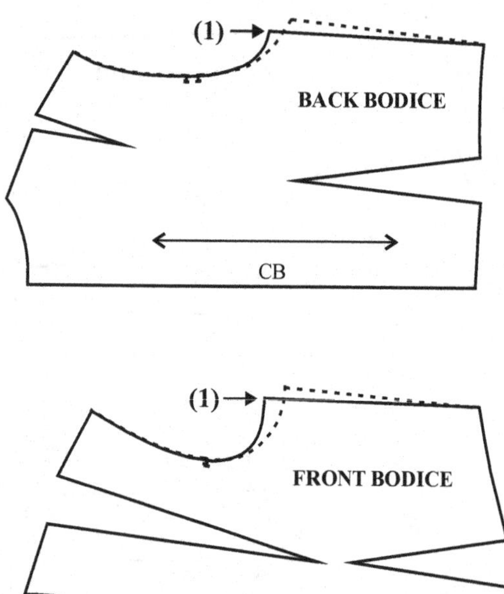

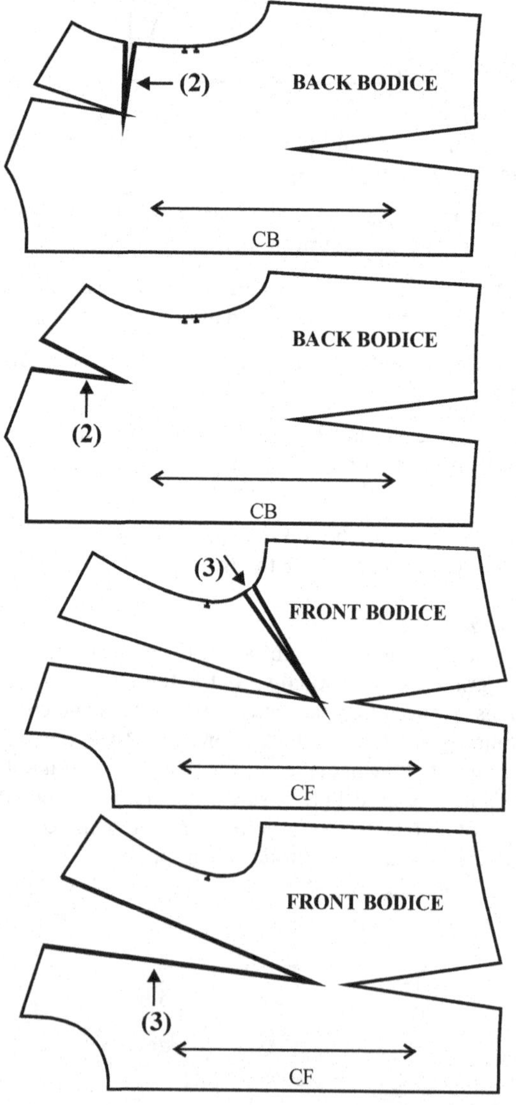

Modification for a lower 'cut-away' round neckline

A 'cut-away' neckline requires fitting more closely to the upper chest area. To prevent this lower cut neckline from gaping it has to be tightened.

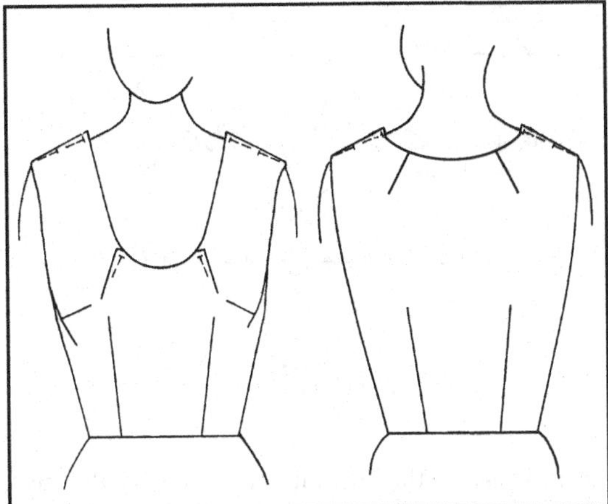

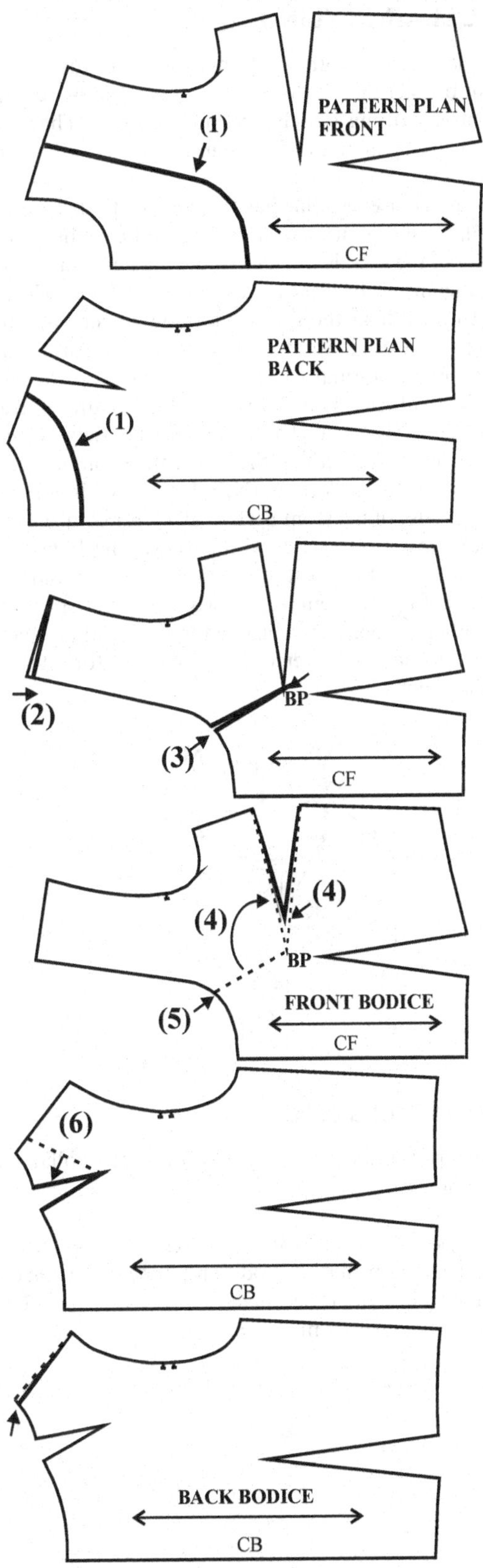

(1) Plan the curve of the new neckline on both the back and front bodice.

FRONT BODICE

(2) At the new neckline lower the front shoulder (approx. 0.5 cm to 1.0 cm) tapering to nothing at the armhole.

(3) A small dart (approx. 0.5 cm or 1.0 cm) is inserted into the neck. The apex should finish at the BP so that this dart can be combined with the underarm dart. (The underarm dart apex has also to be extended to the BP.)

(4) Combine the new neck dart with the underarm dart. Then shorten the underarm dart (approx. 2 cm to 3 cm) from the BP for a smoother fit.

(5) The new neckline may have to be redrawn into a smoother curve.

BACK BODICE

(6) It is advisable to pivot the back shoulder dart into the neck curve as the length of the shoulder seam has been shortened. The back neck can be further tightened by increasing the width of this dart.

(7) At the new neckline lower the back shoulder (approx. 0.5 cm to 1.0 cm) tapering to nothing at the armhole.

COLLAR STYLING

Collars can be both functional and decorative. They can protect the wearer from adverse weather conditions by fitting closely around the neck. They can also be a focal point of a design and act as a frame for the face.

The **bodice neckline** has to fit correctly before the collar can be constructed. If a normal neckline at the base of the neck is too high at the centre front (CF) it is uncomfortable for the wearer. When the neckline is cut too wide at the sides or too low at the back the collar will poke way from the neck. The correct fitting for the normal neckline is where the centre back (CB) is sufficiently high to cover the seventh cervical, known as the nape. The neckline fits closely around the back of the neck to the top of the shoulder seam known as the neck point (NP). It is cut slightly lower at the base of the front neck to allow for comfortable neck movement. Styles with cut-away necklines are discussed in the previous section. It is necessary to decide if the garment is to be worn beneath or on top of other garments. If worn on top the neckline has to be widened and lowered to allow room for another collar beneath.

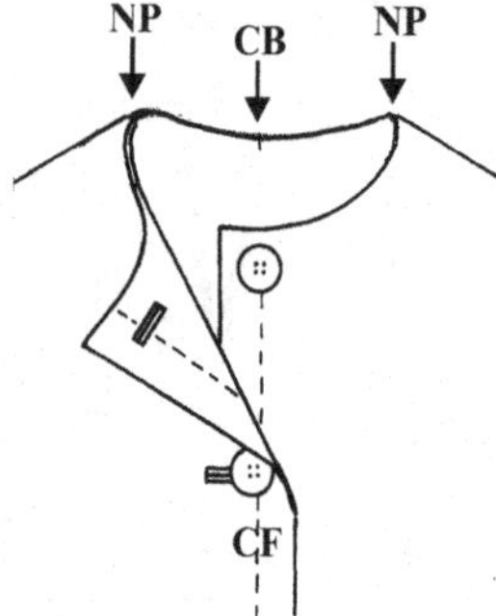

Sections of a collar

The terms which relate to the sections of a collar are as follows:

Neck edge where the collar is attached to the bodice. It is important that the bodice neckline is the correct shape. The collar neck edge and bodice neckline should be of equal measurement.

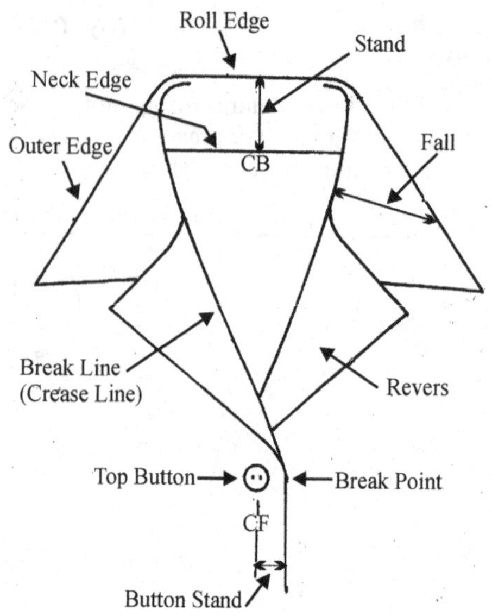

Collar stand is the section that rises up the neck column.

Roll line is where the collar stand rolls over to become the collar fall.

Collar fall is the section between the roll line and outer edge.

Outer edge is the outer seam line of the collar.

Revers are the section of the faced front bodice that turns back on itself to form a V-shape at the neck gorge.

Break line or **crease line** is where the revers fold back.

Break point is the position at the lower end of the break line.

Top button position is important as it determines where the front bodice will fold back at the break point to become the revers.

Button stand is the extension beyond the CF when the buttons are fastened.

Collar patterns

There are generally three pattern pieces to a collar: the top collar, under collar and interlining:

Top collar is the visible surface of the collar fall. The top collar is generally cut slightly wider than the under collar to conceal the outer edge seam (see 'seam allowances' below).

Under collar acts as a facing to the top collar. All the patterns constructed in this section will be for the under collar.

Collar interlining generally interlines the top collar but in some circumstances the under collar. Fusible interlining can be cut 0.3 cm smaller than the top collar to reduce bulk, but should be caught at the stitching line.

Seam allowances depend on the weight and thickness of the fabric. For medium weight fabric generally 0.6 cm is sufficient for the enclosed outer edge. To conceal the outer edge seam of the completed collar an extra 0.3 cm is added to the top collar, tapering to nothing at the neck edge. This forces the seam to roll under. For pointed corners no allowance is needed at the point. The adjacent straight lines are slightly bevelled. This makes the lines straight when the collar is stitched and turned through to the face side. The pointed seam corners can be cut off to reduce bulk. The neck edge seam allowance may vary between 0.6 cm and 1.0 cm according to the method of assembly. When the CB is a mirrored line the neck edge and outer edge should always be at 90° so that there is a smooth line running from the left to the right side. The neck edge of the collar has to be notched to match the bodice neckline, generally at the CB and NP.

Grain lines are positioned according to the style and fabric design. When the roll line is positioned on the bias grain the collar will roll softly around the neck. This is more suitable for flat and semi-stand collars. A firm roll line requires a warp grain, so the grain line is placed on the length of the collar. This is more suitable for high standing collars. For medium and lightweight fabrics the top and under collars have similar grain lines. For thick bulky fabric the under collar can be cut on the bias.

The design of the fabric influences the style of the collar. For example, the positioning of stripes or checks is important so that the left and right side of a collar and revers match. This is controlled by the positioning of the grain line and computer match point coding (see the section on Placement strategies for fabric type and matching in Part 5).

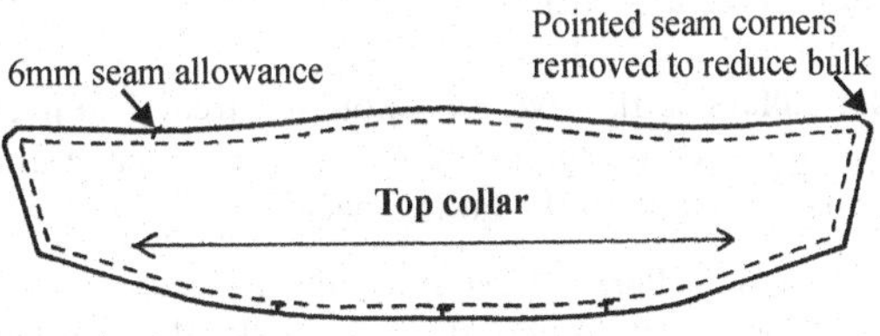
6mm seam allowance
Pointed seam corners removed to reduce bulk
Top collar

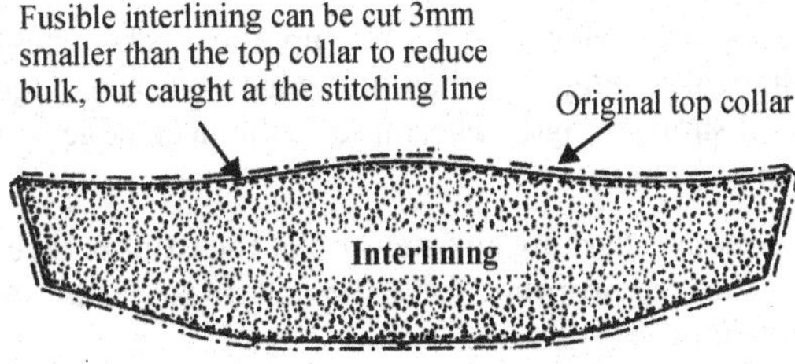
Fusible interlining can be cut 3mm smaller than the top collar to reduce bulk, but caught at the stitching line
Original top collar
Interlining

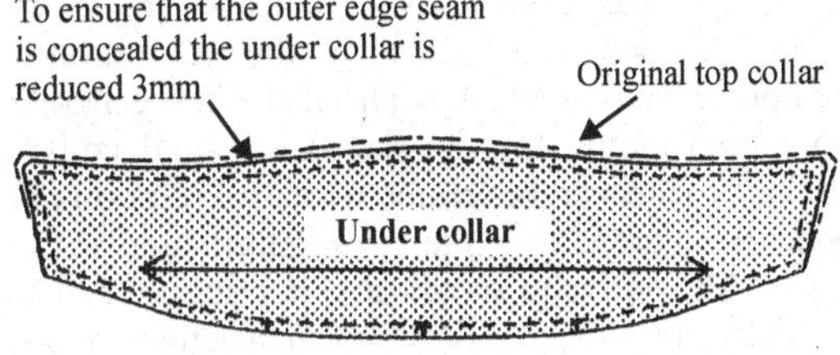
To ensure that the outer edge seam is concealed the under collar is reduced 3mm
Original top collar
Under collar

Types of collars

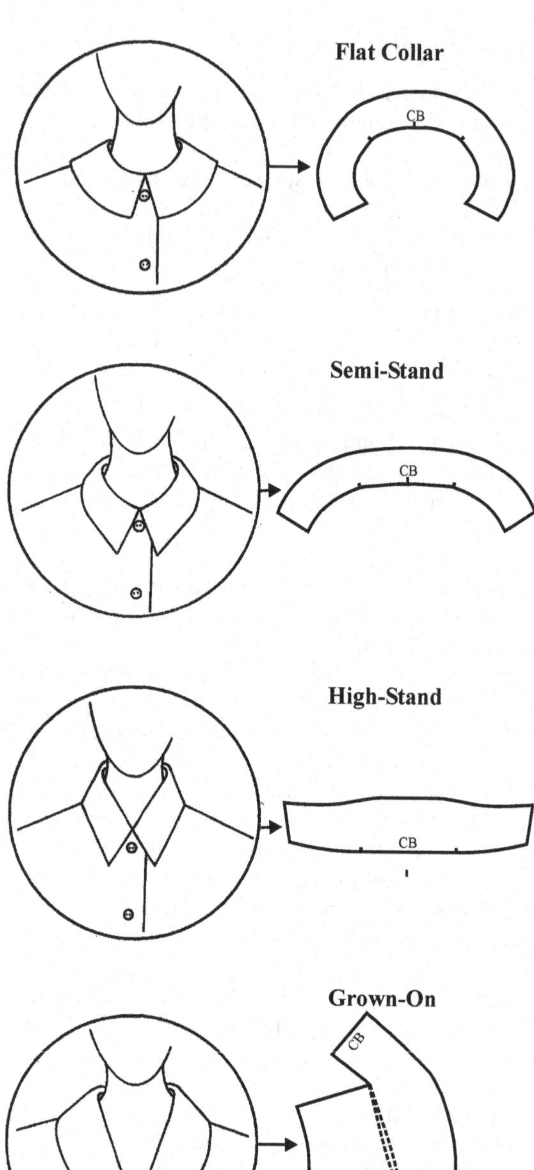

Collars can be grouped into four main categories: flat, semi-stand, high-stand and grown-on. There is a definite relationship between the length of the outer edge, height of the stand and curve of the neck edge. If a flat collar has the outer edge reduced in length by suppression it sits nearer the neck. This forces the collar to stand higher up the neck. Also the neck edge will become less curved.

Flat collars lie flat over the shoulder area. The neck edge and outer edge of the collar are concave and similar in shape to the bodice neckline.

Semi-stand collars roll part way up the neck, mainly at the back. The outer edge and neck edge are less curved than the flat collars but still concave, and the outer edge is shorter.

High-stand collars stand high up the neck. These collars are straight as the outer edge and neck edge are of similar length. The curve of the neck edge can even be convex.

Grown-on collars are an extension of the front bodice. It is more appropriate for these collars to be flat or semi-stand.

Methods of constructing collar patterns

There are several methods of constructing collar patterns. They can be created by the adaptation of the bodice block neck area. This method is generally more appropriate for flat and semi-stand collars. Straight collars can be created by the adaptation of a rectangle representing the neck length and collar width. Alternatively, standard collar shapes can be drafted from a prescribed set of instructions. Complex styled collars that rely on the drape of the fabric can be draped and then digitised into the computer.

Adaptation for a flat collar

This flat collar is adapted from the neckline of a bodice block. The outer edge is slightly tightened to roll the collar sufficiently up the neck to conceal the neck edge seam. When the flat collar is cut in one piece the straight grain line can be positioned along the CB mirror line. Flat collars can have a back opening instead a front one, or can be cut in two pieces giving a right and left collar. These can be cut either on the straight or bias grain according to the style and fabric design.

(1) Copy the front and back bodice blocks. Transfer the back shoulder dart temporarily into the armhole and the front dart into the side seam.

(2) Merge the blocks at the shoulder seam with the neck points (NP) matching for the pattern plan.

(3) Draw the collar shape. The outer edge at the CB should be at right angles to maintain a smooth line across the back. In this example the collar width is 6.0 cm. The front collar point is 2.0 cm from the CF. The space between these points will widen when the garment is worn.

(4) Draw a second grain line parallel to the CF to be used as another grade reference line.

(5) Trace the collar pattern from the plan. Reduce the half collar outer edge length 1 cm between the original shoulder line and CF. This method of minus fullness retains a smooth outer edge and neck edge.

(6) Position a notch at the neck point (NP) of the collar to match the bodice shoulder seam.

(7) The CB is a mirror line for the complete collar.

Suggested grading

The multi-grade reference line grading method would enable the front collar to be graded as for the original front bodice neck and the back collar as for the back bodice neck. The grade for the left side will be mirrored.

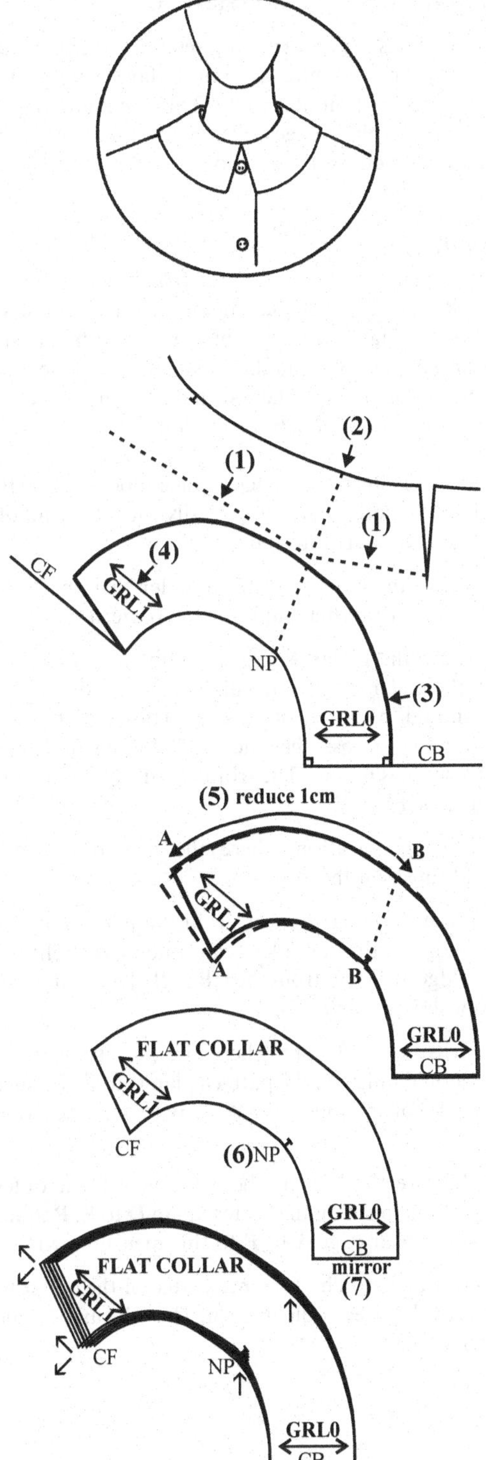

Adaptation for a semi-stand collar

Two methods for constructing semi-stand collars are described below. These are both adaptations of the bodice neck and shoulder area. The first method is by reducing the outer edge to the required length. The second method is by spreading the neck edge to match the bodice.

METHOD 1

The semi-stand collar is adapted from the neckline of a bodice block in a similar method to the flat collar. The outer edge is reduced in length so that it sits nearer the neck edge and the collar stands up the neck column. Generally the straight grain line is positioned along the CB mirror line.

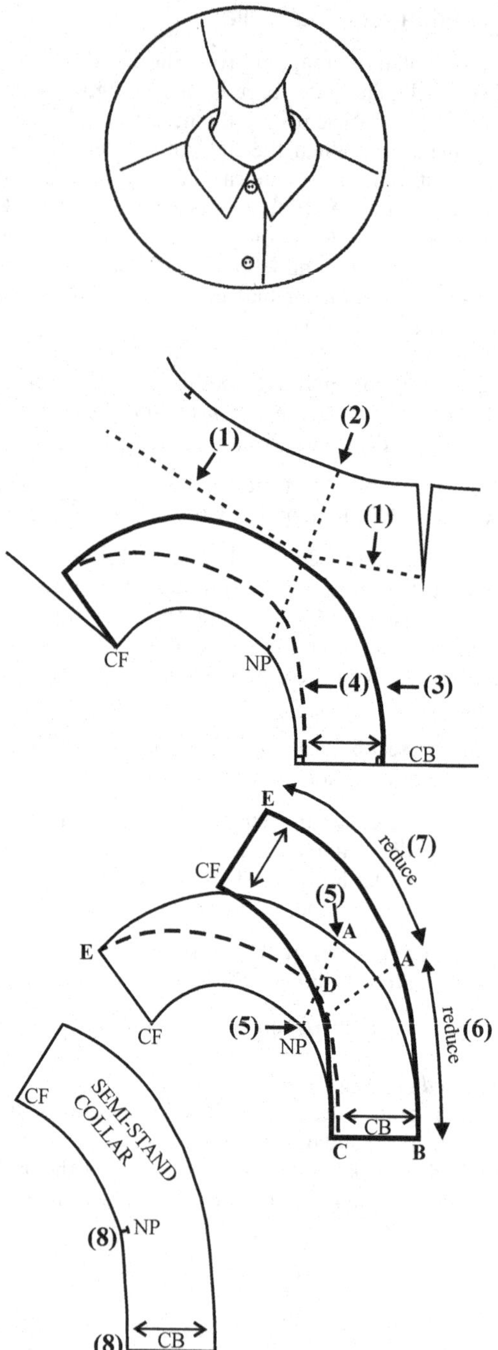

(1) Copy the front and back bodice blocks. Transfer the back shoulder dart temporarily into the armhole and the front dart into the side seam.

(2) Merge the blocks at the shoulder seam with the neck points (NP) matching for the pattern plan.

(3) Draw the collar shape including the width of both the collar stand and fall. The outer edge at the CB should be at right angles. This maintains a smooth line across the back. In this example the collar width is 6.0 cm. The front collar point is 1.5 cm from the CF.

(4) Plan the position where the new outer edge should finish on the bodice.

(5) Trace the collar pattern from the plan. Split the outer edge at the shoulder seam line at **A**. Split the outer edge at 2.0 cm from the CB at **B**. This maintains a smooth line at the CB.

(6) Measure the length where the new back outer edge should finish on the pattern plan **C** to **D**. Reduce the back collar outer edge **A** to **B** to this measurement.

(7) Measure the length where the new front outer edge should finish on the pattern plan **D** to **E**. Reduce the front collar edge **A** to **E** to this measurement.

(8) Position a notch at the neck edge of the collar to match the bodice shoulder seam. The CB can become a mirror line.

METHOD 2

The outer edge, collar-stand, roll line and fall can be planned with tape lines on a sample of the bodice placed on a workroom stand. These positions are then measured and planned using the computer pattern design system. The neck edge has become very short on the pattern plan. To produce the collar pattern the neck edge has to be spread to match the neck measurement of the bodice.

(1) Copy the front and back bodice blocks. Transfer the back shoulder dart temporarily into the armhole and the front dart into the side seam.

(2) Merge the blocks at the shoulder seam with the neck points (NP) matching for the pattern plan.

(3) Plan the position of the collar outer edge.

(4) Plan the width of the fall from the outer edge to the roll line. The back roll line is copied and offset from the back neckline. Taper this line to the CF.

(5) Measure the distance between the planned roll line and the bodice neckline. Then construct the stand width the same distance from the roll line by copying and off-setting the back neckline and tapering to the CF.

(6) Trace the collar shape from the pattern plan. Measure the new collar neckline from the CB at **A** to the CF. From **A** mark a point at **B** that is $\frac{2}{5}$ of the new collar neckline. (This is the same proportion of the back bodice neckline, $\frac{2}{5}$ to the front bodice neck of $\frac{3}{5}$).

(7) Construct a line from the junction of the shoulder seam on the outer edge at **C** to the new back neck width at **B**. Split the neck edge at **B** and outer edge at **C**.

(8) Measure the back bodice neckline from the CB to NP, then spread by adding fullness to the back collar neck edge by the same amount **A** to **B**. The junction of the CB and neck edge and outer edge should be maintained at right angles for a smooth line.

(9) Measure the front bodice neckline from the CF to NP and spread the front collar neck edge by the same amount **B** to **D**. Correct the lines if they are not a smooth curve.

(10) Position a notch at the neck edge of the collar to match the bodice shoulder seam at **B**.

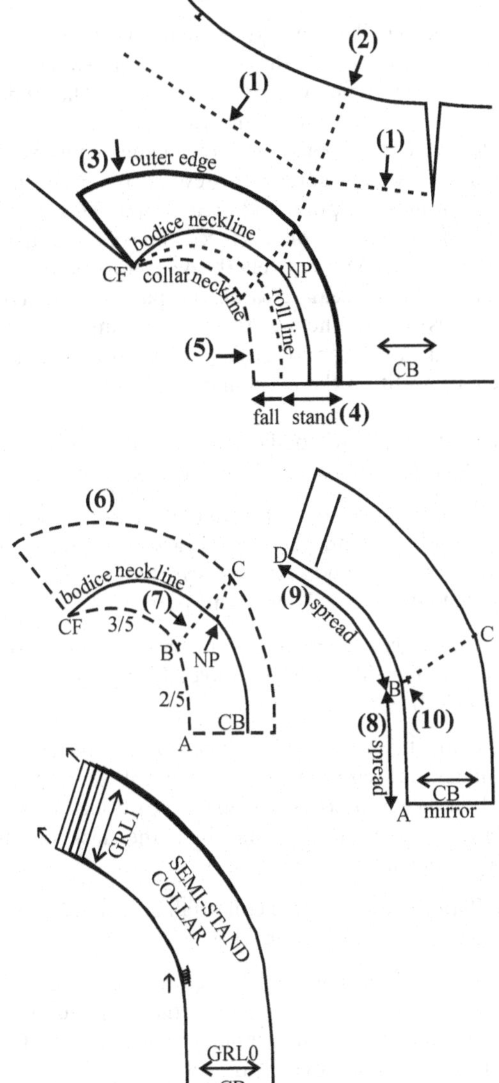

Suggested grading

Measure the amount of the neck grades on the back and front bodice. Add a second grade reference line for multi-grade line grading, parallel to the front collar. This method of grading would enable the front collar to be graded the length of the front bodice neck grade and the back collar the length of the back bodice neck grade.

Adaptation for a grown-on collar

This type of collar is an extension of the front bodice that continues around the back neck to a seam at the CB. It is also known as a shawl collar. The under collar is an extension of the front bodice; the top collar is cut in one piece with the facing. To improve the sit of the neck gorge a fish-eye dart is positioned between the neck point and break point. This dart is concealed when the collar is faced and rolls over from the break line. Alternatively the dart suppression can be included in a seam creating a separate under collar. This avoids the weakness of cutting into the corner at the NP and is easier to sew. It also makes the pattern fit on the fabric more economically.

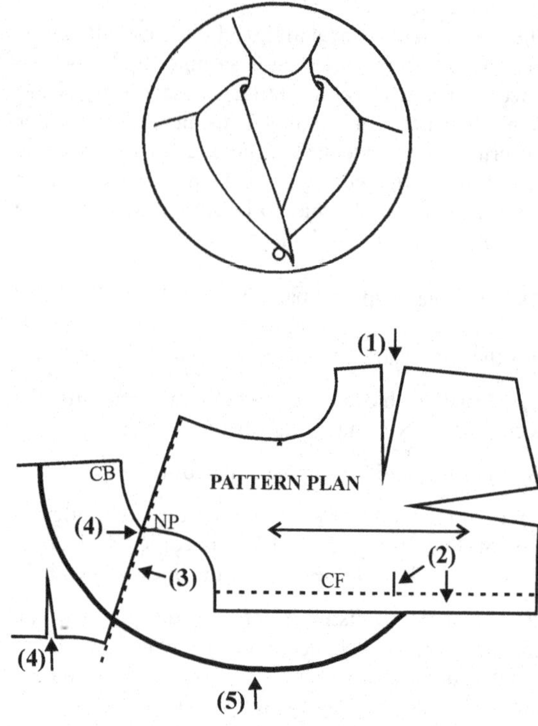

(1) Copy the front bodice block and transfer the shoulder dart temporarily into the side seam.

(2) Plan the position of the top button and the width of the button stand. (For details about the width of button stands and facings see the section on production patterns later in this part)

(3) Construct a line that is a continuation of the shoulder line from the armhole through the neck point (NP).

(4) Copy the back bodice block and transfer the shoulder dart temporarily into the armhole. Flip the back bodice over so that it can be positioned on the extended front shoulder line with the neck points (NP) matching.

(5) Plan the width of the collar fall and stand. (This example is 9.0 cm at the back.)

(6) Trace off the new front bodice including the back collar. Split the outer edge line on the front 2.0 cm to 3.0 cm from the extended shoulder line at **A**. The CB at the collar outer edge is point **B**.

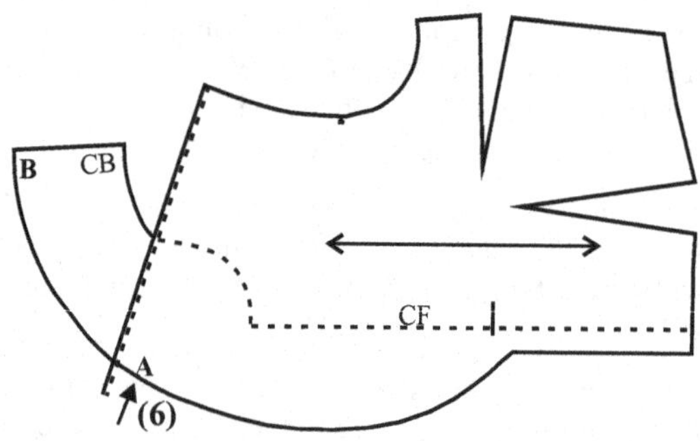

(7) Using the minus fullness command reduce the length of the back collar (between **A** and **B**) by 4 cm to 6 cm according to the required height of the collar stand. (This example is reduced by 6 cm.)

(8) Construct a **fish-eye dart** either side of a construction line from NP meeting the CF line. This dart is 1.0 cm at the mid-point. Drill holes are positioned 1.0 cm from the end of the dart and at the mid point as a guide for stitching.

(9) Alternatively a seam can be positioned which incorporates the dart and creates an under collar. The NP position has to be notched on the under collar.

(10) The facing pattern does not have to include the fish-eye dart. The facing becomes the visible surface of the collar. To conceal the outer edge seam, approximately 0.3 cm should be added to the facing outer edge from the break point to the CB seam.

Suggested grading

The multi-grade reference line grading method would be the most appropriate. The front section of the collar would be graded as for the front neck. The back collar is graded from a grade reference line parallel with the CB. The grading increment for the back collar would be the same as the back neck width grade. (The collar width of this example has remained the same for all the sizes.)

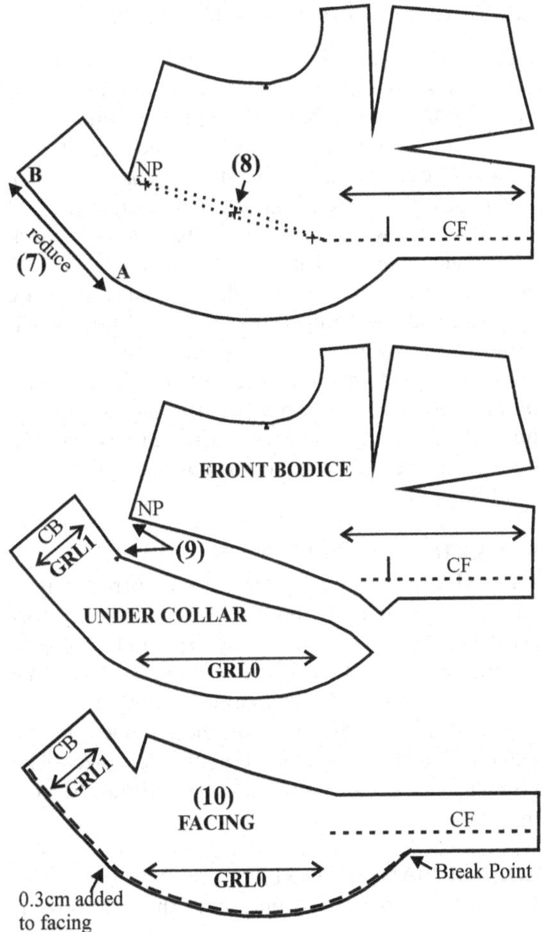

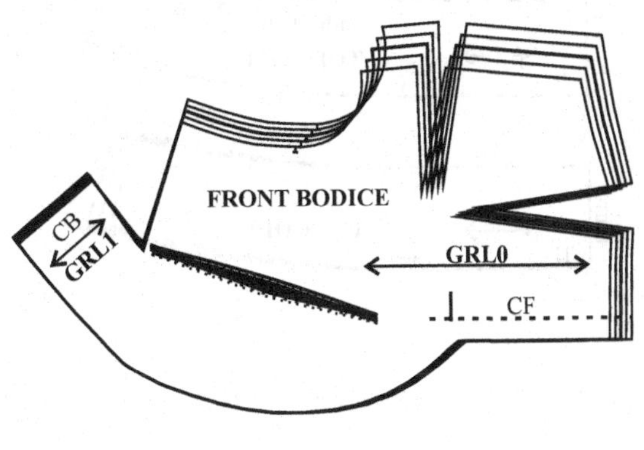

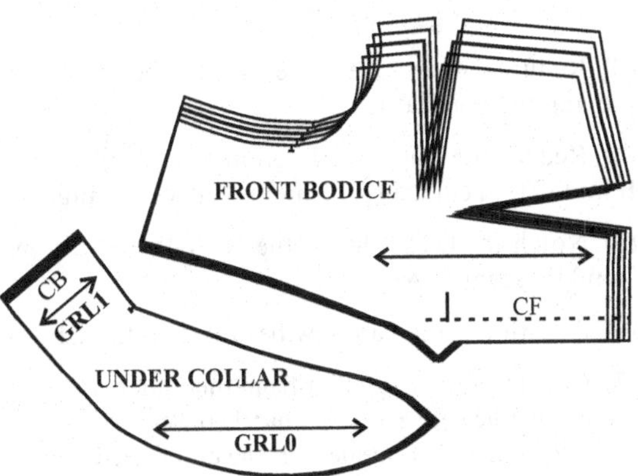

Construction of a high-stand collar

High-stand collars sit better around the neck if the bodice neckline is pitched slightly forward. This group of collars can be based on a rectangle that is modified to conform to the shape of the neck. This is because the girth at the base of the neck is larger than further up the neck column. Several examples are described below, starting with a basic collar band. This collar band can be adapted to create a convertible collar, a two-piece collar that has a separate stand and fall. Separate drafts are given for the high-stand roll collar and shirt style collar. The grain line for these high-stand collars can be positioned either parallel to the CB or at right angles to it, depending on the fabric design. Where a sharp crease is required on the roll line it is better to position the grain line at right angles to the CB.

NECKLINE MODIFICATION

To enable high-stand collars to sit correctly it is advisable to pitch the bodice neckline slightly forward. Lower the centre front neck 0.5 cm tapering to nothing at the middle of the front neck. Raise the centre back neck 0.5 cm tapering to nothing at the neck point. The overall neckline measurement has to remain the same as the original neck size. The neckline at CF and CB should be maintained at a right angle.

COLLAR BAND CONSTRUCTION

This collar band example buttons at the centre front.

(1) Construct a rectangle for the under collar equal to half the bodice neckline by the required width.

(2) Measure the bodice back neck edge from the CB to the NP. Split the collar neckline at the NP position at **A** and top edge line at **B**.

(3) Reduce the top edge at the CF a further 0.5 cm by moving the point at **C**.

(4) Reduce the length of the front top edge between **B** and **C** by 1 cm using the minus fullness command.

(5) Notch the neck edge at the NP position for the shoulder seam at **A**.

(6) A button stand can now be added to the CF.

(7) Construct the top collar from the under collar by increasing the width approximately 0.3 cm so that the top edge seam is concealed on the completed collar.

Suggested grading

Using the computer measure the amount that the half bodice back and front necklines have been graded. The collar is then graded from the mirrored CB according to these measurements.

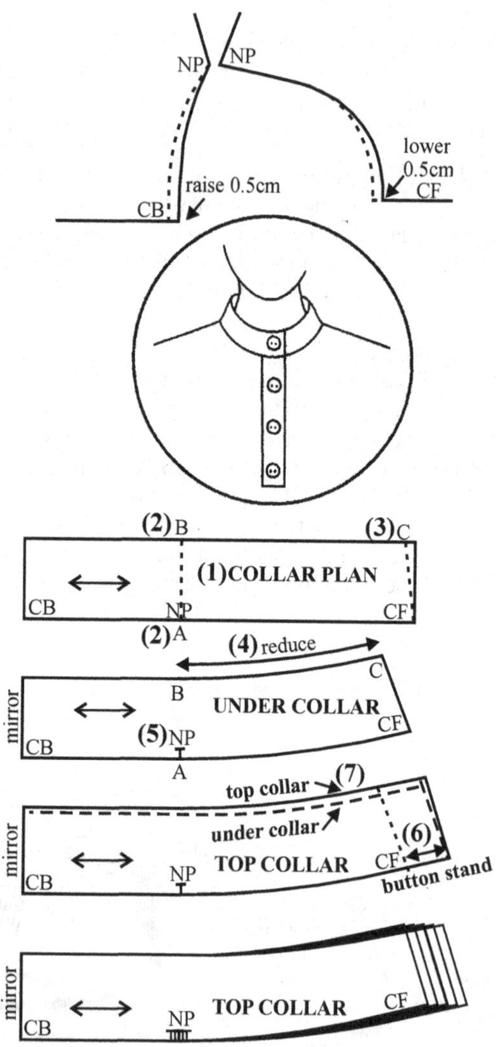

Adaptation for a convertible collar

This type of collar is known as a convertible collar as it can be worn either buttoned up to the neck or opened to form a collar and revers. It can be constructed as an adaptation of the collar band by extending the top edge to form the collar fall. The measurements quoted refer to the illustrated style as guidance.

(1) Copy the collar band under collar, excluding the CF button stand. Construct a guideline at right angles to CB that is the width of the collar fall from the top of the band (minimum of 0.5 cm wider than the stand at the CB to cover the neck seam when worn). Square to the CF.

(2) Draw the outer edge according to the style.

(3) This shaped outer edge of the collar will require a top and under collar pattern. The top collar should be increased slightly in width (approx. 0.3 cm) to roll the seam under and conceal it when worn.

(4) A straight outer edge of the collar can become a fold so that the top and under collar are cut as one pattern piece.

Suggested grading
Use the same method as for the collar band.

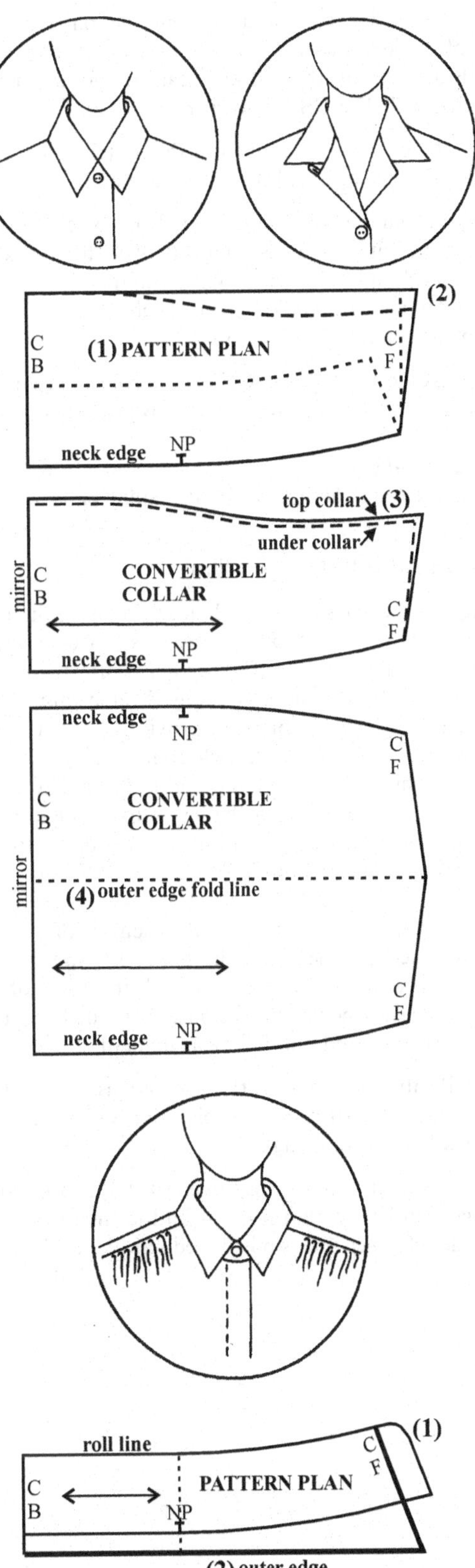

Adaptation for a two-piece collar

This two-piece collar has a separate stand and fall. This enables the collar roll line to fit more closely to the neck because the stand is cut separately. The outer edge of the collar fall has to be increased in length to lie over the shoulder area.

(1) Copy the collar band under collar, including the button stand. The button stand can be shaped.

(2) Add the collar fall width from the roll line and draw the outer edge.

(3) Trace the collar band from the pattern plan. Notch the CB and NP on the neck edge. Notch the CB and CF on the roll line seam for joining to the collar fall. The CB is a mirror line.

(4) Trace the collar fall from the pattern plan. Split the roll line at **A** and the outer edge at **B**.

(5) Widen the collar outer edge between **B** and **C** so that the outer edge sits smoothly over the shoulders. (This example is 1.5 cm.) Wider collars will require a greater increase according to where the outer edge finishes on the shoulder area.

(6) Notch the CB and CF on the roll line seam for joining to the collar band. The CB is a mirror line.

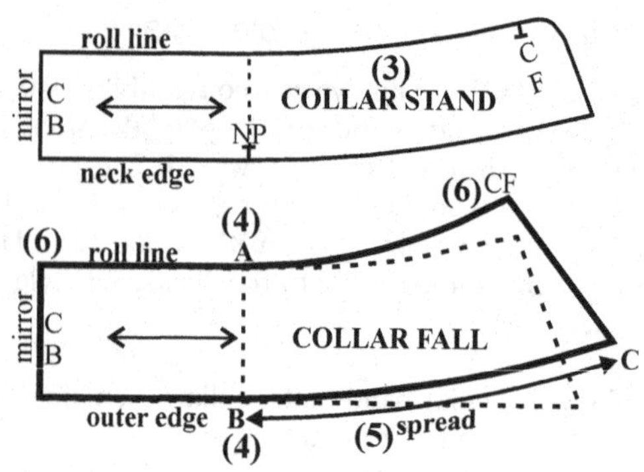

Suggested grading
Use the same method as for the collar band.

Draft for a high roll collar

High roll collars can mould closely around the neck because they are cut on the true bias of woven fabric or in a stretch fabric. For woven fabric there generally has to be a neck opening. Stretch fabric does not require a neck opening providing that there is sufficient elasticity to stretch comfortably over the head. This style has the top and under collar cut in one piece. The example below is for a high stand collar in woven fabric with a left side buttoned opening.

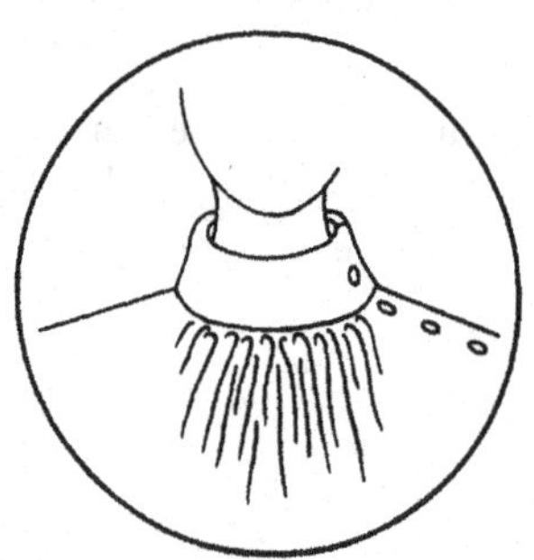

(1) Construct a rectangle with a length of the full bodice neckline measurement (minus 1.0 cm to 3.0 cm according to the fabric bias stretch in excess of the neck measurement) plus the button stand. The width is twice the stand and fall measurement.

(2) Position notches on the neck edges for the CF, NP and CB, distributing the bias evenly (this example has a left side opening).

(3) The grain line is positioned at 45° to the outer edge fold line so that it is cut on the tube bias. The grade reference line is positioned parallel to the neck edge.

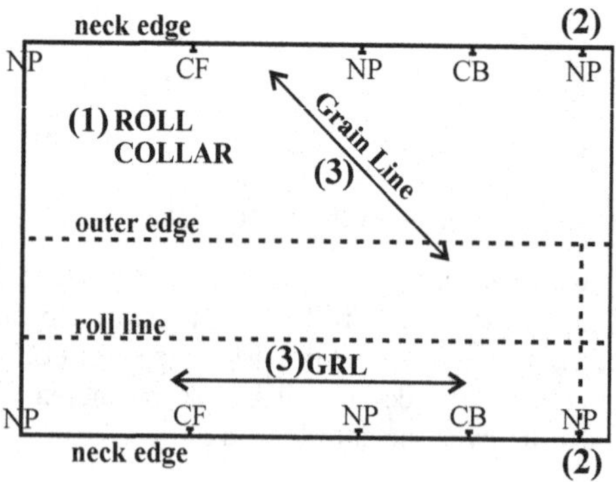

Suggested grading
The neck edge grade is equal to that of the bodice neckline. The computer is programmed to rotate the collar pattern on to the 45° grain line when lay planning.

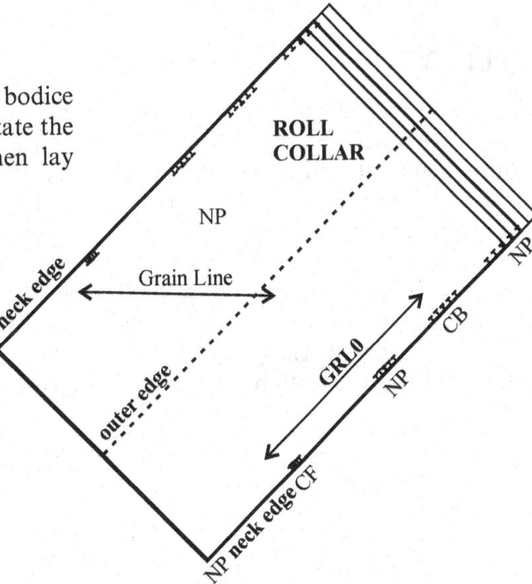

Draft for a shirt collar

This example is obtained by a direct drafting method for half a shirt collar, that is then mirrored at the centre back.

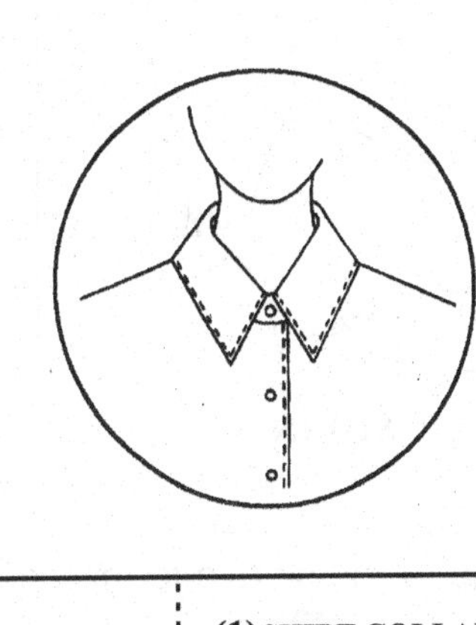

(1) Construct a rectangle with the length of 21.5 cm (half the garment neck 20 cm plus a button stand 1.5 cm) by the required width (this example 8.5 cm).

(2) Measure from the CB half the back neck width (8 cm) for the NP position. Measure from the CB half the neck length for the CF (20 cm).

(3) Draw the neck edge curve above the lower line:
1.5 cm at the CB
1.0 cm at the neck point
0.75 cm at the CF
1.5 cm for the button stand.

(4) Position the CF line 3.5 cm up from the lower right corner and square in 2.5 cm. Draw the button stand width 1.5 cm parallel to the CF line and curve.

(5) Draw the outer edge and front collar point according to the required style. In this example the front collar outer edge has been lowered 0.5 cm.

(6) Trace the shirt collar from the draft. The CB is a mirror line.

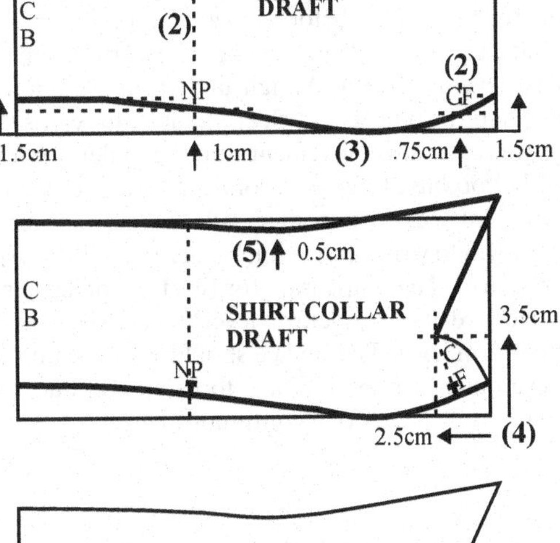

Suggested grading
Use the same method as for the collar band.

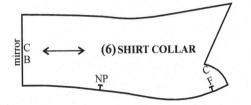

SLEEVE STYLING

Sleeves are an important style feature and contribute to the silhouette of the garment. Sleeve styles can be considered in three main categories. The first are those that are stitched into the bodice armhole, known as set-in sleeves. They can be loose fitting by not conforming to the shape of the arm or can fit more closely following the arm contour. The second are known as raglan sleeves and have a part of the bodice attached to the sleeve head. The underarm has a similar fit to the set-in sleeve. The third category is those that are cut in one with the bodice, often classed as kimono styles. These have to drape loosely under the arm to allow for arm movement. The raglan and kimono sleeves are better suited to soft and moulding fabric. All these types of sleeves can be adapted from the basic sleeve block patterns. Whatever the style they must be comfortable to wear and not restrict arm movement. However there are many variations to these styles. These can be interpreted for computer use from books for manual pattern construction.

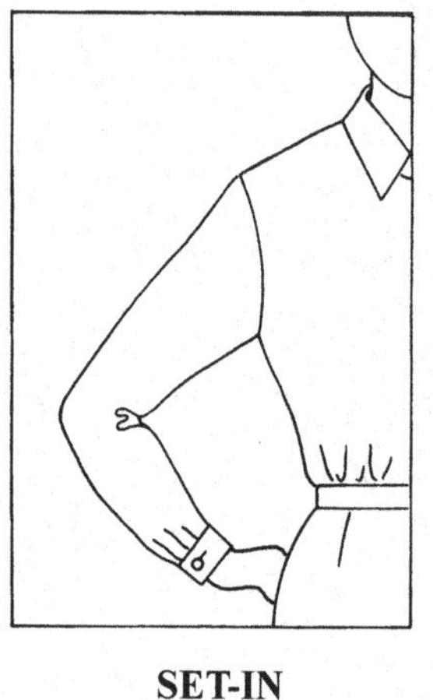

SET-IN

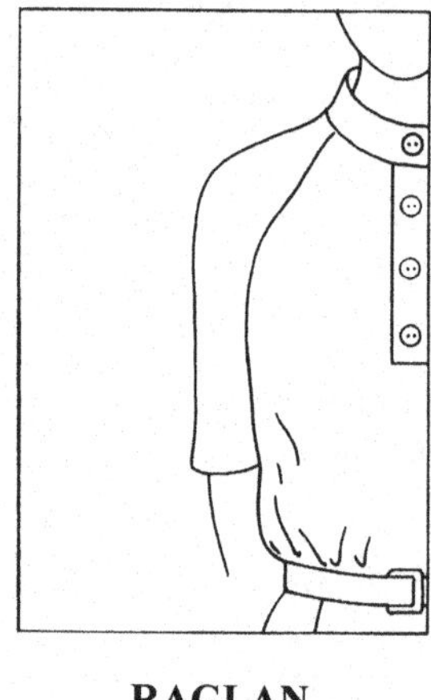

RAGLAN

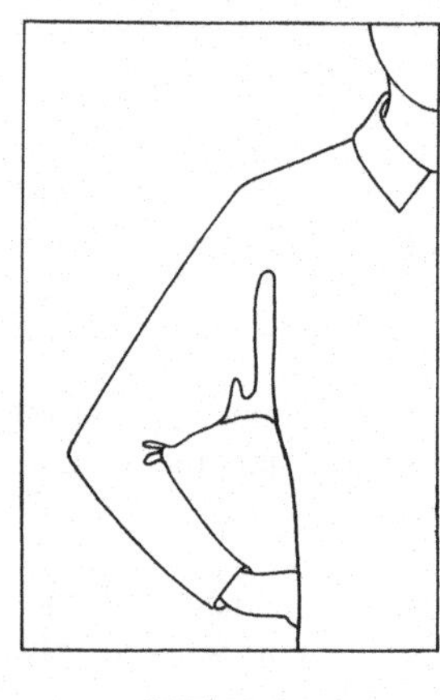

KIMONO

Relationship of the sleeve to the bodice

The fit of the block pattern armhole and sleeve head has to be correct before it can be used for style adaptation. The height of the underarm from the waist level is crucial. A high underarm gives longer length to the bodice and sleeve underarm seams that allow greater arm movement. But the underarm must not be too high to give discomfort. A low underarm seam shortens the bodice side seam and sleeve underarm to wrist seam, which can restrict the raising of the arm. To compensate for this the bodice has to be widened at the underarm level and the sleeve at the upper arm level. The bodice shoulder length and the sleeve outer arm length have to be correct and considered together as one continuous length.

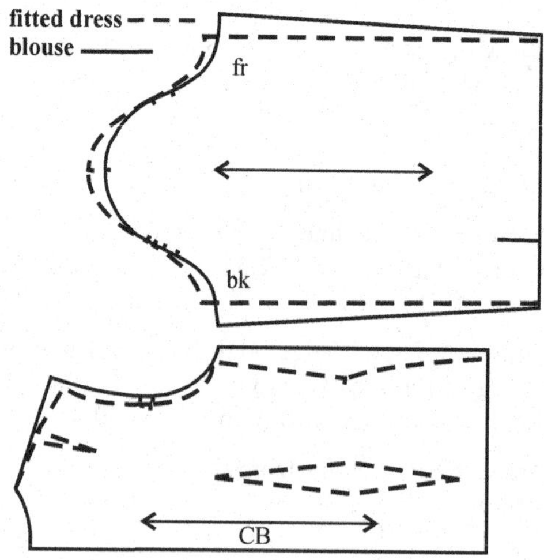

Sufficient 'ease allowance' has to be added to the width of both the bodice and sleeve for arm movement. This is especially so for woven fabric. The position and amount were explained in Part 1, 'Size chart formulation' and illustrated in Figure 1.4. The crown of the sleeve also has extra measurement that is **'eased in'** to the armhole for moulding around the top of the arm. This is controlled by the balance notches positioned at the middle of the back and front armholes and for the shoulder seam position. These notches also control the hang of the sleeve so that it conforms to the forward bend of the arm. This is known as the pitch of the sleeve. The positioning of the sleeve grain line also helps to maintain the correct hang. Once the fit of the bodice and sleeve has been approved the correct measurements, balance notches and grain line have to be maintained when the block patterns are adapted to a new style.

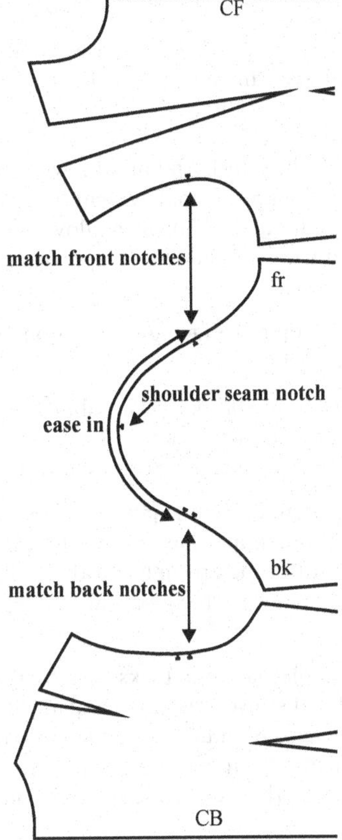

Sleeve lengths

All the different categories of sleeves can also vary in length.

Cap sleeve just covers the top of the arm.

Short sleeves end approximately midway between the elbow and underarm level.

Elbow length sleeves are considered more comfortable if they finish just above the bend of the elbow. This prevents them creasing at the forearm when the elbow is bent.

Three-quarter length sleeves are generally considered as three-quarters of the outer arm length.

Long sleeves refer to those that finish at the wrist bone. This length is generally established when the elbow is bent at right angles. It therefore appears a little longer when the arm is straightened.

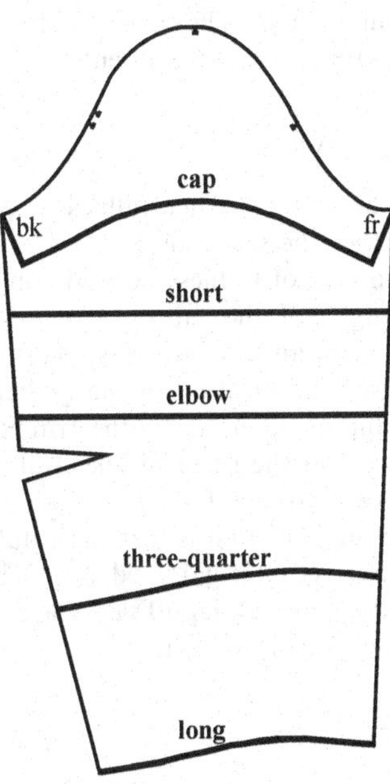

Blouse sleeve with buttoned cuff

This is an adaptation of the blouse straight sleeve block. The sleeve can be gathered or tucked into the cuff.

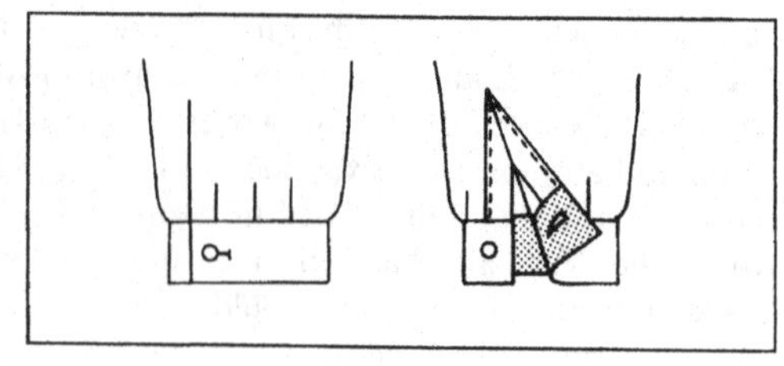

(1) Construct the cuff by taking the wrist measurement and adding approximately 3 cm or 4 cm for the buttoning of the cuff. A further allowance for the button stand has to be added, of approximately the button diameter.

(2) In this example the facing is cut in one with the cuff and is interlined.

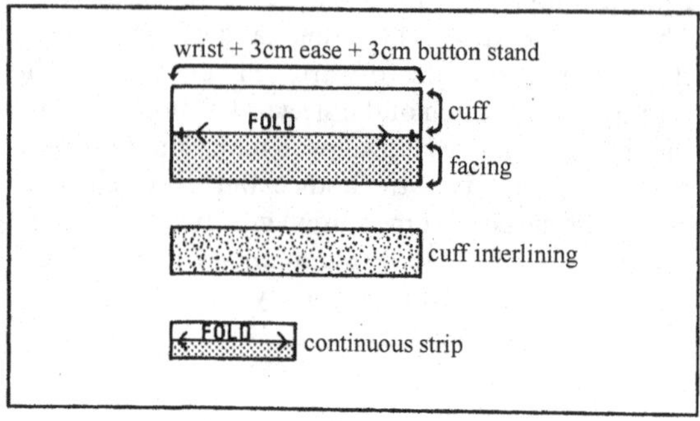

(3) The basic blouse block sleeve is shortened by half the cuff width. The wrist can be straight or curved down at the back arm line for a more pouched effect.

(4) A slit opening of 7 cm is positioned on the back arm line. This example has a continuous strip of twice the length of the opening and the width of the button stand. The strip facing is cut in one with the strip.

(5) This example has four tucks of 2.5 cm each so that the cuff and sleeve wrist are of equal length. The width of the back continuous strip has to be included in this calculation. (If necessary the width of the sleeve wrist can then be reduced at the underarm seam.)

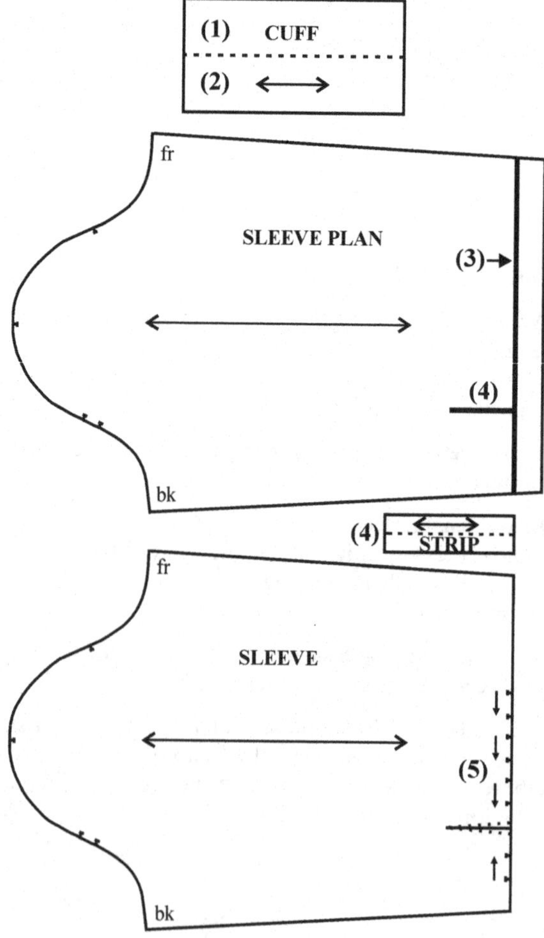

Blouse sleeve grading

The blouse sleeve can be graded as for the original sleeve block. The cuff length grades as for the wrist girth measurements. The cuff width generally remains the same for adult garments.

Puff sleeve styles

Puff sleeves are created when fullness is added then gathered so that the sleeve puffs out away from the arm. The amount of fullness depends on the thickness and weight of the fabric. The ratios of the gathers are the same as those suggested for the gathered skirt. The position of the gathers is more appropriate on the outer part of the arm between the back arm line and the forearm line. Fullness under the arm can cause discomfort. Two examples of puff sleeves are given. The first is for a half puffed sleeve where the lower edge is gathered into a band. The second example is for a full puff sleeve with gathers at the lower edge and sleeve head.

Half puff sleeve

(1) The sleeve band is constructed with a length equal to the biceps measurement plus 2 cm ease allowance. The width varies according to the style. The facing and band are cut in one.

(2) Shorten the sleeve to the required length, in this example at the mid biceps.

(3) Split the sleeve perimeter line at the head and the lower edge at the back arm line, forearm line and central line.

(4) Add the required fullness at the lower edge between the central line and back arm line (on this example twice the original width). Use the front underarm seam as the **hold line** to maintain the sleeve on the correct grain line.

(5) Add the required fullness at the lower edge between the central line and front arm line. Use the back underarm seam as the **hold line**.

(6) Extra length can be added to the lower edge of this sleeve to enable it to puff away from the arm.

(7) Add notches at the lower edge of the sleeve for the gathering position.

(8) Retain the original sleeve head notches so that they match the normal armhole without any gathers.

Half puff sleeve grading

This sleeve can be graded using the multi-grade reference line method. Grain reference line **GRL0** is as the original sleeve grain line. This is used for grading between the back arm line and the forearm line as the original sleeve. **GRL1** on the front underarm is as the original grain line before the sleeve was pivoted. **GRL2** on the back underarm is as the original grain line before the sleeve was pivoted. Using grade reference lines enables the grade to maintain the correct circular curve caused by the added fullness. When the sleeve is gathered it will set into the armhole like the original sleeve. The sleeve band grade is according to the arm width grade.

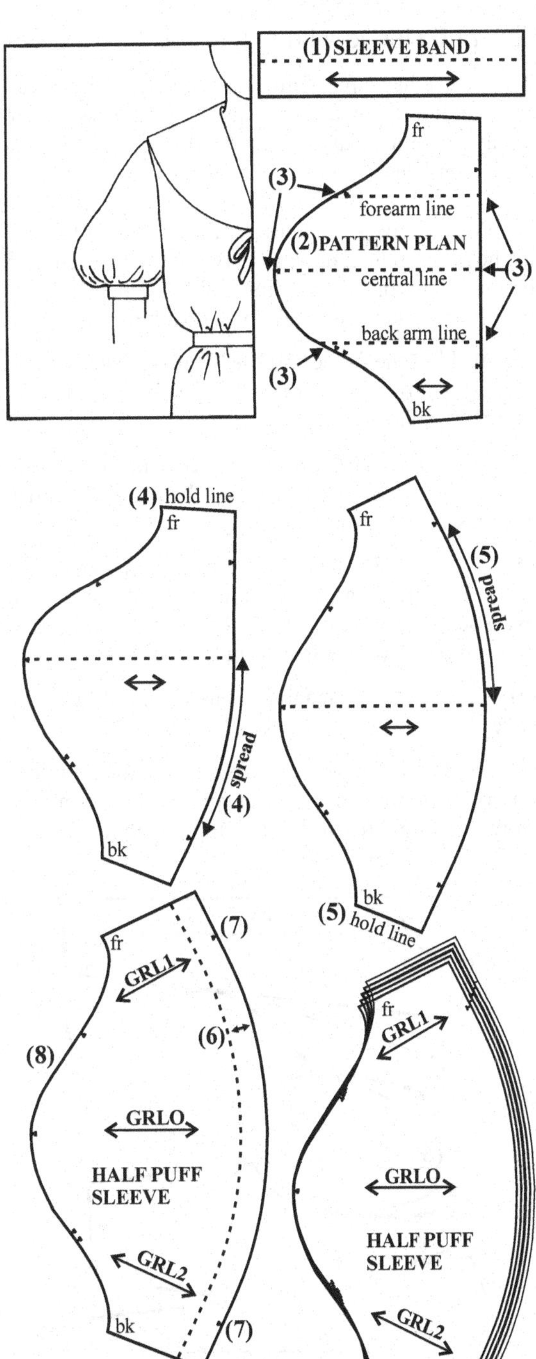

Full puff sleeve with frill

This sleeve has equal fullness at the sleeve head and lower edge. (The example has the fullness added equal to the measurement between the forearm and back arm lines.)

(1) Position five slash lines parallel to the centre line, at the forearm line, midway between the forearm and centre lines, the centre line, midway between the centre and back arm lines, at the back arm line.

(2) Add $\frac{1}{5}$ of the fullness at the front slash lines 1 and 2, then $\frac{1}{10}$ at the centre line 3. Use the back underarm seam as the hold line.

(3) Add $\frac{1}{5}$ of the fullness at the back slash lines 5 and 4, then $\frac{1}{10}$ at the centre line 3. Use the front underarm seam as the hold line.

(4) To enable the sleeve to puff out away from the arm add extra height to the crown of the sleeve and extra length to the outside of the lower edge (the amount varies according to the style).

(5) A frill can be added to the length according to the style.

(6) To enable the sleeve to rise above the shoulder reduce the shoulder length 1 cm to 2 cm from the shoulder seam tapering to the mid armhole.

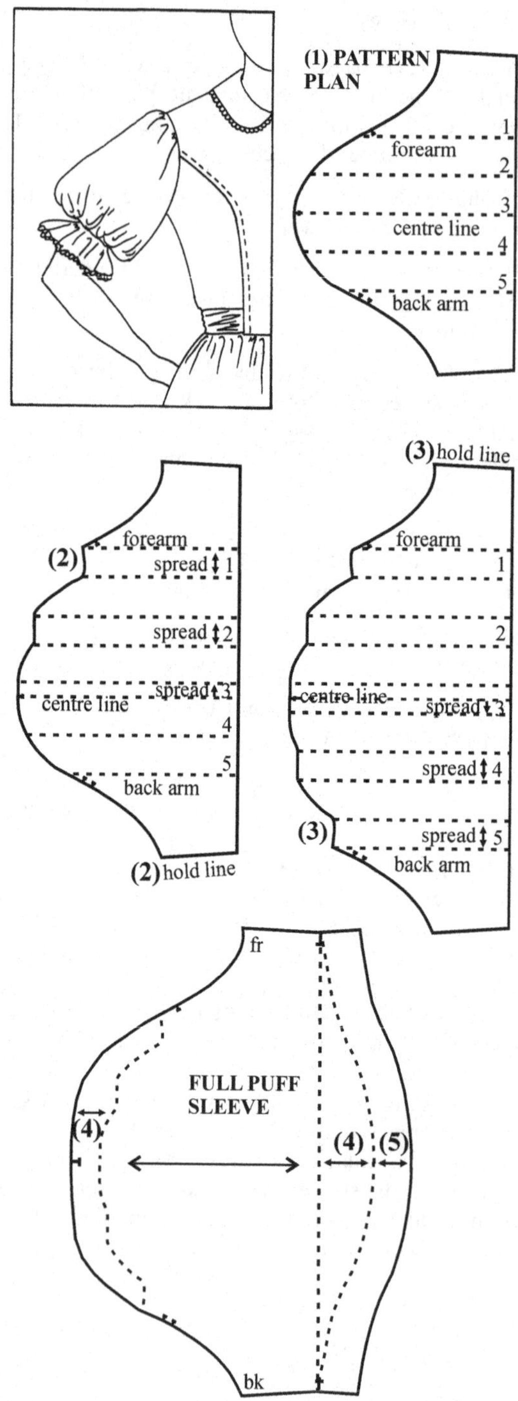

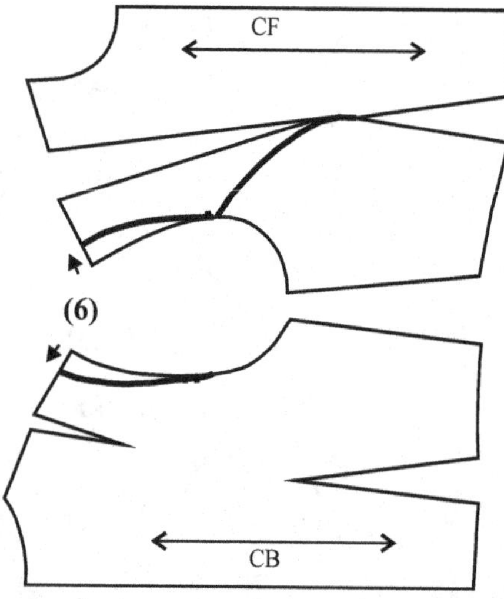

Full puff sleeve grading

This sleeve can be graded as for the original sleeve block.

Raglan sleeve

A raglan sleeve can be an adaptation of a set-in sleeve. Therefore the fit of the raglan under the arm is similar. The style feature of the raglan is obtained by joining the yoke section of the bodice to the sleeve head. The shoulder seam can form a dart or a seam that extends from the shoulder to the lower edge.

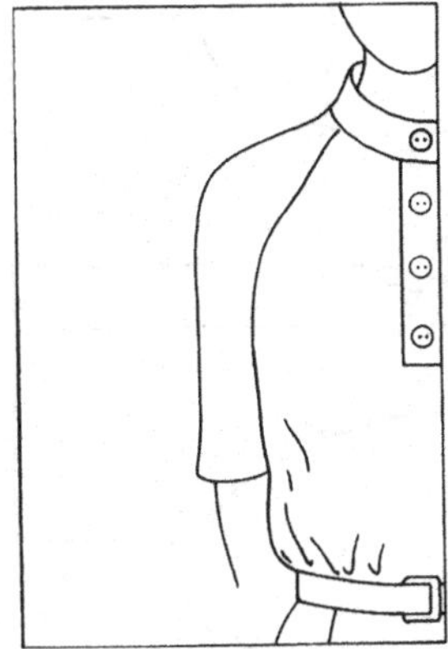

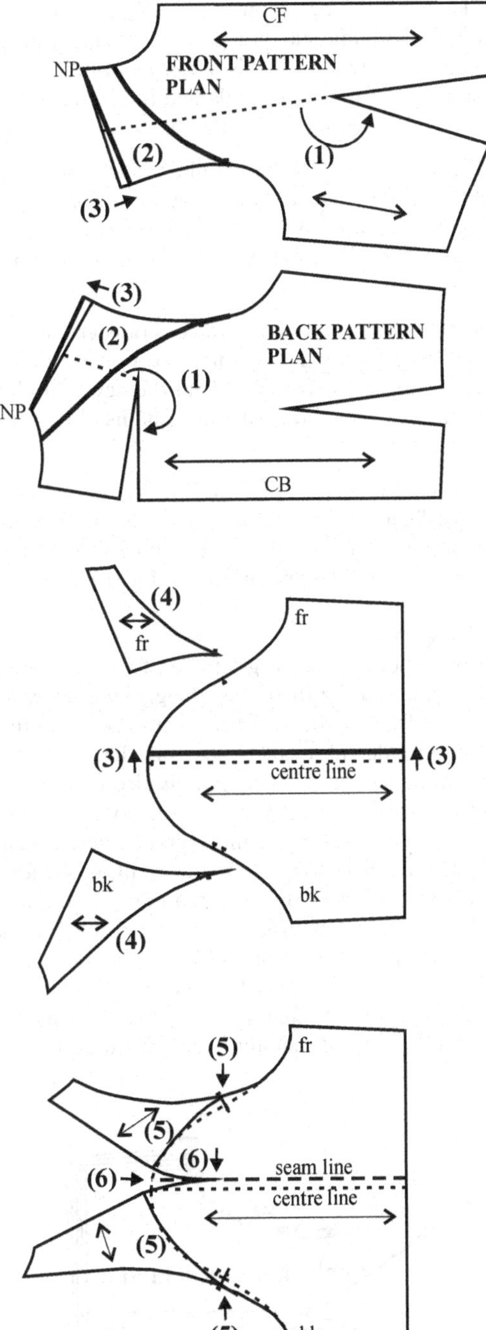

(1) Copy the back and front bodice and sleeve. Transfer the back shoulder dart temporarily into the CB. Transfer the front shoulder dart into the waist dart.

(2) Plan the bodice yoke seams to finish approximately at the mid armhole balance marks.

(3) Reposition the shoulder seam **1 cm** forward at the armhole tapering to nothing at the neck point NP. Position the sleeve centre line **1 cm** forward to match. Plan the sleeve length and waist seam according to the style.

(4) Trace off the back and front yoke sections. Retain the original grain lines for grading.

(5) Position these yoke sections at the sleeve head approximately **1 cm** from the mid armhole balance marks, touching the sleeve head at the front shoulder seam and **0.5 cm** from the back shoulder seam. Construct lines that blend the yoke seam into the underarm seam. Transfer the balance notches on to the new seam lines.

(6) Extend the shoulder seams to form a dart which finishes at the balance mark level and **1 cm** forward from a central line (for a two-piece raglan sleeve continue a seam which extends to the lower edge **1 cm** forward from the central line).

(7) Trace off the new front bodice. Include the original mid armhole balance notch that matches the sleeve. (The waistline of this pattern can been styled for a gathered waistline, explained in the section on bodice styling earlier in this Part.)

(8) The front leg of the shoulder dart should be slightly shorter than the back leg. Position notches **5 cm** from the neck so that the back leg eases into the front (for two-piece raglan sleeve notch at the end of the dart).

(9) Trace off the new back bodice. Transfer the back shoulder dart into the original position so that it opens at the new seam line. Blend the new seam line to a smooth curve and delete the remains of the back shoulder dart.

(10) The new back bodice and sleeve seam line is notched **5 cm** from the neck and at the original mid armhole. This enables the back bodice to ease into the sleeve to give shaping for the shoulder blade area.

Raglan grading

The back bodice can be graded as for the original bodice blocks using the CB as the grading reference lines. The front bodice has been flipped over so that common grade rules can be used as the back. The multi-grade reference line method is also used on the front as explained in the section on bodice styling. The sleeve is graded by the multi-grade reference line method. **GRL0** is used from the armhole balance notches to the lower edge for grading this section as for the original block. **GRL1** is drawn parallel to the original front bodice CF line. **GRL2** is drawn parallel to the original back bodice CB line. This enables the shoulder length grading increment to be in the correct direction for matching the graded bodice.

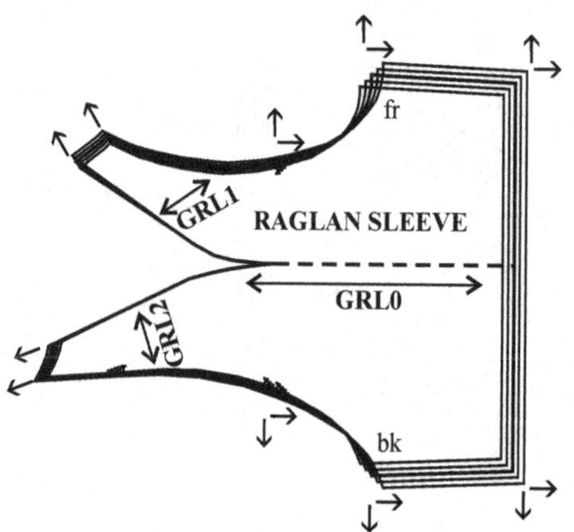

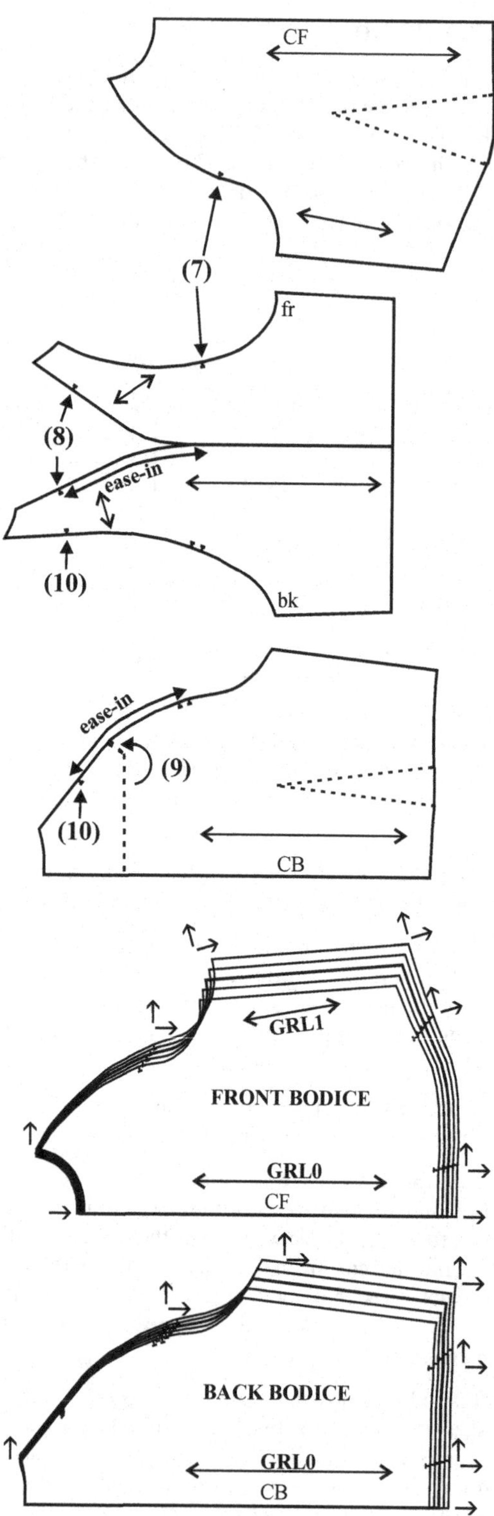

Kimono styled sleeve

There are many variations of kimono styled sleeves. Originally they were based on the traditional Japanese kimono that is loose fitting. The style described below is fairly loose fitting without a gusset and adapted from a one-piece dress block (the illustrations are to the hip level). This kimono style is more suitable for knitted or stretch fabric. Modern close-fitting kimono styles in woven fabric require a gusset inserted under the arm for movement. For this type of style refer to books on manual pattern construction.

Copy the back, front and sleeve blocks for a loose fitted dress.

(1) Transfer the front shoulder dart into the armhole.

(2) Transfer the back shoulder dart into the armhole.

(3) Reposition the bodice shoulder seam 1 cm forward at the armhole, tapering to nothing at the neck NP.

(4) Position the sleeve outer arm seam forward 1 cm from the sleeve centre line. Otherwise the sleeve tends to slip backwards when worn.

(5) Construct a line on the sleeve at the underarm width.

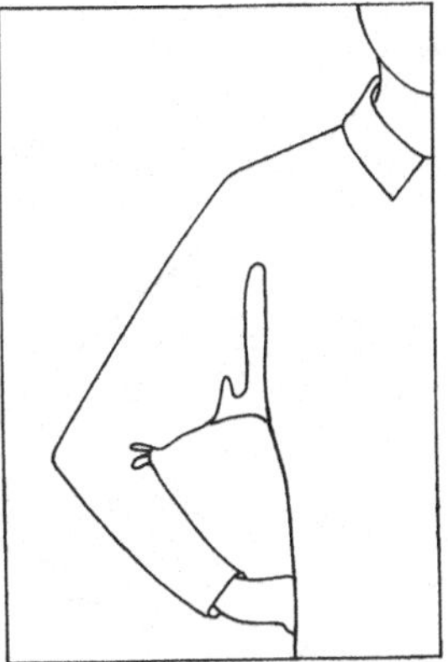

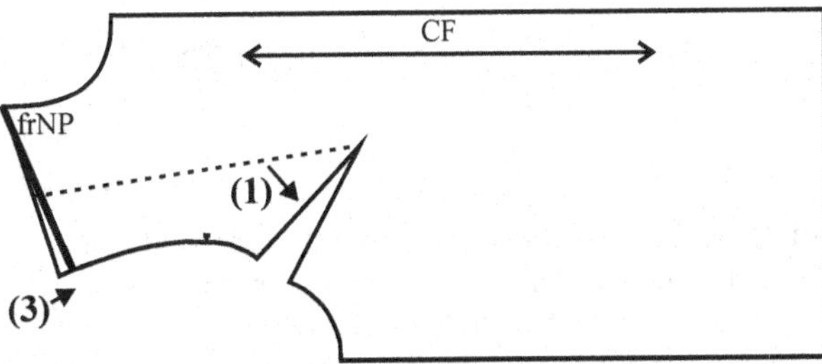

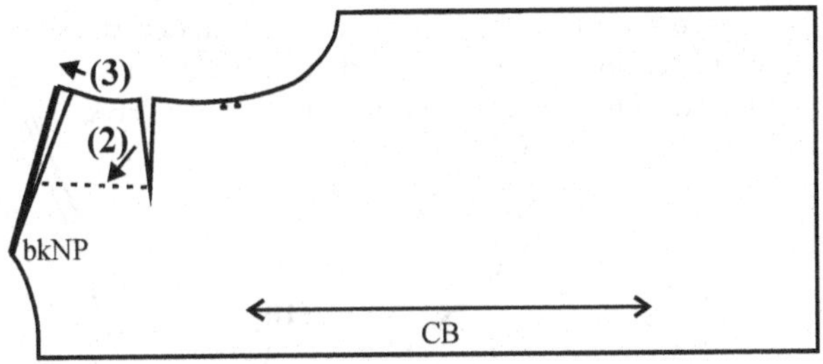

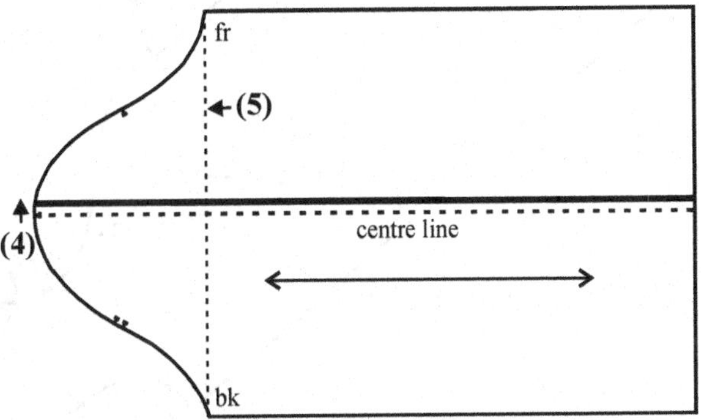

(6) Join (merge) the back and front blocks along the shoulder.

(7) Construct a line from the bkUP to the frUP.

(8) Position the sleeve matching the underarm line (5) to the bodice and position on line (7), and matching the sleeve seam line (4) to the shoulder line (6). (If required, widen the sleeve at the upper arm line to equal the length of line (7).)

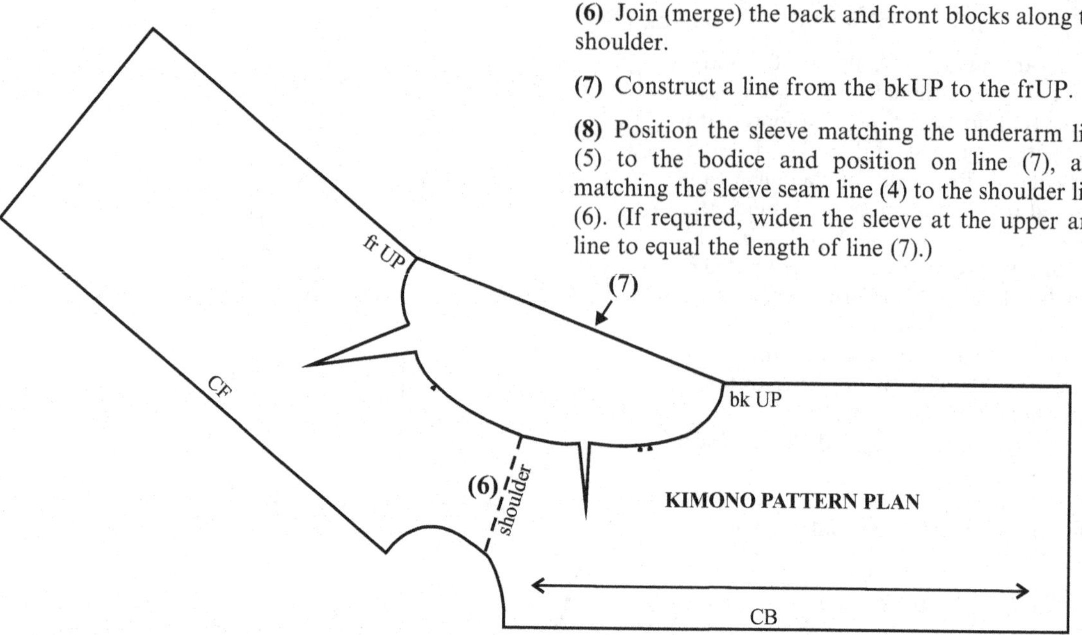

(9) Extra length can be added to the sleeve at the wrist if required. This compensates for the loss of length and is equal to half the amount of overlap between the sleeve the bodice at the end of the shoulder.

(10) Curve the new underarm seams on the back and front according to style. Check that the underarm seams match in length. (The underarm seams may require adjusting in shape as the angle of the back and front sleeves may differ.)

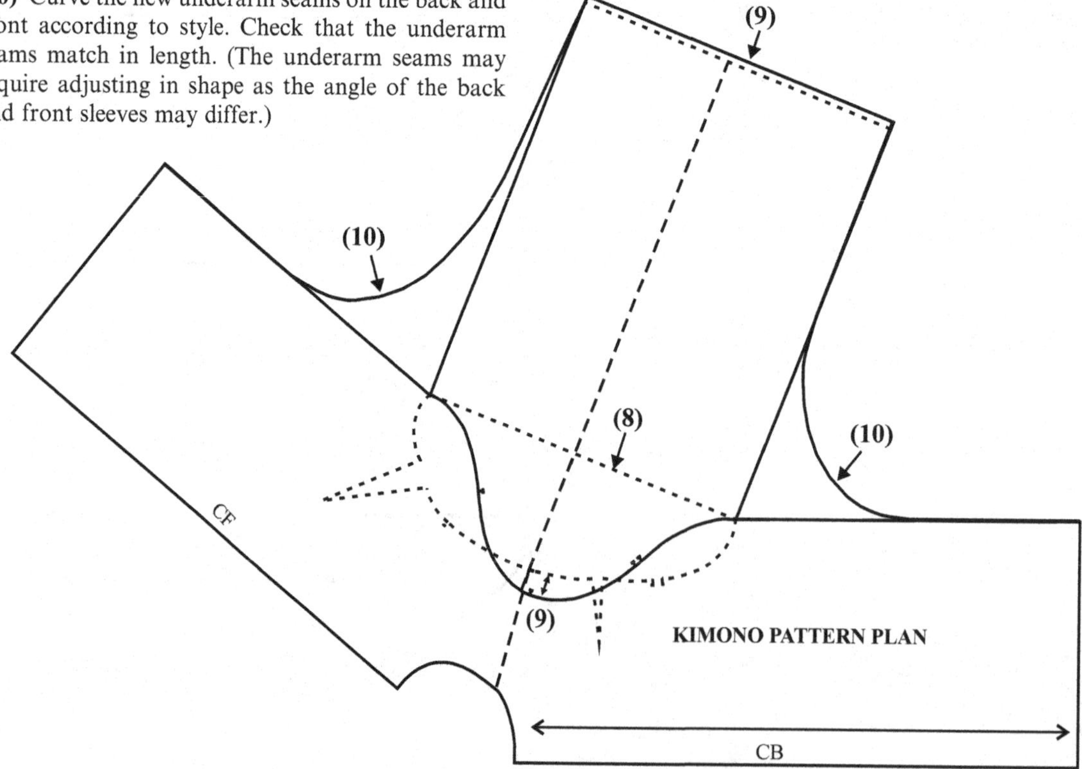

(11) Trace off the new back and front kimono blocks. The front is then turned over and pivoted so that the CF is positioned on the straight grain.

(12) Notch both the back and front outer arm seam at the original upper arm line for matching when stitching.

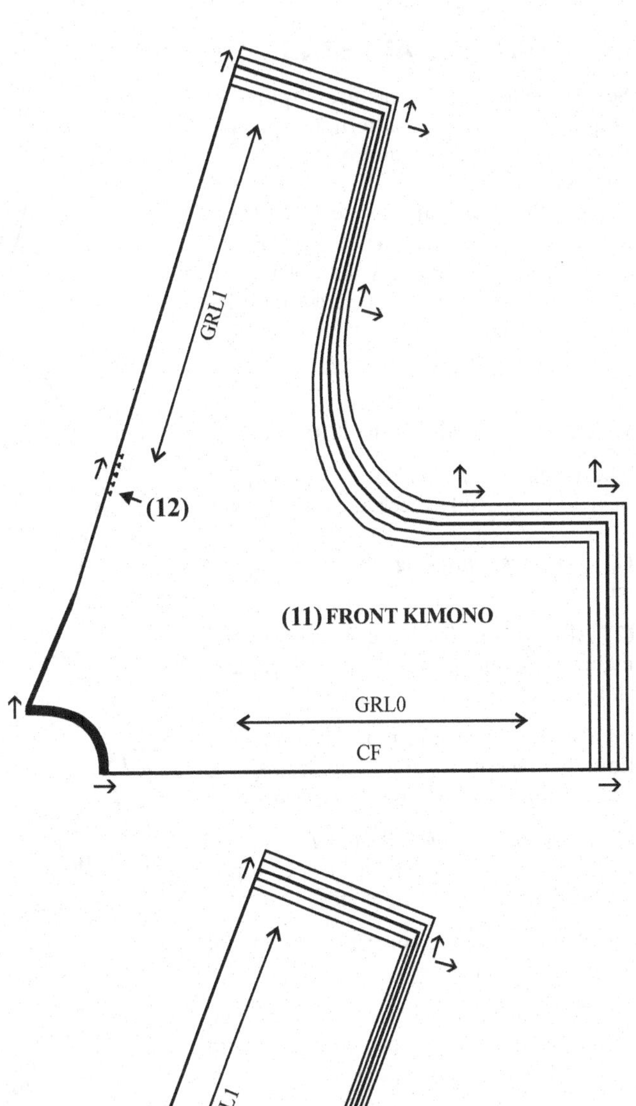

Kimono sleeve grading

The multi-grade reference line grading method can be used for kimono styles. In this example the bodice neck, shoulder length, waist width and bodice length can be graded as for the original bodice block by using the grade reference line **GRL0** (parallel to the CB and CF). The sleeve grade reference line **GRL1** is drawn parallel to the original sleeve grade reference line (parallel to the sleeve central line). This enables the sleeve to be graded as for the original sleeve block.

It is now possible via the CAD media to record each procedure of modification made to the pattern. This can be applied in the development of new patterns, thus saving time. These procedures can be placed in a data library and applied automatically for future new pattern modifications.

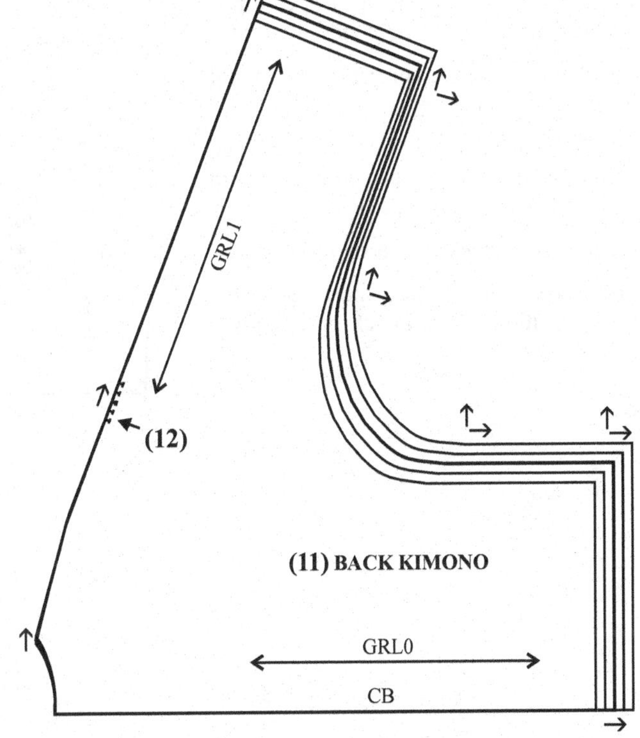

PRODUCTION PATTERNS

Seams, hems and facings are required to complete a garment pattern for production. The pattern technologist has to be aware of all the factors that influence the completion of the patterns, such as the end use of the garment, the design, fabric, production methods and cost. The perimeter of the patterns constructed so far has been the stitching line, that first has to be checked for accuracy.

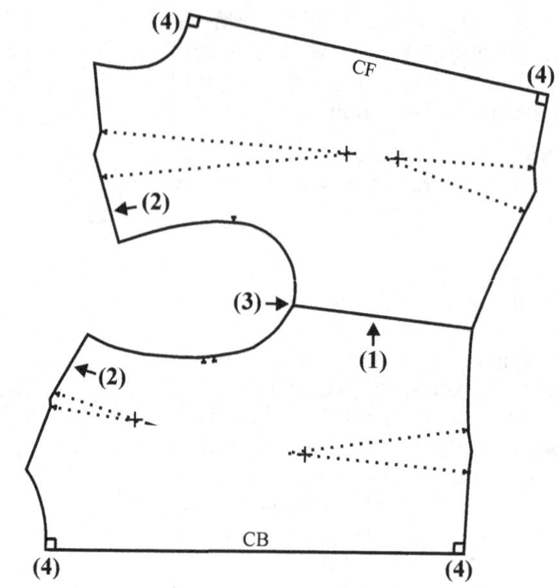

Checking the stitching line

The **stitching lines** of each pattern piece have to be carefully checked to ensure that:

(1) *Adjoining stitching lines* match in length.

(2) All *straight lines* are truly straight, including when the darts are folded, e.g. shoulder, side seam.

(3) *Curved lines* run smoothly and continue at the end of adjoining seams, e.g. neck, armhole, waist and flared hems.

(4) The *corners at the CB and CF* are at 90° so that the neck and waist lines continue smoothly from left to right side.

(5) *Darts* fold in the correct direction and have an underlay.

(6) *Drill holes* are correctly positioned within the underlay (approximately 0.5 cm to 1 cm from the dart apex).

(7) The *grain line* is correctly positioned.

(8) *Notches and balance marks* are correctly positioned and match on adjoining seams.

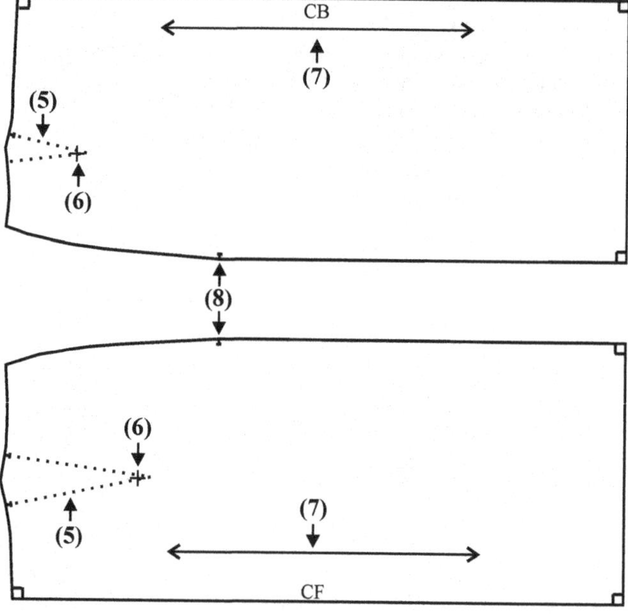

Seam allowances

The seam allowance is the distance from the stitching line to the perimeter of a cut garment part. The adding of seam allowances within a computer pattern design system is both quick and accurate. The same width can be added to the total pattern, or varied seam widths can be added to selected lines. Existing seams can be altered or removed. It is useful to display the stitching line when checking the pattern measurements, but they are not required for marker making and cutting garments. The amount of seam allowance can vary greatly according to:

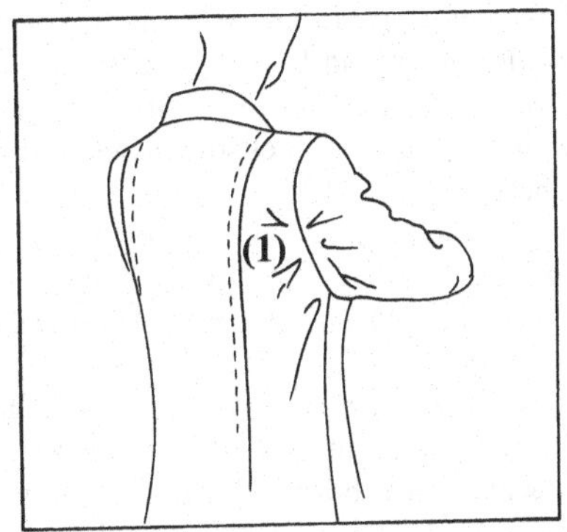

(1) The *position of the seam* and how much stress it has from body movement, e.g. an armhole and sleeve head (minimum allowance 1 cm).

(2) The *curvature of the seam* influences the seam width, e.g. the curved seam of a panelled bodice requires a narrower seam where convex and concave curves have to be joined together (maximum of 1 cm).

(3) *Enclosed seams* require a narrow seam allowance, e.g. the faced neck and armhole (maximum of 0.7 cm).

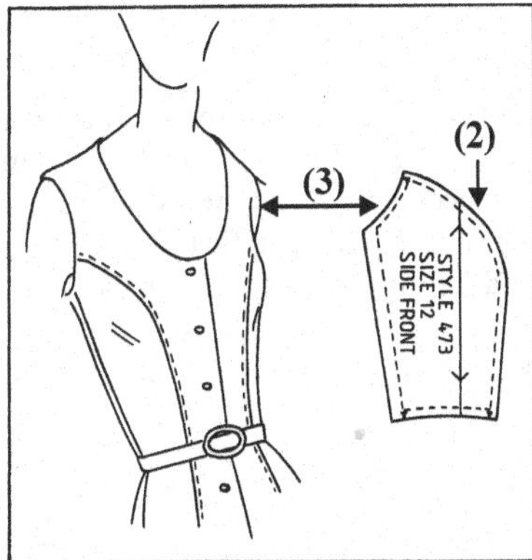

(4) The *type of machinery* required for stitching the seam influences the seam allowance width. For example, a flat felled seam where a twin-needle machine with a folder attached is used to enclose both seam allowances by interlocking opposing folded edges beneath two parallel rows of stitching through all plies; and a decorative seam with bias cut piping which encloses a cord that is inserted between flat and frilled fabric at the same time by an overlocking machine with a folder attachment and differential feed.

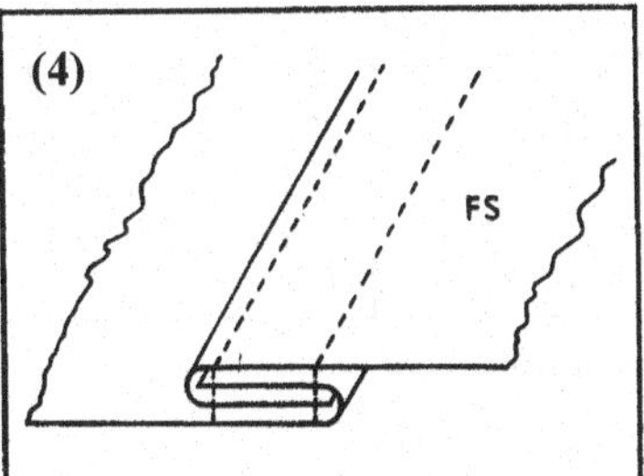

Flat Felled Seam

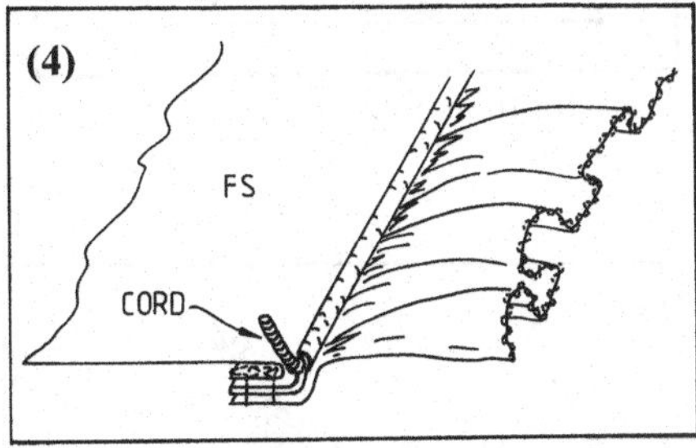

Decorative Seam

(5) *The type of fabric* often determines the width of the seam allowance and how it is neatened:

- e.g. *loosely woven fabric* requires a wide seam allowance for plain lockstitch machine and is neatened by overlocking
- e.g. *sheer fabric* can be neatened by a French seam which is constructed so that a narrow seam is enclosed within a larger one to produce a clean finish
- e.g. *thick fabric* can be neatened by being bound with *bias binding* (cut at 45° to the warp and weft); this is suitable for both straight and curved seams
- e.g. *stretch fabric* can be overlocked by enclosing the raw edge within the loops of the stitch to form the seam; it can have 3 or 4 threads and in this example it has 5 threads which includes a safety chainstitch.

(6) A *wide seam allowance* is required for inserting a zip fastener, e.g. *skirt placket* (1.5 to 2.5 cm). Wide seam allowances known as **inlays** are also required where a garment is specifically constructed for altering and letting out (1.5 to 2.5 cm).

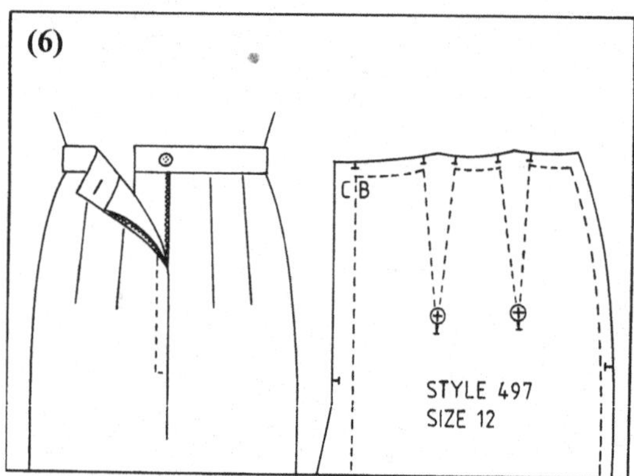

Wide seam allowances

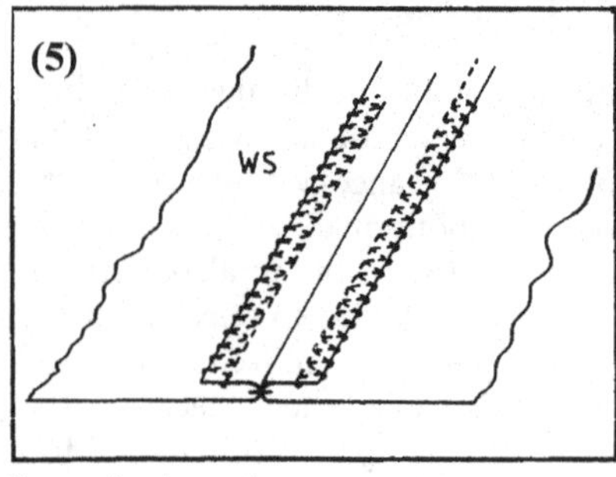

Seam for loosely woven fabric

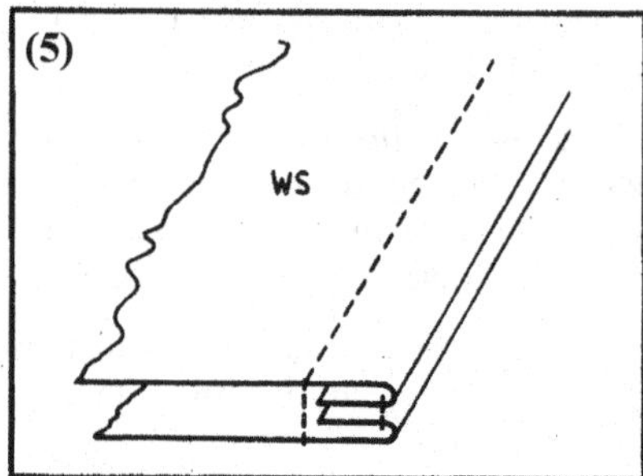

Seams for sheer fabric

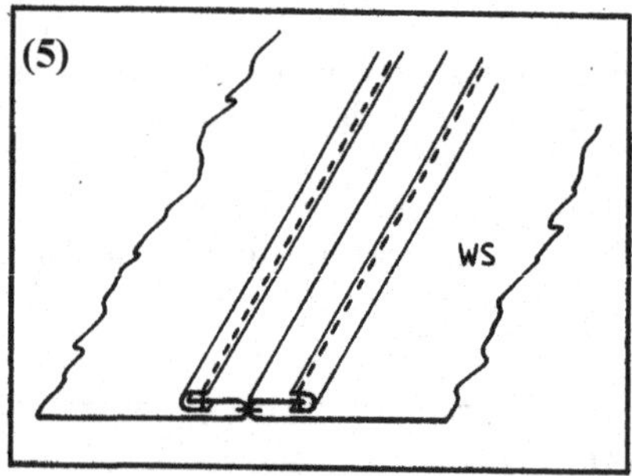

Seam for thick fabric

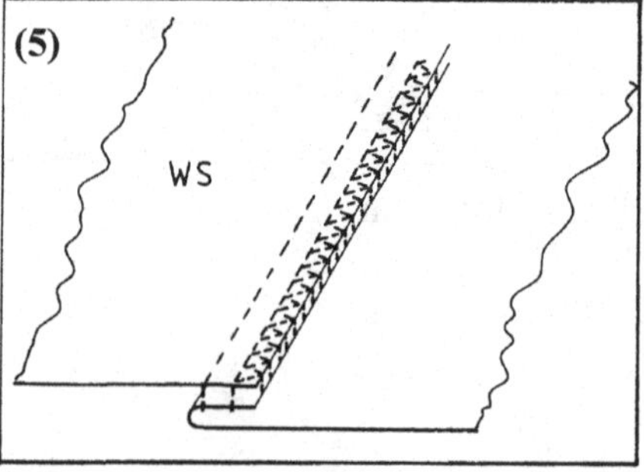

Seam for stretch fabric

Seam corners

The type of corners at the intersection of two perimeter lines can be varied according to the method of production, the type of fabric and whether notches are required. The selection and method of changing these types of corners will vary according to the different computer systems. The following examples are a selection of the most often used corner types.

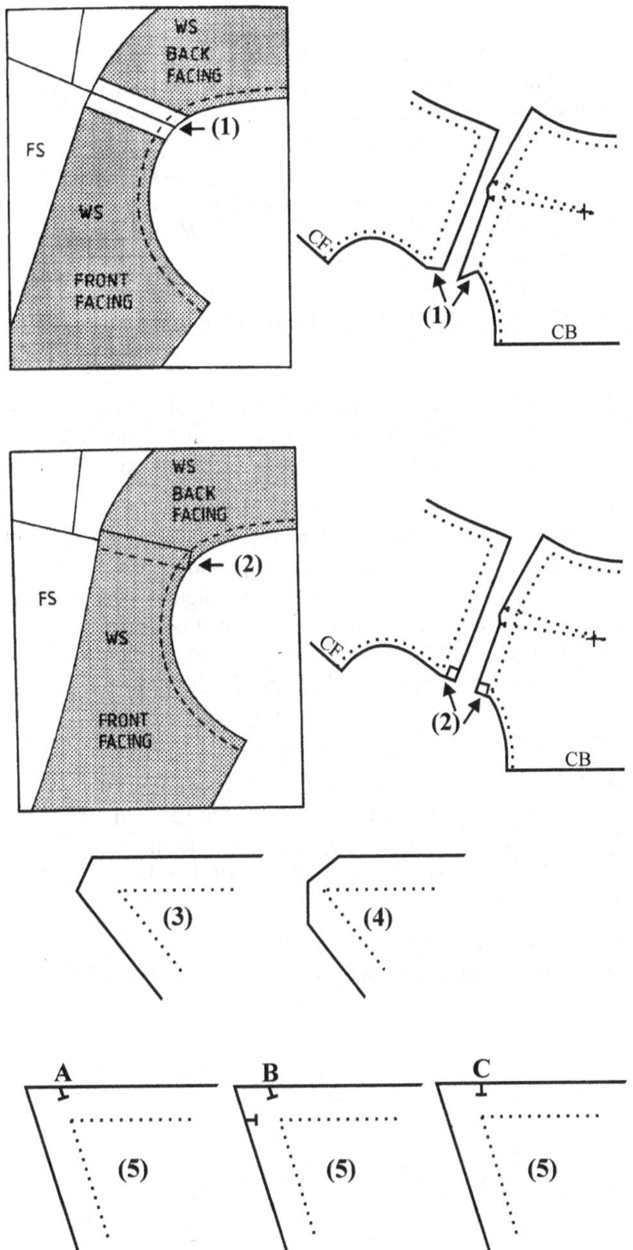

(1) *Mirrored corners* are used where a seam is pressed open and the contour of the intersecting perimeter line is continued. This is achieved by selecting the seam that has to fold back from the stitching line. In some cases it can be more difficult to match the stitching lines when stitching, or the sharp point may be difficult for cutting a quantity of garments (e.g. the neck edge at the shoulder seam).

(2) *Square corners* can be used where the seam is pressed in one direction. This is also easier for matching identical contours when stitching, and easier for cutting a quantity of garments. However, care has to be taken that the corner is held by the intersecting seam (e.g. the neck and shoulder seam).

(3) A *mitred corner* is clipped off straight and perpendicular to the corner to remove excess fabric.

(4) A *shaped corner* removes more excess fabric by clipping off a double mitred corner.

(5) For *notching at corners* it is advisable to restrict the notches to one non-standard seam allowance, as example **A**. The notching of both sides of the corner, as example **B** can cause weakness and fabric fraying. Some computer programs generally insert the notches at right angles to the pattern edge, as example **C**. However, these notches can be angled in the direction of the stitching line as in examples **A** and **B**. (For types of notches see 'Notch parameter table' in the section on Pattern preparation for digitising in Part 2).

Hems

The shape of the hemline influences the width of the **hem allowance**. A straight hem lies flat when folded back so the hem can be quite wide. Curved hems do not lie flat when folded back and therefore the hem width has to be reduced. The corners at each end of the hem also have to be considered so that the turned up hem mirrors the intersecting seam. Where the hemline has a more complex shape and will not fold back, a separate facing will have to finish the edge. The following are some typical examples of hems that fold back:

(1) A *straight hem* allowance can be wide (4 to 6 cm) where the intersecting seams at each end of the hem are at right angles to the hem line, e.g. a straight skirt.

(2) A *straight hem with mirrored corners* is required for the hem where the intersecting seams at each end of the hem are not at right angles, e.g. short sleeve. The hem allowance has to mirror the angle of the intersecting seams. The hem allowance can be added as a wide seam allowance with mirrored corners when using the computer. Alternatively the pattern can be mirrored at the hemline and split at the hem width.

(3) *Flared hems* will vary according to the amount of curve that allows the hem to turn back flat (2–4 cm). When the hem does not lie flat the allowance has to be reduced. The corners should mirror the intersecting seams.

(4) *Circular hems* are very curved and therefore require only a very narrow hem allowance (1 cm to 2 cm).

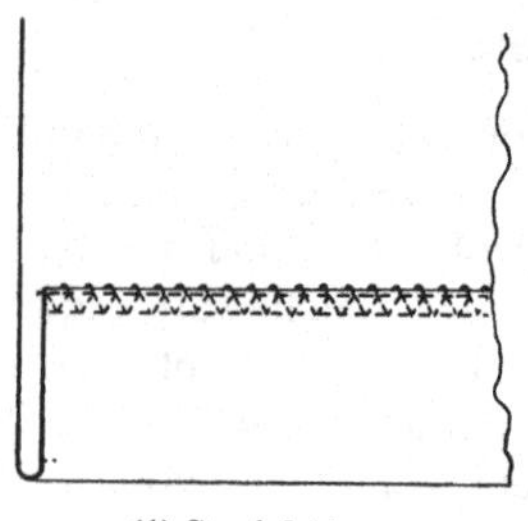

(1) Straight hem

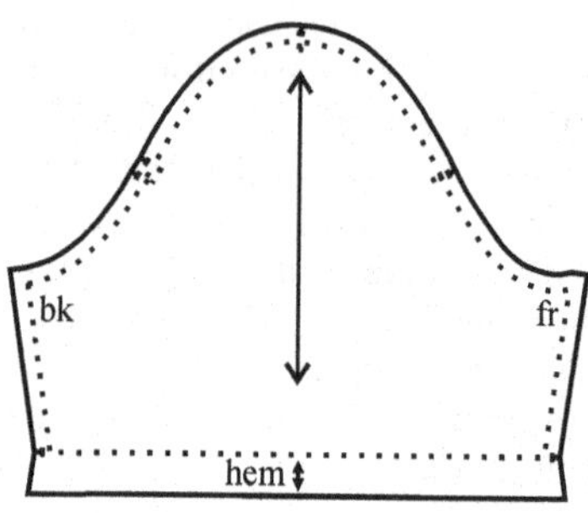

(2) Straight hem with mirrored corners

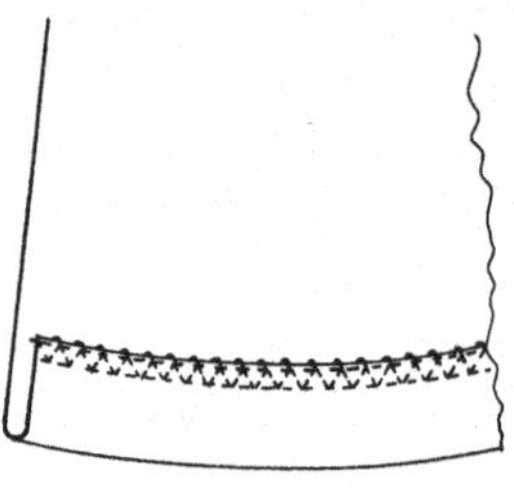

(3) Flared hem

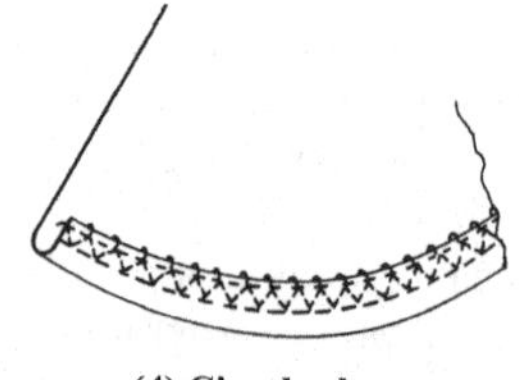

(4) Circular hem

Facings

A **facing** neatens the edge of a garment. It can be an extension of a garment part or cut separately. It is generally concealed by turning underneath to the inside of the garment. It can also turn back to be revealed on the face side. The following are some examples of different types of facings:

(1) For *facing a buttoned opening* the button size has to be known and the buttons positioned on the fitting line. The **button stand** is the extension beyond and parallel to the fitting line. The two button stands overlap the garment to prevent the garment gaping. The width of the button stand is calculated as either the diameter of the button or the radius plus 1 cm. The buttonhole length is calculated as the diameter of the button plus 0.2 or 0.3 cm according to the shape and size of the button shank (BS 3866 1997 Holes and Shanks for Buttons). The width of the facing is calculated from the edge of the button stand to approximately 2 cm beyond the end of the buttonhole. The facing is generally interlined to stabilise the buttons and buttonholes.

(2) *Extended facings* are also known as 'grown-on' facings as they are an extension of a garment part. This example is of a collarless bodice with a CF buttoned opening. The button stand has first to be added to the CF and the width of the facing calculated. The facing is planned on the pattern parallel to the edge of the button stand to 2 cm beyond the buttonholes. To face the neck edge the facing width curves to the shoulder seam at approximately 5 cm from the neck edge. The pattern can then be mirrored and split along the edge of the facing. A 0.5 cm seam allowance can be added to the cut edge of the facing for overlocking or neatening. For production a fusible interlining pattern can be made for reinforcing the button and buttonholes. This interlining finishes from the fold edge of the button stand. The interlining seam allowances can be reduced 0.2 cm to reduce bulk but are still caught by the stitching.

Grading the extended facing uses the same grading reference line as for the bodice. The grade rules for the facing NP and shoulder will be the opposite from the bodice. The interlining can be graded either as the bodice or facing.

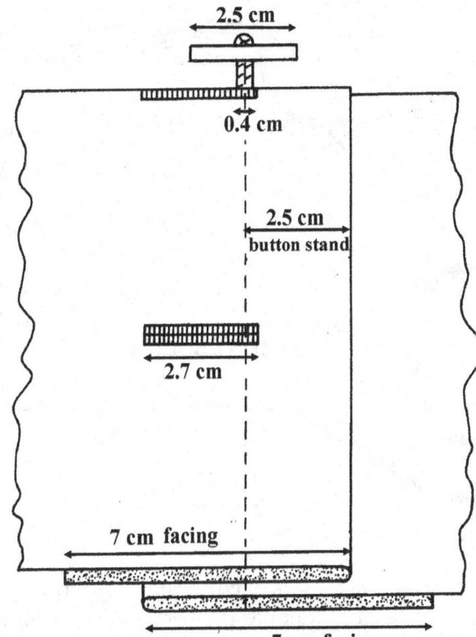

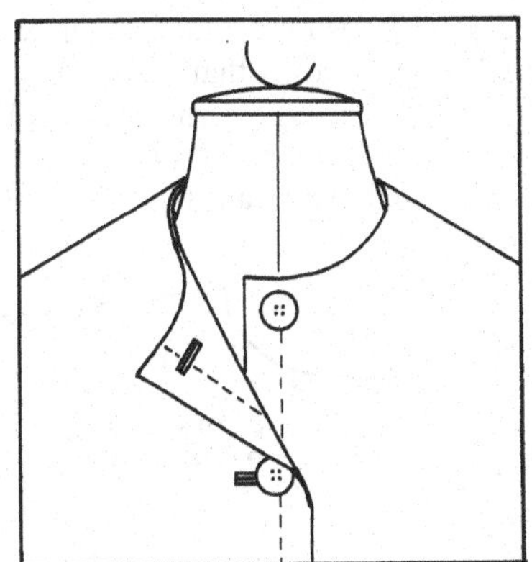

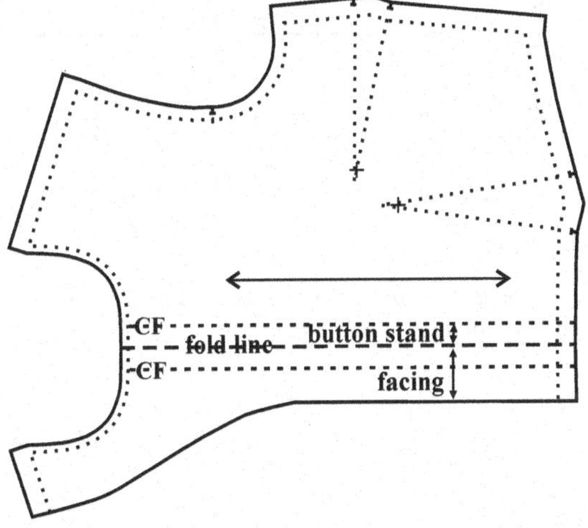

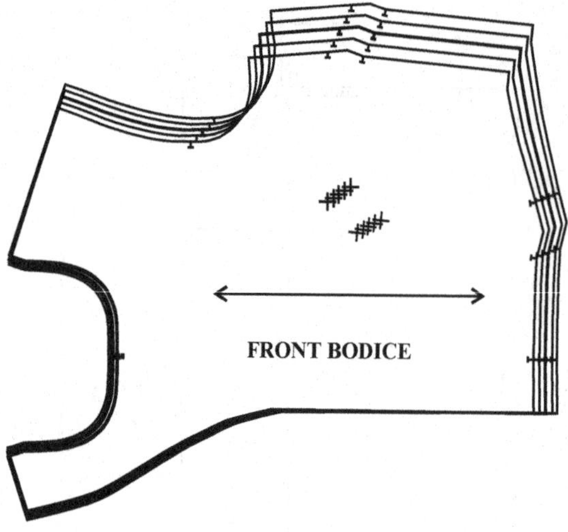

(3) *Separate facings* are cut separately from the garment part, for example the back neck of the collarless bodice. This facing is planned on the back bodice pattern the same width as the front facing at the shoulder. The CB of the facing finishes deeper than the CF neck edge so that the neatened lower edge is not visible when the garment is displayed on the hanger. This is to improve the 'hanger appeal' when it is displayed in a retail store.

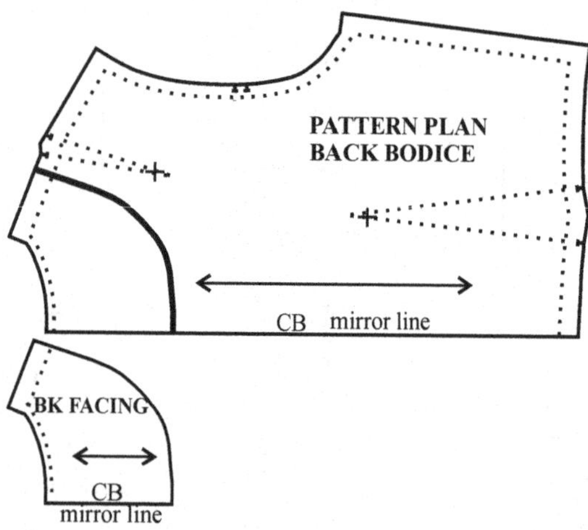

Grading the separate facing is identical to grading the back bodice.

(4) *Facing shaped revers* requires separate facings. The area of the revers that turns back on to the face side has to be slightly larger than the fronts so that the seam at the edge of the revers is concealed by rolling underneath. This facing is increased (approximately 0.3 cm) at the point of the revers tapering to nothing where the collar joins and where the crease line breaks at the first button. When facing straight lines it is advisable to bevel the stitching line slightly outwards. This has the effect of the seam appearing straight when the facing is turned through on to the face side, otherwise the seam can have a curved appearance. The interlining of the facings and fronts will vary according to the type of garment and method of production.

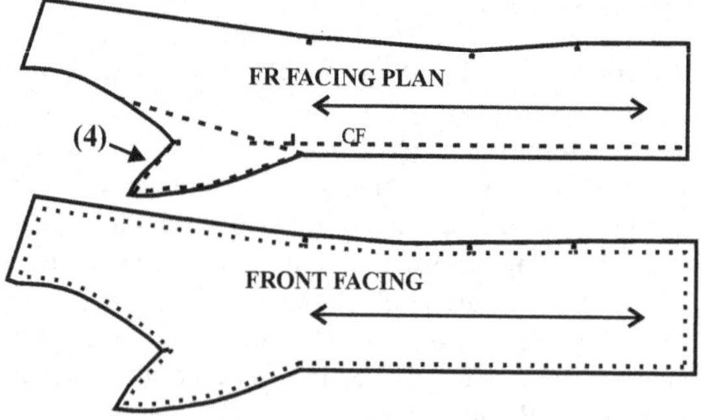

Grading the shaped revers facing is identical to the fronts.

(5) A *combined neck and armhole facing* is required for this example with a low-cut neckline and sleeveless armholes. This gives a neater finish than separate neck and armhole facings. The facing pattern is planned by the front bodice and side panel being positioned or merged on their stitching lines between the armhole and BP (before seams are added). The width of the facing at the front is 2 cm beyond the end of the buttonholes to the edge of the button stand; at the underarm it is 4 cm. These lines are blended together above the bust prominence. The facing can be reduced in size to ensure that rolling to the inside of the garment conceals the seams. To achieve this the facing pattern can be reduced (approximately 0.3 cm) at the shoulder, side seam and at the top of the armhole.

Grading the combined neck and armhole facing requires two grade reference lines. The CF and neck can be graded according to the front bodice **GRL0**. The facing armhole and side seam have a grade reference line according to the side bodice **GRL1**. This is because the facing is constructed with all the dart suppression transferred to the waist.

CAD systems will now create facings automatically using the intelligent data library stored. This can be applied to linings and interlinings and reduces time in pattern development.

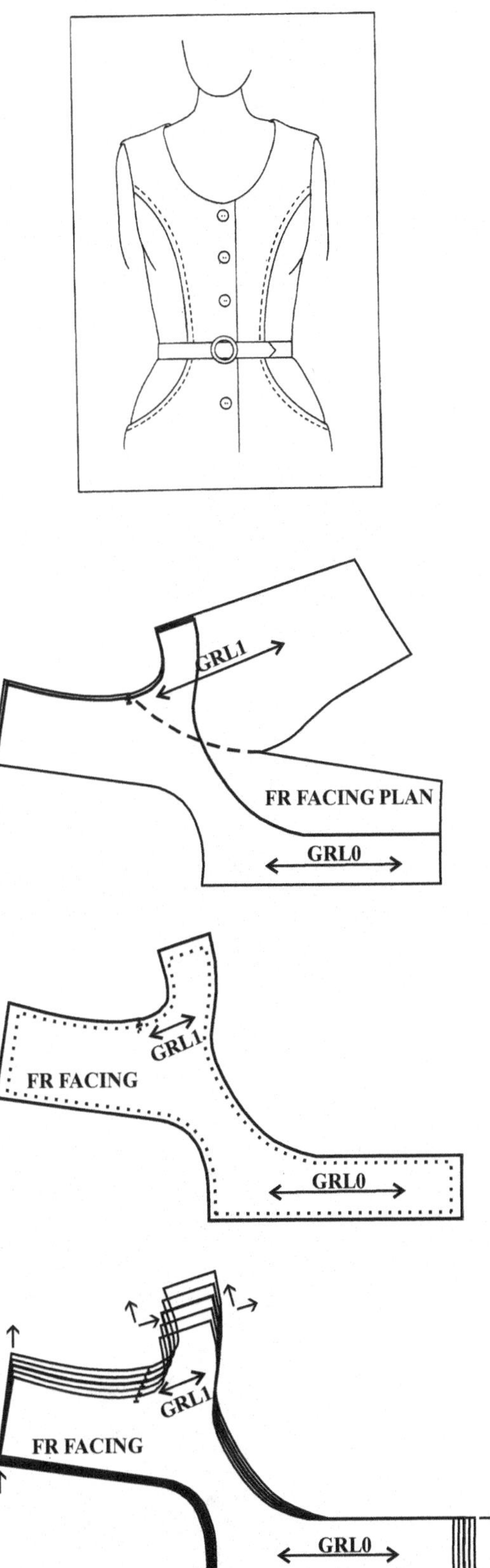

Approval of the sample garment and graded patterns

The original patterns developed for a garment style are in the sample size, which is generally the centre of a size range. These patterns have to be tested by cutting and making a sample garment. This garment has to be approved for the design, size and fit, fabric behaviour, method of assembly and cost. The pattern and garment may require altering and correcting. Once the style has been accepted the graded patterns have to be tested. Generally this is by cutting and making two garments in each size and colourway for further approval before a quantity of garments can be put into production.

Part 4
Pattern modification for garment size and fit

To set up and use computer pattern alteration systems or made-to-measure systems requires a good basic understanding of the variation in figure shapes, and the appropriate pattern modification. This part covers:

- Assessing figure shape and garment fit
- Pattern alteration according to wearer's:
  - bone structure
  - posture
  - body size and contour
- CAD technology for customisation.

The subject of garment fitting is very large and requires a book devoted entirely to the subject. In this section only some guidance for a few typical examples is given, which hopefully will encourage the reader towards further study of garment fit (Bray 2003; Liechty 1992). Most computer pattern alteration systems and made-to-measure systems modify a standard size pattern within a size range stored in the computer memory.

Several of the 3D computerised body measuring systems can be linked to the pattern alteration or made-to-measure systems. For those who do not have access to these computer body measuring systems there is still the alternative of using manually taken measurements and inputting them into a system. Patterns can also be manipulated to the new measurements within a pattern design system.

ASSESSING THE FIGURE SHAPE AND GARMENT FIT

The fit of a garment is affected by variations in the wearer's:

- Bone structure
- Posture
- Body size and contour.

There are many factors which contribute to figure variation such as occupation and lifestyle, growth and maturity or ethnic origin. To assess this variation an average figure shape has to be defined. Body measurement surveys are the most reliable source but are not always up-to-date and available for the segment of the retail market for which the garments are being manufactured. The average measurements quoted in the size charts in Part 1 were based on small-scale surveys as no up-to-date national survey data was available. These body measurements will be used as a standard to illustrate the assessment of the variations in figure shape and pattern alterations.

Identifying and understanding the cause of a fitting fault gives a clear indication of how to correct it. First, when assessing a garment on a figure the areas that fit correctly have to be identified as these must not be interfered with when correcting a fault in another area. Most faults can be identified by unwanted creases or folds and uneven hemlines. The cause of the fault is often found at the top of a fold or at the end of a crease. These can be diagnosed as:

- Incorrect lengths
- Incorrect widths
- Incorrect suppression.

The examples of fitting faults and corrections described below are a few of those most frequently encountered. They can also be applied to the reverse fault, e.g. square or sloping shoulders. It is advisable, where possible, to position the alteration lines on the straight warp and weft grains. This is to prevent the patterns from becoming skewed off grain when modified.

VARIATION IN BONE STRUCTURE

Much of the figure shape depends on the bone structure of the skeleton. This determines the lengths of the torso and limbs, the slant of the shoulders, the size of the rib cage and the width of the pelvis, the shape of the legs and the stance.

Height

The variation in height can be divided into three categories based on the small-scale surveys detailed in Part 1. The average heights in the centre of each category are:

- Short 155 cm
- Medium 165 cm
- Tall 175 cm.

Figure 4.1 shows that it is in the limb lengths that most of the variation occurs when the eight-head theory is applied. This classic theory of dividing the body height by the length of the head has proved to be very useful when proportioning garments to the average figure. The survey found that there was only a slight difference between women's head lengths. These head lengths had no relationship to the women's heights. It was also found that the proportions of the taller women, whose heights range between 170 cm and 180 cm, conform closely to the eight-head theory. This is illustrated in Figure 4.1 with the underarm level at the second head length, the waist at the third and the hip at the fourth. The knee is a little above the sixth, giving greater length in the lower leg. The central size of the medium height of 165 cm averaged seven and a half head lengths, and the short height of 155 cm averaged seven heads. There is only a slight difference in length between the shoulder and underarm level as the variation takes place more between the underarm and waist. There is a little variation between the waist and hip levels as most variation in the leg length is between the hip level and ground. The arm lengths also vary proportionally to match the torso. The variation in length measurements for a bodice, sleeve, skirt and trouser are based on the height variations of tall 175 cm, medium 165 cm and short 155 cm. For those who do not have the exact height quoted their differences can be a percentage of the alteration measurement. Reference should be made to the actual body measurements for those who may have the proportional variation of a long torso and short legs or long legs and short torso. The neck length variation between the three height categories is approximately 1 cm. The positions on the patterns for length alteration are illustrated in Figure 4.2 for both an increase and decrease in height:

(1) Bodice underarm to waist ±3 cm
(2) Sleeve underarm to elbow ±2 cm
(3) Sleeve elbow to wrist ±1 cm
(4) Skirt and trouser waist to hip ±1 cm
(5) Skirt and trouser hip to knee ±2 cm
(6) Skirt and trouser knee to ankle ±3 cm.

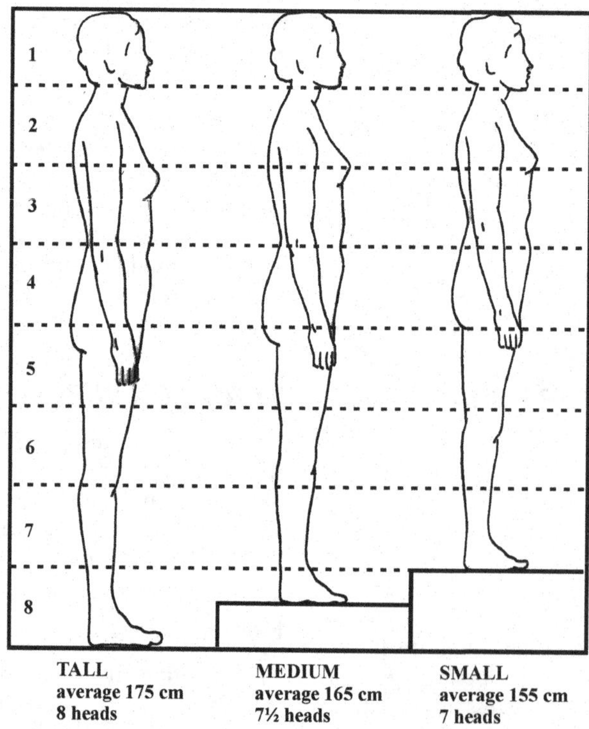

Figure 4.1 Height variations

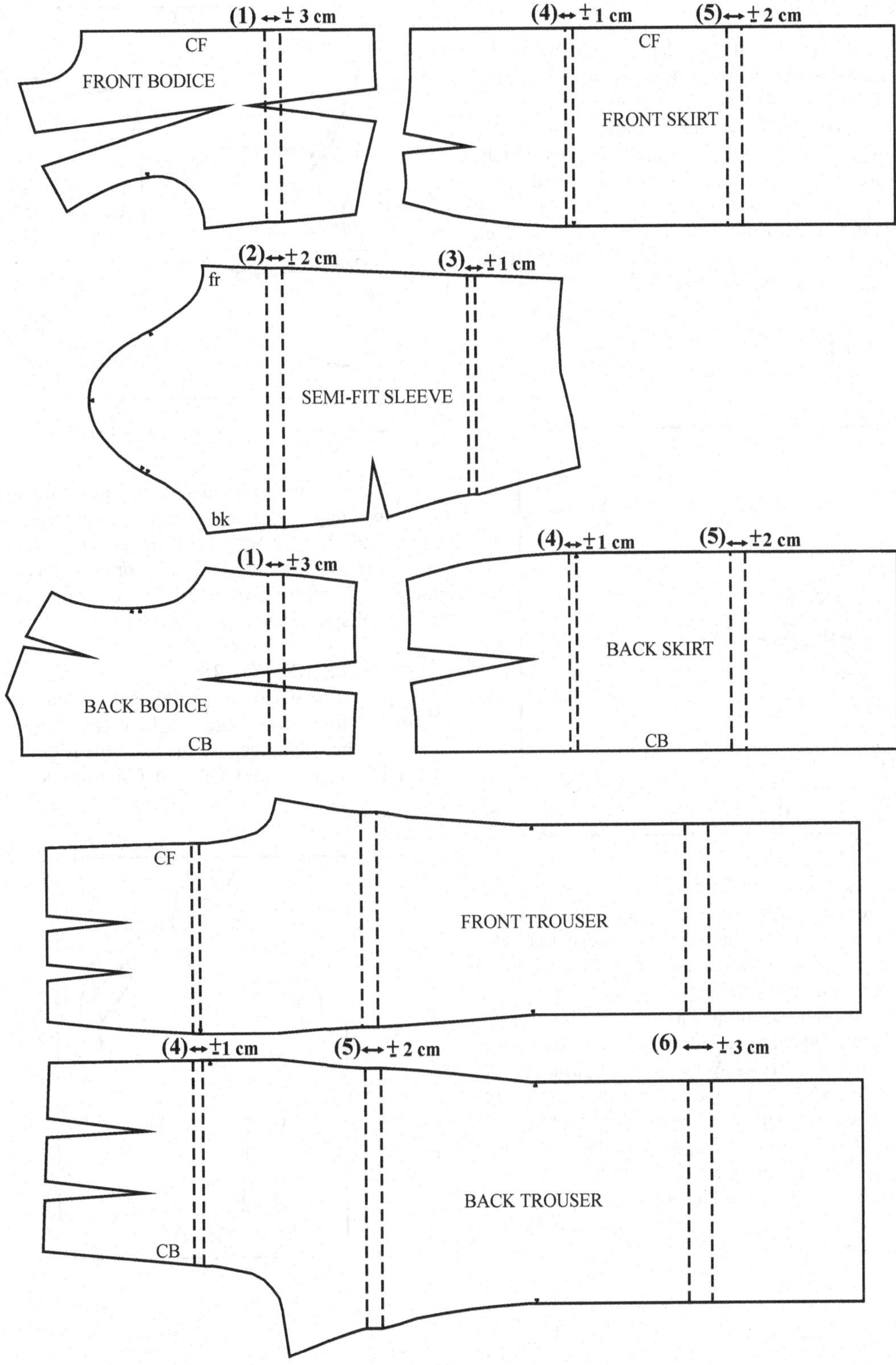

Figure 4.2 Positions of pattern alterations for variation in length measurements

Shoulder slant

The shoulder width and slant depends mostly on the bone structure although the muscles do contribute. Taller women tend to have squarer and wider shoulders than shorter women. From an analysis of 100 young women's shoulder slant, approximately 50% were in the range of 20–24°, 30% within a range of 15–19° and 20% within a range of 25–29°. This gives a total range of 14°. To categorise the shoulder slant for assessment purposes the centre of these three ranges could be considered:

- Square 17°
- Medium 22°
- Slanting 27°.

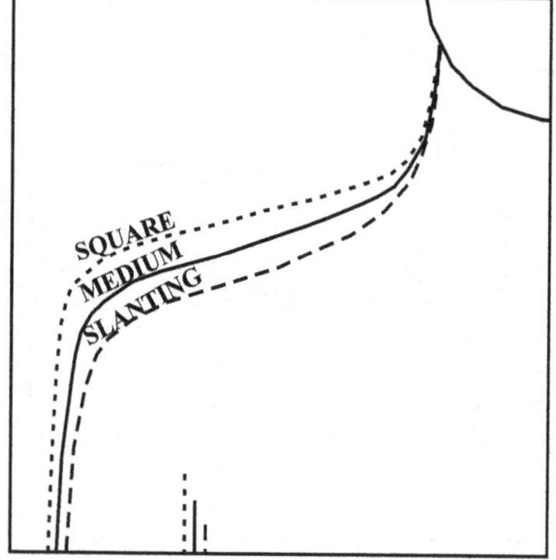

The average shoulder slant of 22° is also supported by the findings of other research. Because a woman's shoulder slant varies, this does not mean that the armhole increases or decreases in size. To retain the same size armhole the underarm level also has to be altered by the same amount. This is an example of the type of information required for defining made-to-measure alteration rules. When sample garments of bodice block patterns with a 22° shoulder slant were fitted on women with a 17° shoulder slant it was found that the shoulder height and the underarm level had to be raised 1 cm on both the back and front. The slanting shoulder of 27° required the shoulder and underarm level to be lowered by the same amount.

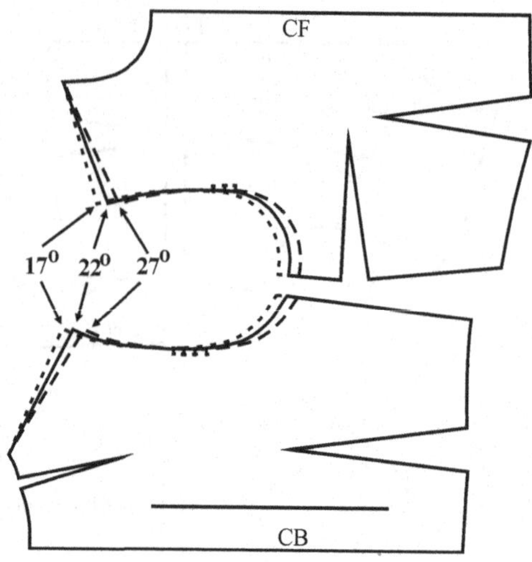

The slant of the shoulders has a major influence on both the hang and fit of a garment. An incorrect slant is detected by unsightly folds in the garment. First, the areas where the garment fits correctly have to be located. Illustrated below are two examples of possible faults and methods of correction.

SLOPING SHOULDERS

The problem of sloping shoulders is identified by diagonal folds from the armhole and side seam towards the neck, because the armhole is collapsing whereas the neck and CB and CF lengths fit correctly.

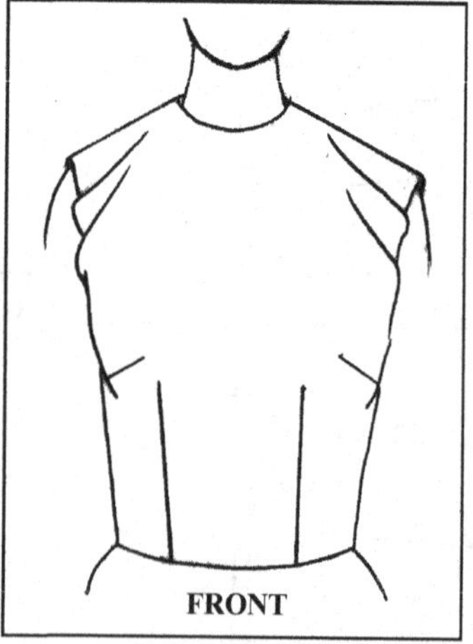

To correct

(1) The back and the front pattern are corrected by lowering the shoulder at the armhole (approximately 1 cm for a 27° slope) tapering to nothing at the NP.

(2) The UP is also lowered by the same amount to maintain the armhole at the same size.

(3) In some cases the across back and across front have to be narrowed slightly, the amount being determined by checking these measurements.

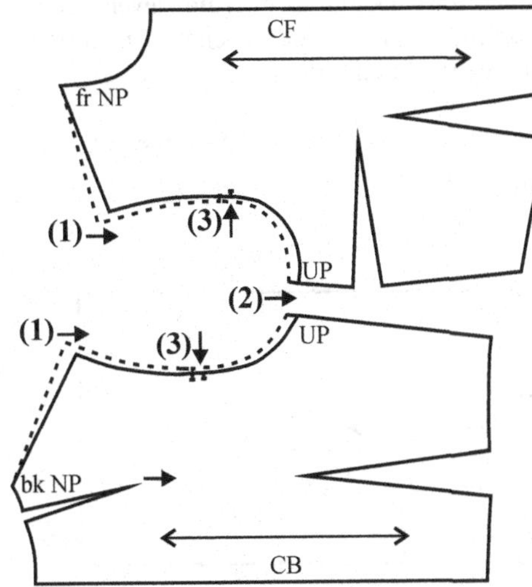

(4) The reverse alteration is made for square shoulders by raising the shoulder and underarm and widening the across back and across front.

(5) In some cases the shoulder length may need increasing.

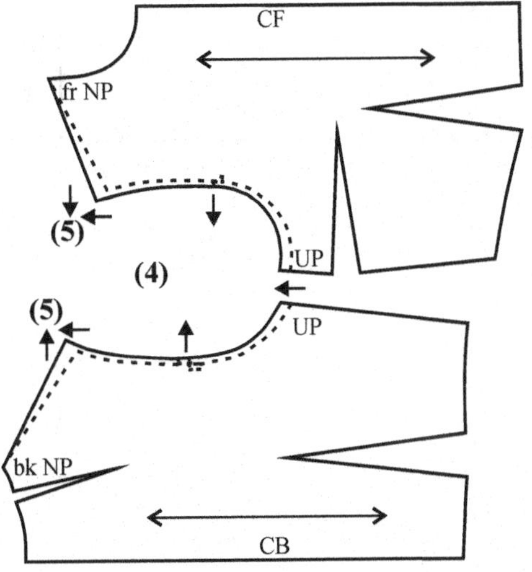

SQUARE SHOULDERS

The problem of square shoulders is identified by creases running horizontally below the neckline and across from the ends of the shoulders. The neckline may also ride up the neck column whereas the armhole and side seam fit correctly.

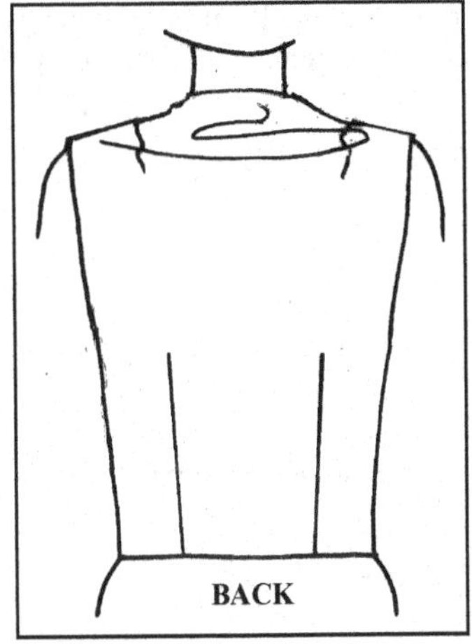

To correct

(1) Lower the back and front necklines and the NP the required amount.

(2) Taper the altered shoulder line to nothing at the armhole.

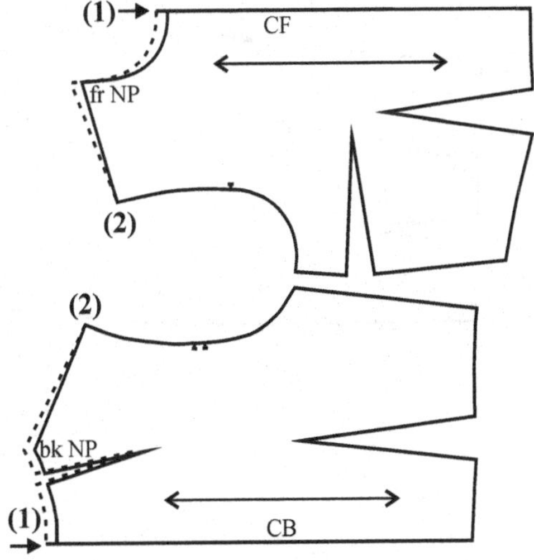

Length of the upper and lower torso

Women of the same height can have a proportional variation in their torso lengths. They can have a long torso and shorter legs or the reverse. This variation is caused by their bone structure. The bodice and skirt alterations would be the same as for the height variation but according to the individual's measurements. For trousers the length of the pelvis can affect the inside leg measurement, making it too high or too low.

CRUTCH LEVEL TOO HIGH

The problem of the crutch level being too high is diagnosed by the CB and CF waist levels being pulled downwards, and shortness in the total crutch length causing discomfort at the trouser fork. The level of the crutch from the waist can be estimated by the inside leg measurement being subtracted from the outside leg minus 1 cm to 2 cm for movement and comfort.

To correct

(1) Lower the back and front crutch level at the fork. This also reduces the inside leg length (the outside leg measurement remains the same as the original pattern).

The reverse alteration can be done for a crutch level that is too low.

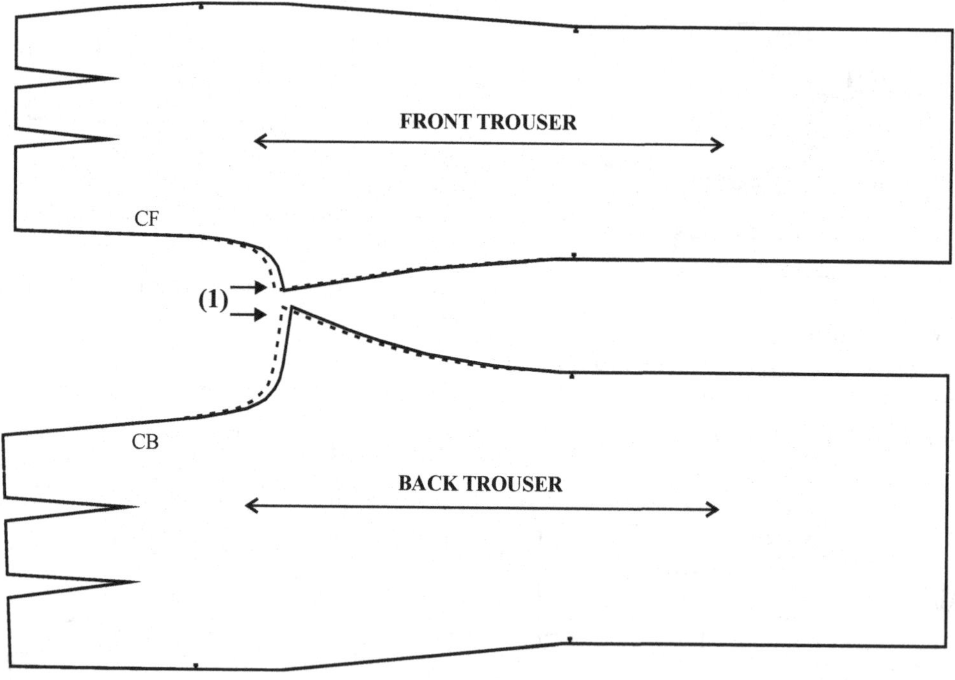

Stance

The **stance** of the figure influences the hang of the trouser legs. When the wearer stands or walks normally, approximately with heels 12 cm to 15 cm apart, the central crease line should appear in the centre of the foot. Also the inside and outside leg hang equally either side of the leg. The variation in stance is according to the angle of the leg bones and the width of the pelvis. Women with a wide pelvis tend to have a closed stance where the legs touch together at the thighs and knees when they stand normally. Women with a narrow pelvis tend to have an open stance where the legs are set apart and curve away at the knees and calves.

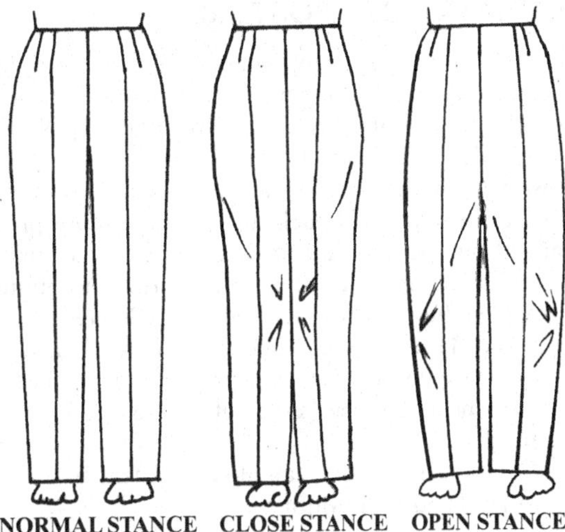

CLOSED STANCE

The problem of a closed stance is diagnosed by the trouser legs hanging close to the inside of the leg, the crease line swinging towards the outside leg, and the outside leg hanging away at the knee and ankle.

To correct

(1) At the knee and ankle levels, on the back and front trouser patterns, move the inside and outside leg seams and crease line towards the inside leg. (The amount is to the new position of the crease line in the centre of the foot.)

(2) Taper from the knee level to the crutch level both the inside and outside leg seams.

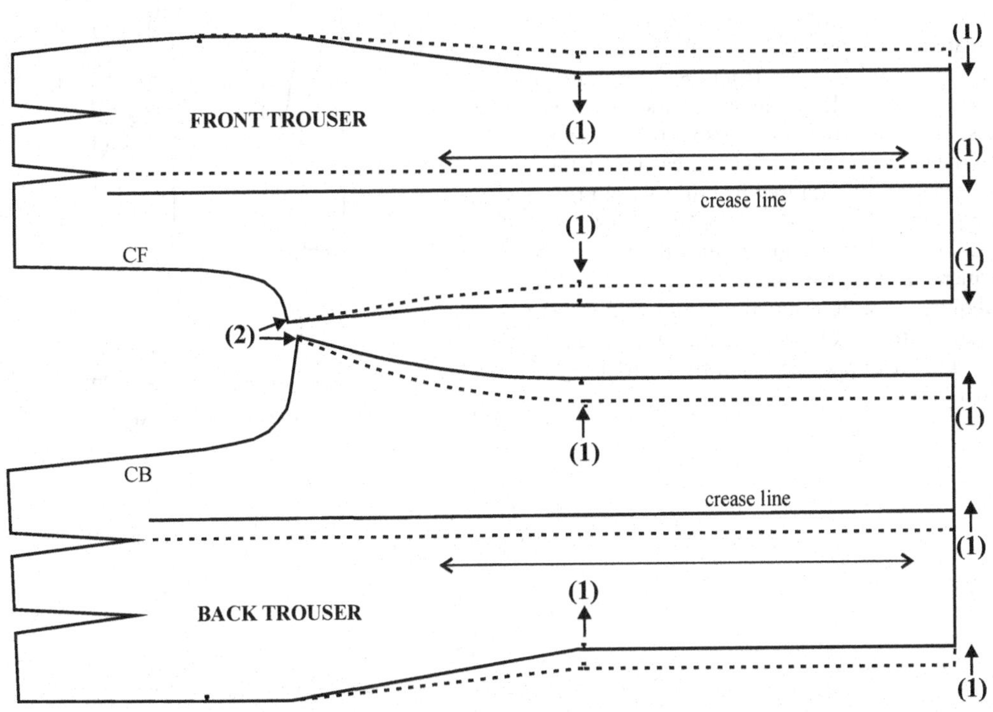

VARIATIONS IN POSTURE

The posture is generally determined in the way a woman stands. This influences both the prominence and the hollows of the contour. If the normal posture is considered fairly erect, in comparison a stooping posture would have a forward neck, a hollow chest and a more rounded back, the stomach more prominent but the seat flatter. A very erect stance may have the hip and waist area as the norm but a more prominent bust, flatter back and a straight erect neck column. The 2D and 3D computerised body-scanning systems plus the comparison of body measurements can be very useful in assessing the figure shape.

Variations in posture can affect garment balance, waist level and neck width.

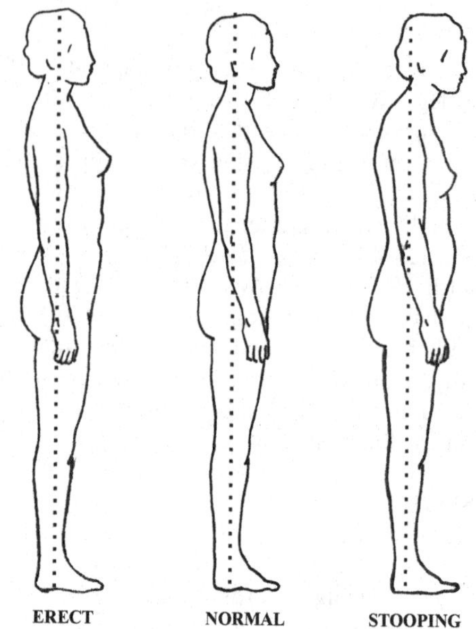

Garment balance

The correct **balance** of a garment is where the hemline is parallel to the ground and the CB, CF lines and side seams are perpendicular to the ground. There are several possible causes of incorrect balance. It could be the wearer's posture, either stooping or erect, or the variation in body size and contour due to the increase or decrease in prominence of the bust, seat or stomach. These alter the relationship between the back and front widths and lengths, also the amount and position of the garment suppression.

There are two main areas that control the garment balance. For garments worn from the upper torso it is the area between the shoulders to the bust level. For those worn from the lower torso it is between the waist to the hip level. Faults that appear below these control areas are often caused by faults within them, for example an uneven hem or crooked side seam.

The following three examples alter the balance of the garment in different ways. To know which to apply requires careful observation in identifying the fault. The methods of pattern alterations can be reversed and so applicable to either a stooping or erect posture.

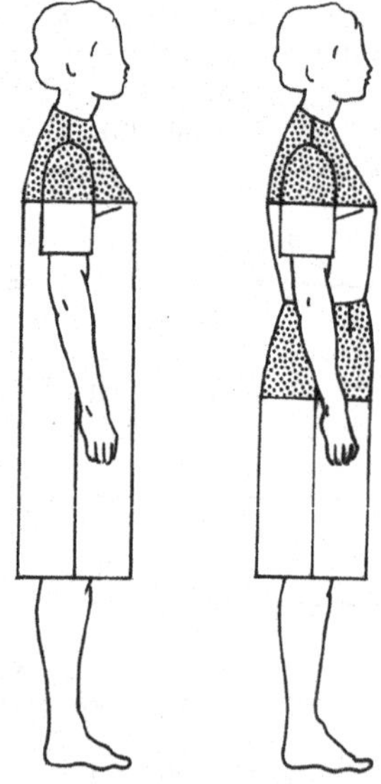

SHORT BACK LENGTH
The problem of a short back length is caused by a slightly stooping posture and diagnosed by the garment pulling up at the upper back and diagonal drag folds from the shoulder blades. The front neck rises up the neck column and the back neck is too low. The back hem appears too short and hangs away from the body at the CB. The lower side seam slants backwards. The upper front also appears too long and the CF front hem pulls close to the body. The amount of length alteration is estimated by measuring the difference between the CB and CF hem levels from the ground.

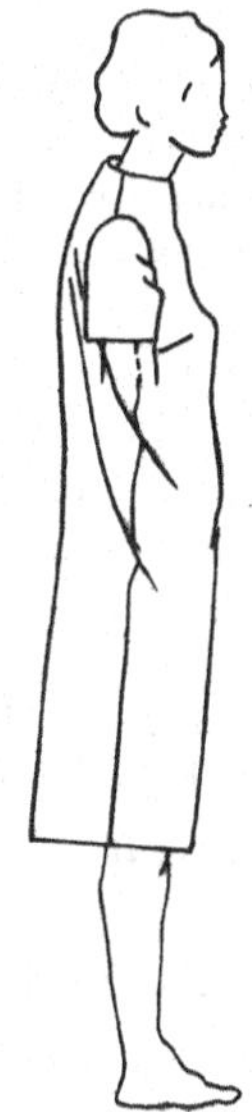

To correct
This correction alters the back and front lengths and repositions the shoulder seam and neckline. It maintains the original side seam length, armhole and bust suppression.

(1) The front length is decreased by subtracting parallel fullness of half the difference between the CB and CF hem levels, from the CF to just above the mid armhole balance mark.

(2) The back length is increased by adding parallel fullness of half the difference between the CB and CF hem levels, from the CB to just above the mid armhole balance mark.

(3) The shoulder notch at the top of the sleeve has to be moved forward to match the increase in the back armhole length and the decrease in the front armhole length.

The reverse alteration can be undertaken for a slightly erect posture.

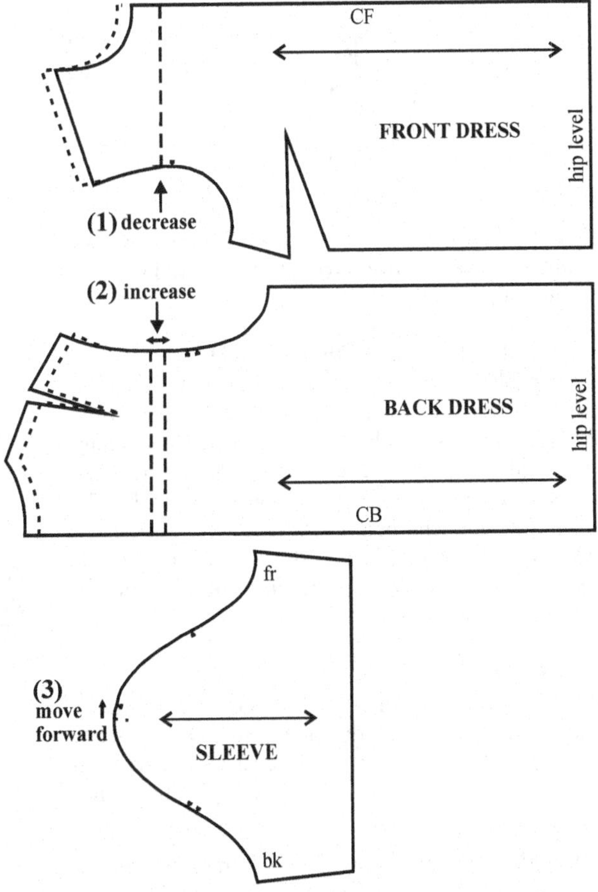

SHORT FRONT LENGTH

The problem of a short front length is caused by an erect posture and diagnosed by the garment pulling up at the front with diagonal folds from the bust to the side hip. The front hem is short and hangs away from the body at the CF. The back and side lengths are correct but the CB hem pulls close to the body and the lower side seams slant forwards. This erect posture often gives a more prominent bust that requires an increase in bust suppression.

To correct

(1) Lengthen the front at the bust level by adding parallel fullness from the CF to the dart apex.

(2) Re-align the wider dart to the required level (the dart may require closing temporarily to redraw a smooth side seam that matches the back).

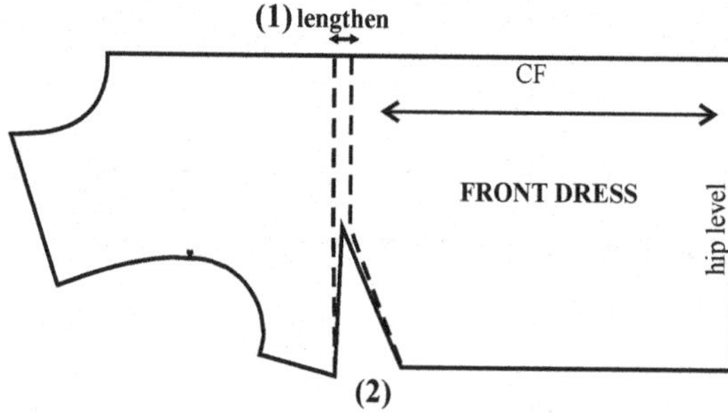

(Where there is no side seam dart the extra bust suppression can be transferred to the other positions of bust suppression.)

LONG BACK AND SIDE LENGTH

The problem of a long back and side length is caused by the flatter back of an erect posture and diagnosed by creases across the back waist continuing as diagonal folds to the bust. The front length is correct but the hem hangs away from the body at the CF. The back and side lengths are too long causing the CB hem to pull close to the body and the lower side seams to slant forwards. The bust suppression will also have to be increased.

To correct

(1) Increase the width of the bust side dart to match the reduced length of the side seam. (If necessary move the dart to the required level and close temporarily to redraw a smooth side seam that matches the back.)

(2) Reduce the back length by subtracting parallel fullness at the bust level from the CB to the side seam.

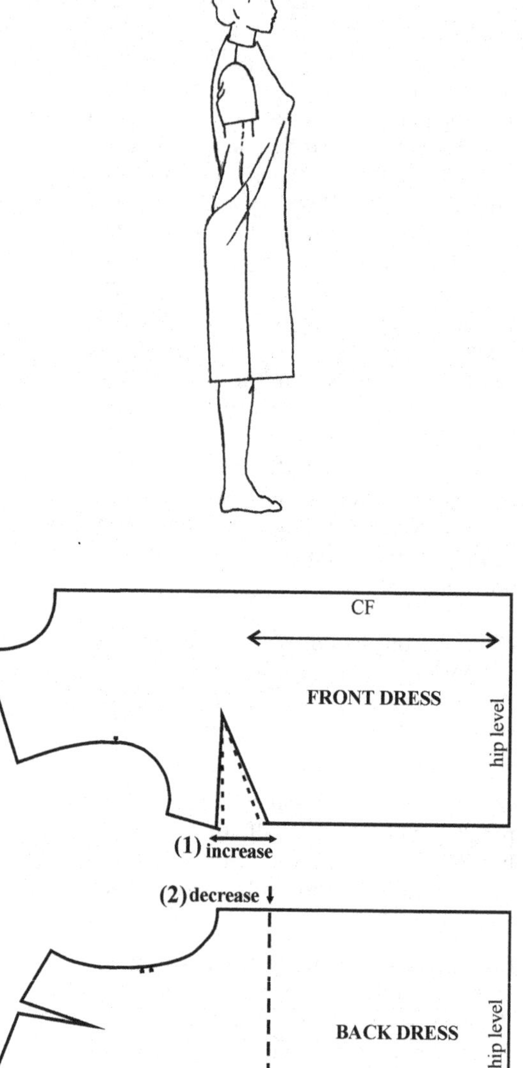

(Where there is no side seam dart the extra bust suppression can be transferred to the other positions of bust suppression.)

Skirt waist levels

Waist levels are rarely parallel to the ground. The average waist level is generally higher at the sides than at the CB and CF. The variation in the waist level at the CB and CF is mainly due to the posture of the wearer that is reflected by the forward or backward tilt of the pelvis.

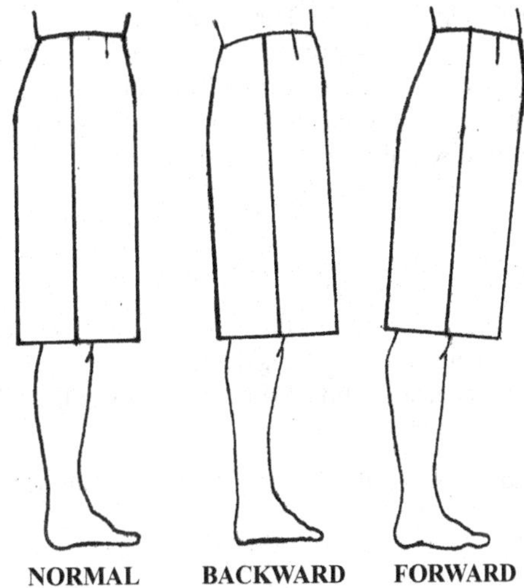

SHORT FRONT LENGTH
The problem of a short front length caused by a backward tilting pelvis is diagnosed by the side seam swinging forward as the CF hem rises up and hangs away from the legs. The CB skirt drops lower and hangs closer to the legs.

To correct
This correction alters the back and front lengths at the waist levels to maintain straight CB and CF. The side seam length and shape remain the same as the original. The amount of length alteration is estimated by measuring the difference between the CB and CF hem levels from the ground.

(1) Raise the CF waist level half the difference between the CB and CF hem levels and shape to the side waist according to the new waist level. (Occasionally the side waist level may require altering.) If the width measurement of the new front waist seam is less than the original, decrease the width of the dart suppression at the waist.

(2) Lower the CB waist level half the difference between the CB and CF hem levels and shape to the side waist according to the new waist level. (Occasionally the side level may require altering.) If the width measurement of the new back waist seam is greater than the original, increase the width of the dart suppression at the waist.

The reverse alteration can be undertaken for a forward tilting pelvis.

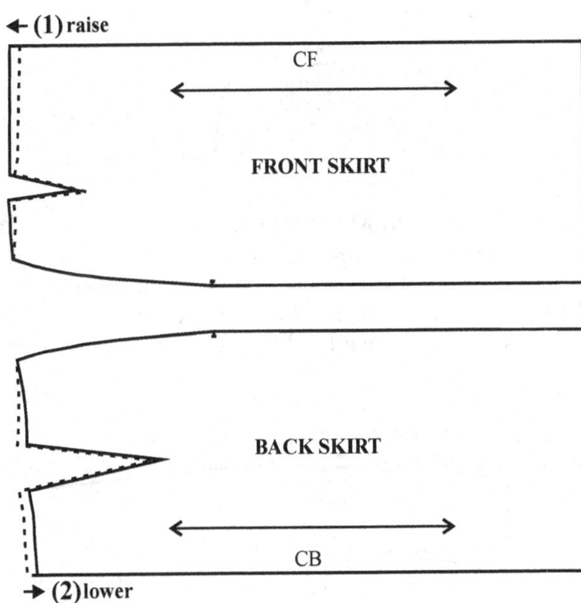

Trouser waist level and seat angle

The back **seat angle** on trousers refers to the slant and length of the seam from the CB waist to the hip level. This can vary due to the posture of the wearer, as reflected by the forward or backward tilt of the pelvis.

It is critical that this angle and length is correct for the wearer as it affects the balance between the back and front lengths, the forward stride of the legs, sitting comfortably and maintaining the correct waist level. (Active sportswear generally requires a greater seat angle and CB seam length than more formal wear.)

CENTRE BACK SEAM TOO SHORT

The problem of the centre back seam being too short can be caused by the forward tilting of the pelvis. This can be diagnosed by restricted forward leg movement, discomfort when seated and the CB waist level being pulled downwards as the seat angle is too straight.

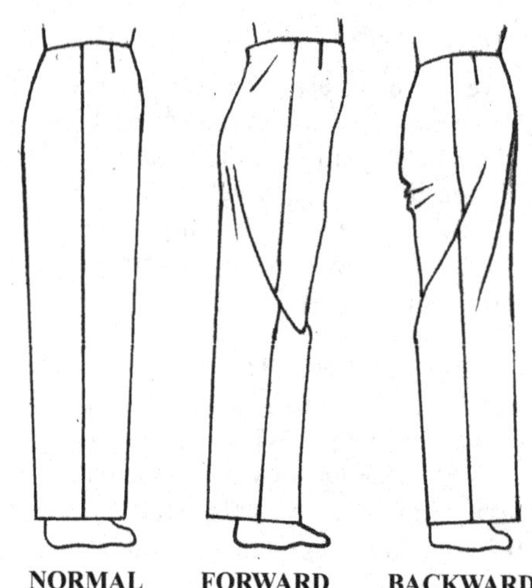

To correct

(1) Insert a wedge to add fullness to the CB hip level, tapering to nothing at the side hip. The required amount is generally estimated by the amount the CB waist level requires raising.

(2) Blend the altered CB seam line and side seam smooth.

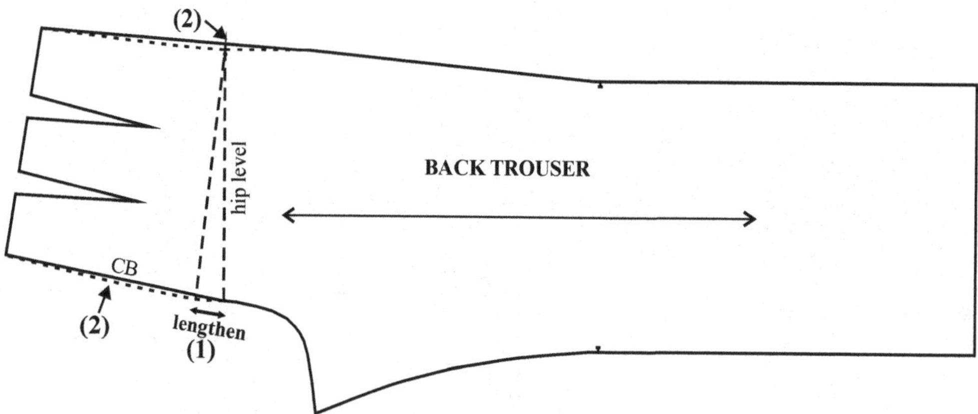

CENTRE BACK SEAM TOO LONG

The problem of the centre back seam being too long can be caused by the backward tilting of the pelvis. This is diagnosed by creases forming under the seat, the back fork appearing too low and the seat angle being too sloping

To correct

(3) Subtract a wedge to reduce fullness at the CB hip level, tapering to nothing at the side hip. The required amount generally estimated by the amount the CB waist level required lowering.

(4) Blend the corrected CB seam line and side seam smooth.

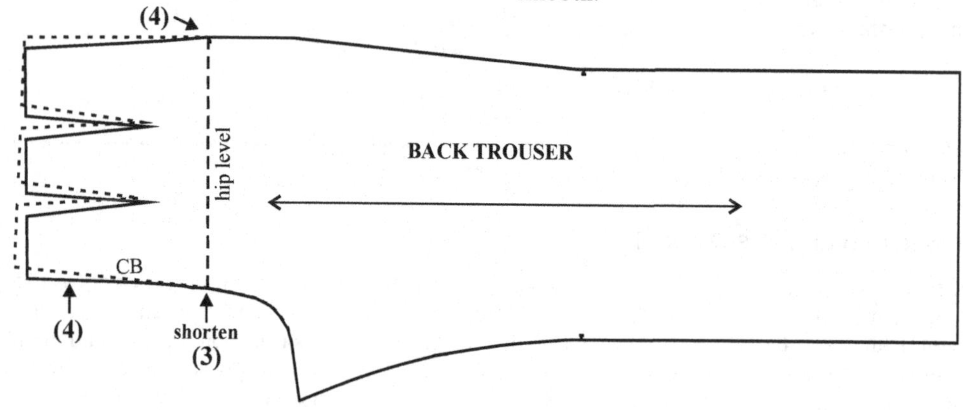

Neck width

The relationship between the front and back neck widths influences the hang at the centre front of a garment that is worn open. A correctly hanging garment should have the centre fronts meeting when the garment is unfastened. First, it has to be assessed whether it is the back or front neck width that is incorrect. The following examples refer to the front neck width where the back neck is considered correct.

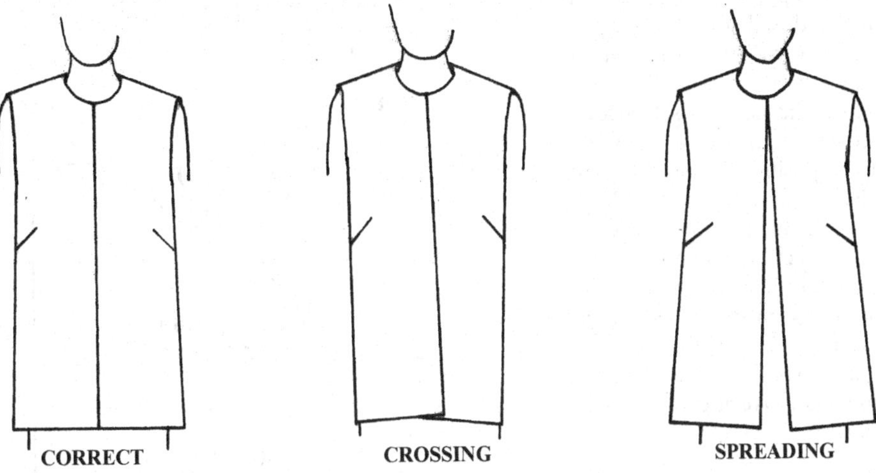

FRONT NECK TOO NARROW
The problem of the front neck being too narrow is diagnosed by the centre front opening crossing over at the hem. This is corrected by the front neck width being widened.

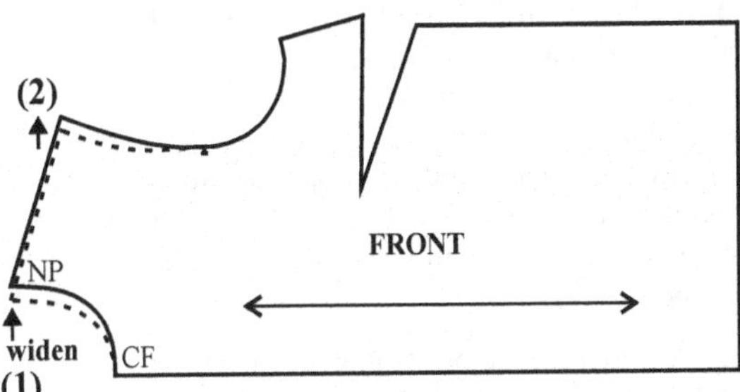

To correct

(1) Widen the width of the front neck at the NP and taper to nothing at the CF.

(2) To retain the shoulder length increase the same amount at the armhole, tapering to nothing at the mid balance mark.

FRONT NECK TOO WIDE
The problem of the front neck being too wide is diagnosed by the centre front spreading open at the hem. This is corrected by the front neck width being made narrower.

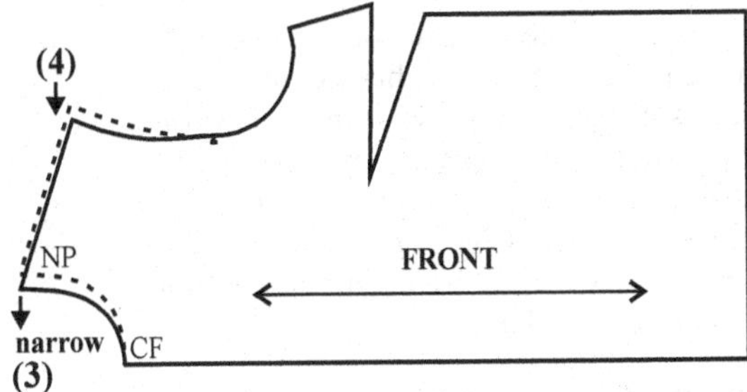

To correct

(3) Narrow the width of the front neck at the NP and taper to nothing at the CF.

(4) To retain the shoulder length decrease the same amount at the armhole tapering to nothing at the mid balance mark.

Sleeve pitch

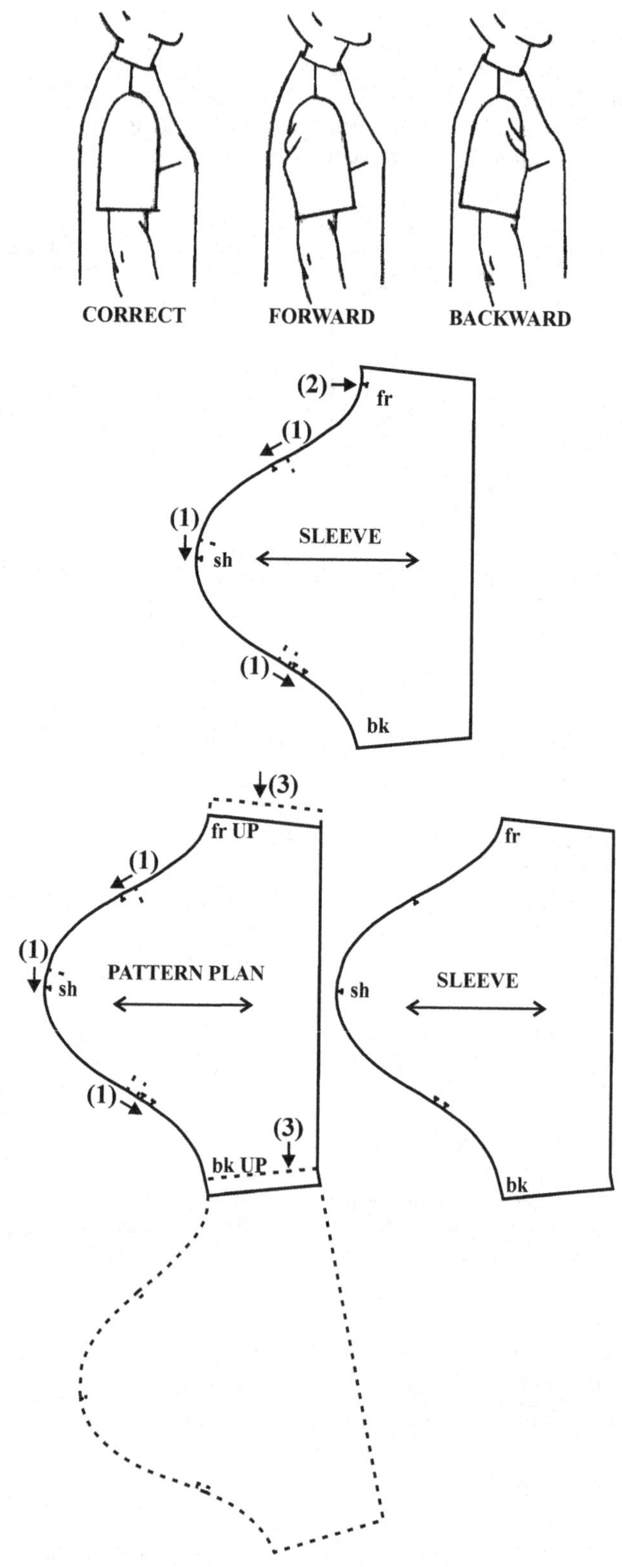

The arm for a normal posture generally hangs vertically from the shoulder to the elbow and then swings slightly forward from the elbow to the wrist, but this can vary according to the posture of the wearer. For a stooping posture the arm may swing forward and for an erect posture slightly backward. The sleeve has to hang in harmony with the arm. The angle or **pitch** to which the sleeve is set into the armhole is controlled by balance marks and the correct positioning of the grain line. First, the armhole has to be fitting correctly before the pitch of the sleeve can be assessed. Incorrect pitch can be altered by re-aligning the balance marks and matching the underarm seams. (For ergonomic reasons a sleeve for active sports or work wear may be deliberately pitched forward as this can give greater forward arm movement.)

SLEEVE TOO FAR FORWARD

The problem of a sleeve pitched too far forward is diagnosed by creases at the back of the sleeve head.

To correct

(1) Move the back and front balance mark notches and the shoulder seam notch towards the back until the sleeve hangs correctly.

(2) A notch can be positioned at the sleeve underarm to match the bodice underarm seam.

(3) Alternatively, where it is necessary to match the sleeve and bodice underarm seams displace the sleeve seam backwards by the same amount as the notches by subtracting from the front seam and adding to the back. (The correct curve of the sleeve head has to be maintained at the junction of the UP.)

SLEEVE TOO FAR BACK

The problem of a sleeve pitched too far backward is diagnosed by creases at the front of the sleeve head. There will also be a restriction of forward arm movement.

To correct
Reverse the alterations described above by moving the front and back balance marks and shoulder seam notch towards the front. The underarm seam position can be notched or displaced forward

VARIATION IN BODY SIZE AND CONTOUR

There are several factors that contribute to variation in body size and contour, such as bone structure, posture and flesh distribution. These variations affect both the girth and length measurement and also the amount and position of suppression.

The contour and proportional variation of the major girths of the bust, waist and hips are influenced by both the bone structure of the rib cage and pelvis and the flesh distribution. Some women require a different size garment for their upper torso from their lower torso. For example they may require a size 12 bodice but a size 14 skirt. First, the average measurements have to be identified. The body measurements quoted below are for a size 12 based on the size charts discussed in Part 1:

Bust 88 cm   Waist 70 cm
Hips 96 cm   Upper hips 90 cm

This gives a ratio, known as the 'drop', of:
Hips minus bust = 8 cm
Hips minus waist = 26 cm
Hips minus upper hips = 6 cm
Bust minus waist = 18 cm

Below are some illustrations of different body proportions.

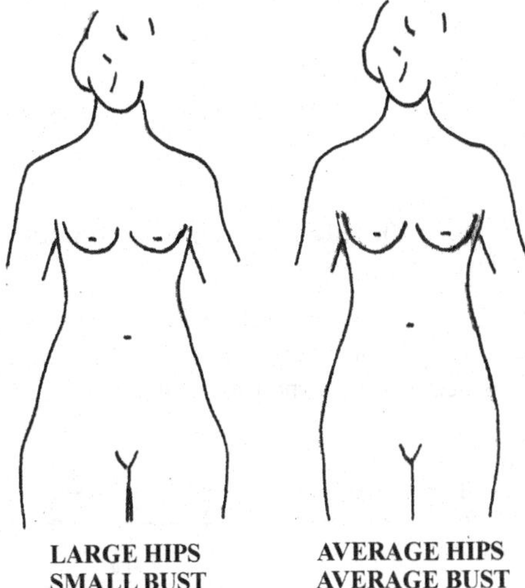

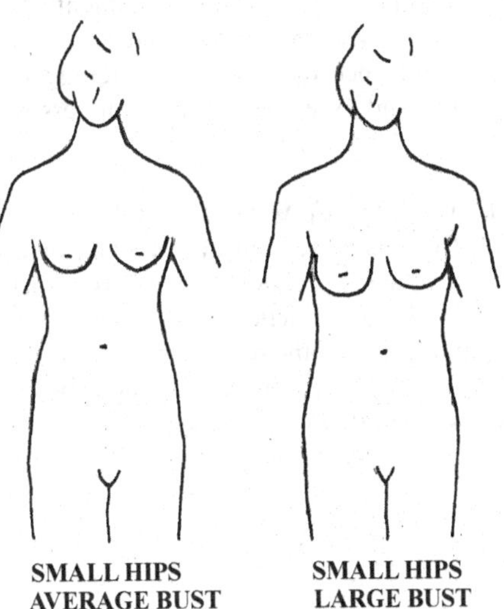

Small adjustment to the waist size

Where there is only a small difference in the waist sizes when joining a bodice of one size to a skirt of another size, the adjustment can be at the waist side seam only. The two examples below are based on a 4 cm increment between the waist sizes. The same adjustment can be made according to an individual's measurements.

To increase the bodice waist of a size 12 to fit a size 14 skirt:
(1) Increase the bodice back and front side seams 1 cm at the waist tapering to nothing at the UP.

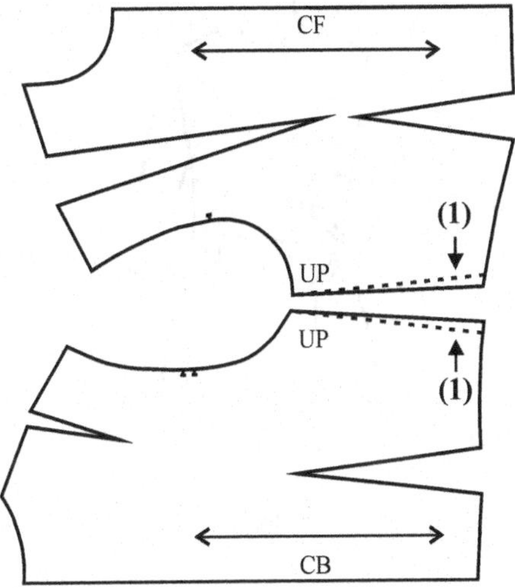

Larger adjustment to the waist size

Where there is a greater difference in the waist girths when joining a bodice of one size to a skirt of another size, the adjustment should be at both the waist side seam and waist darts.

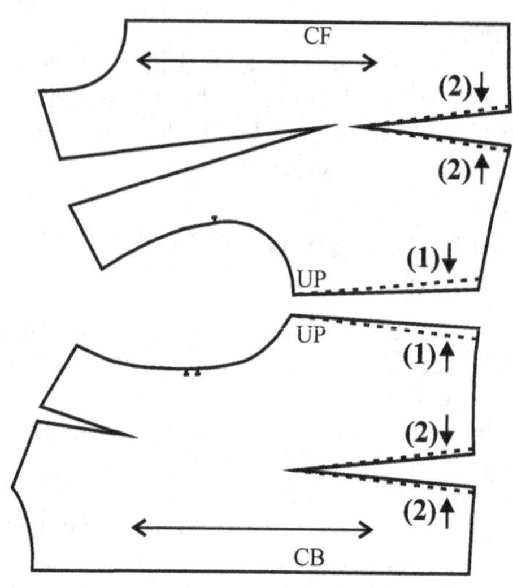

To increase the bodice of a size 12 to fit a size 16 skirt:
(1) Increase the bodice back and front side seams 1 cm at the waist, tapering to nothing at the UP.

(2) Reduce the back and front waist dart suppression equally or according to the contour of the wearer. (For women with a prominent stomach the front waist is generally increased by a further reduction of the dart suppression.)

Incorrect suppression

An incorrect amount of suppression in a garment can cause fitting faults. This creates diagonal dragging folds from a prominence due to the shortening in length in one area and excess length and fullness in another.

OVER SUPPRESSION AT WAIST
The problem of over suppression at the waist on a side seam and under suppression beneath the bust and shoulder blade prominence is diagnosed by diagonal dragging folds from the side waist level to the prominence of the bust, stomach, shoulder blade and seat. This is illustrated on a one-piece dress.

To correct
(1) Widen the side waist, tapering to nothing at the hip level and UP.

(2) Redistribute the suppression to the other darts beneath the bust prominence (BP) and shoulder blade according to the figure shape.

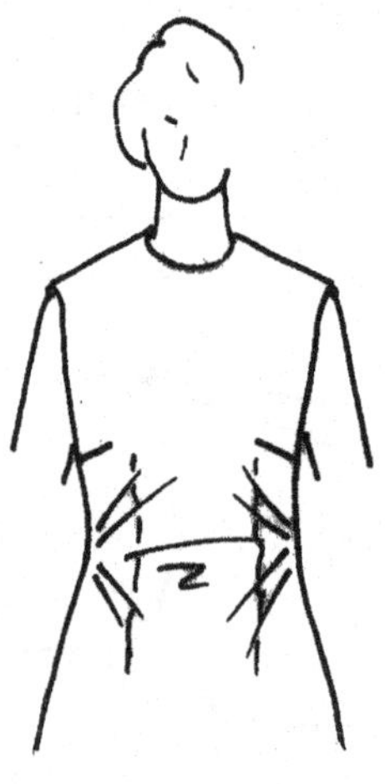

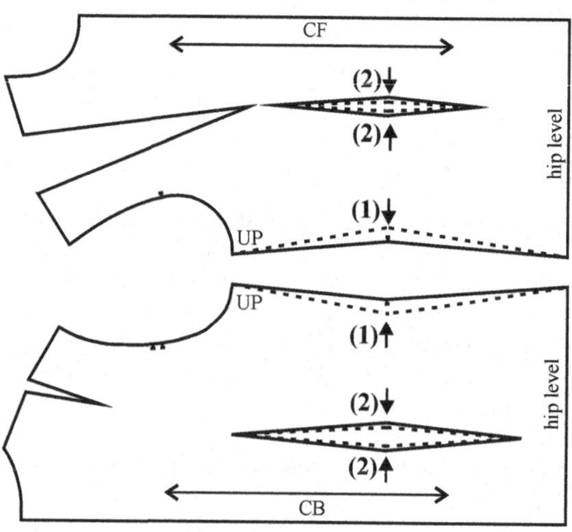

Prominences of the upper torso

The two main prominences of the upper torso are the bust and shoulder blades. Their variation is influenced by the wearer's posture and flesh distribution. A fuller and heavier bust often makes the wearer stand more erect and have flatter shoulder blades, whereas a wearer with a stooping posture may have more prominent shoulder blades and a less prominent bust. These figure variations will affect the garment width and length over the prominences and also the amount of suppression required.

Bust prominence

Some women vary from the average bust size of a B cup fitting. This would require the front bodice to be increased to decreased in both length and width over the area of the bust prominence. The example below for a fuller bust retains the same length at the adjoining seams. The CF length is altered but remains straight as this may be required as a mirror line or a front opening. (For a smaller bust reverse **(1)** to **(6)**.)

FULLER BUST

The problem of a prominent bust is diagnosed by diagonal dragging folds radiating from the bust prominence, shortness of fabric length and width over the bust and insufficient dart suppression.

To correct

(1) Plan the position where the extra fullness will be added by constructing a line perpendicular to the CF at the new bust level and another line from the new bust prominence (BP) to the end of the shoulder.

(2) Reposition the darts to these lines. Extend the dart apexes to the BP.

(3) Add parallel fullness at the bust level between the CF and the dart apex.

(4) Re-align the wider side seam dart to the required level and shorten 2 cm to 4 cm from the BP. (The dart may require closing temporarily to redraw a straight side seam and match the back.)

(5) Add fullness by inserting a wedge at the apex of the waist dart tapering to nothing at the end of the shoulder.

(6) Re-align the wider waist dart to under the new BP and shorten 2 cm to 4 cm. (The dart may require closing temporarily to redraw a smooth waist seam.)

(7) This is a comparison of the original and modified pattern.

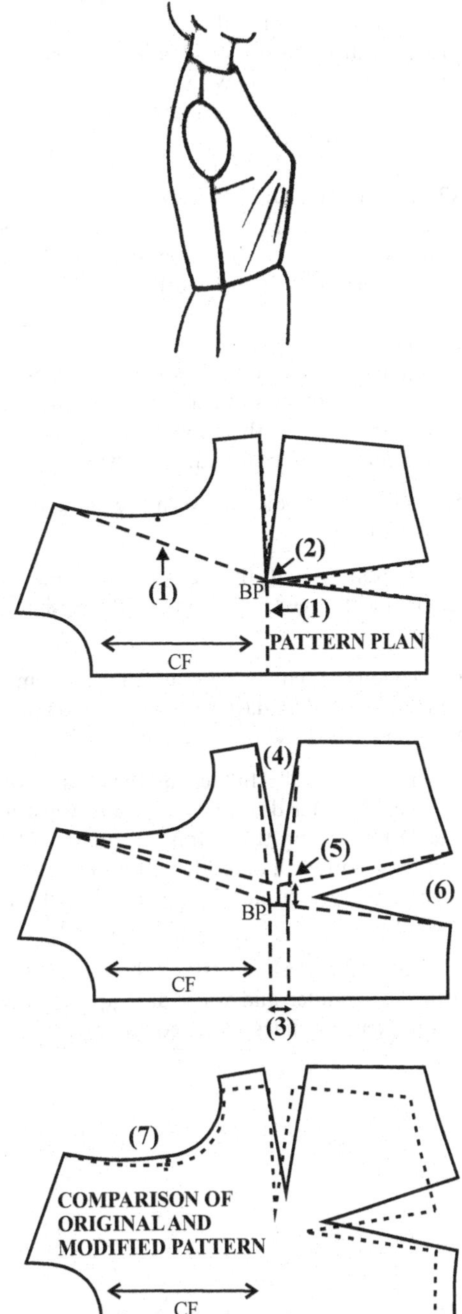

Shoulder blade prominence

The amount of prominence of the shoulder blade influences both the length and width of a back bodice and the dart suppression. For women with very flat backs there is generally insufficient suppression for a shoulder dart. The small amount of suppression at the back shoulder seam can be eased on to the front. Those with prominent shoulder blades will require increased dart suppression. This example for flatter shoulder blades retains the same length at the adjoining seams. The CB length is altered but remains straight as this may be required as a mirror line or a back opening. (For prominent shoulder blades reverse **(1)** to **(6)**.)

FLATTER SHOULDER BLADES

The problem of flatter than average shoulder blades is diagnosed by folds of excess fabric across the back due to too much length and width.

To correct

(1) Plan the position where the surplus fullness will be decreased by constructing a line perpendicular to the CB at the level of the shoulder dart apex, and another connecting the shoulder and waist darts.

(2) Split the back bodice between the shoulder and waist dart apexes.

(3) On section **(A)** at the construction line level reduce parallel fullness between the CB and shoulder dart apex.

(4) On section **(B)** reduce the same amount as minus fullness at the shoulder dart apex tapering to nothing at the armhole at the constructed line level.

(5) Subtract the parallel fullness on the lines between the original shoulder dart apex at **C** and waist dart apex at **D** by moving the points **C** and **D** half the amount to be removed, on both sections **(A)** and **(B)**. The lines will taper to nothing at the shoulder and waist seams.

(6) Merge sections **(A)** and **(B)** at the line (**C** to **D**) between the shoulder and waist dart apexes (check that both the merge lines are of equal length).

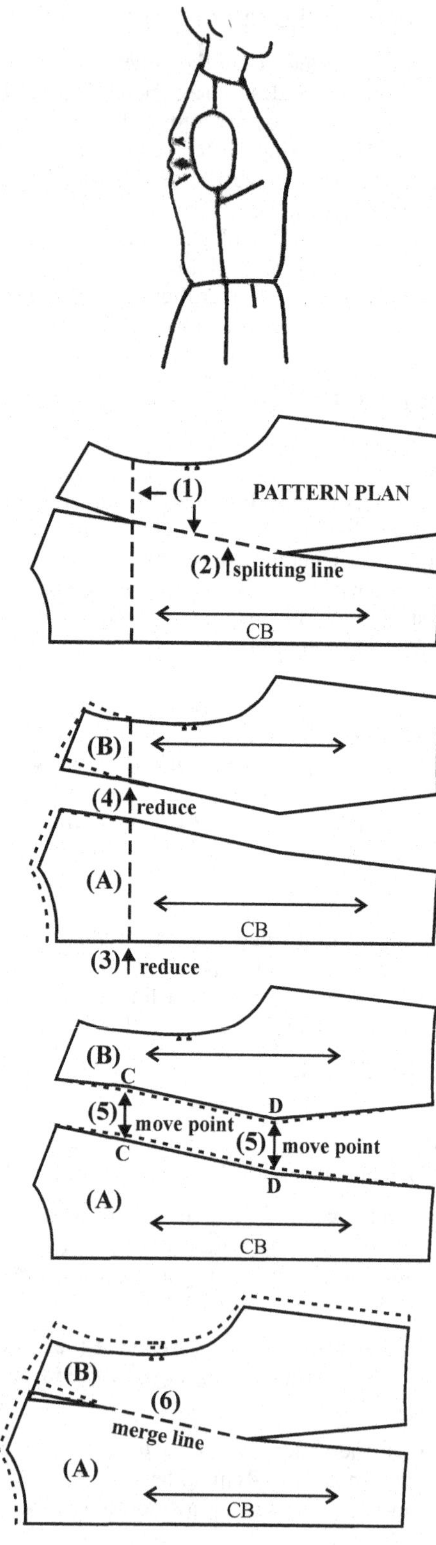

(7) Delete the shoulder dart and ease in the remainder of the suppression on to the front shoulder seam between the notches to give shaping for the shoulder blade prominence.

(8) Extend the shoulder seam at the armhole to the correct shoulder length plus 0.5 cm to be 'eased-in' between the notches.

(9) This is a comparison of the original and modified patterns.

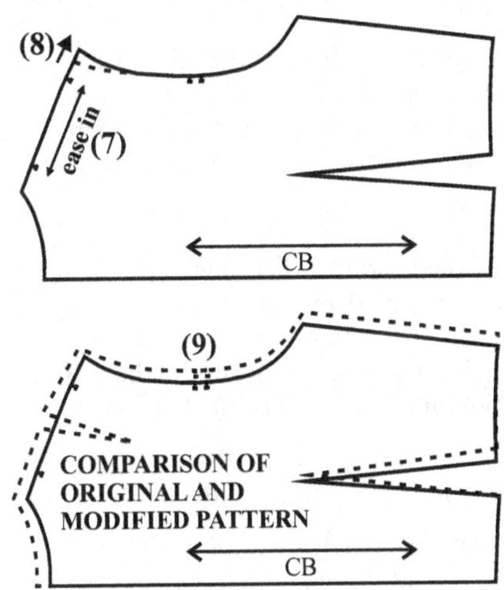

Prominences of the lower torso

The three main prominences of the lower torso are the seat, side hip and stomach. Their variation is influenced by the wearer's posture, the tilt of the pelvis and flesh distribution. Their change in prominence can also affect the waist size and level. The following examples of pattern alterations are given for both the straight skirt and trousers as the methods differ. This is because it is advisable to maintain a skirt CF and CB straight as it is often used as a mirror line or an opening, whereas a trouser CF or CB line can be cut at an angle and incorporates some suppression.

Straight skirt modification for a prominent seat

The prominence of the seat influences the contour of the back. The two following examples explain the pattern modification for a prominent and flat seat.

The problem of a prominent seat is diagnosed by the back skirt riding up causing horizontal folds under the waist, tightness over the prominence of the buttock, and the back skirt is too short and hangs away from the legs at the hem level. The side seam pulls backwards at the hem. The front skirt hem hangs too close to the legs.

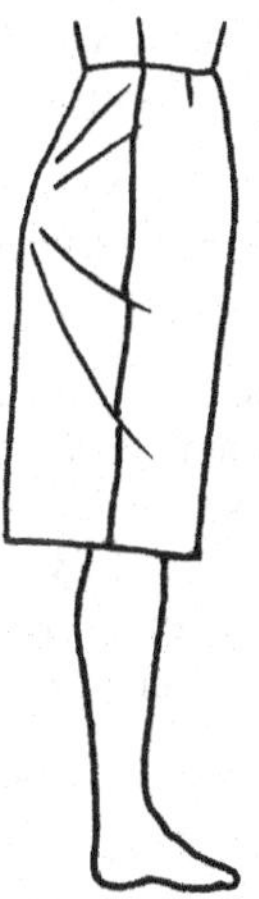

To correct

This correction maintains the same waist and side seam lengths. The CB length is increased but remains straight. The hem increases in width and remains straight to finish parallel to the ground.

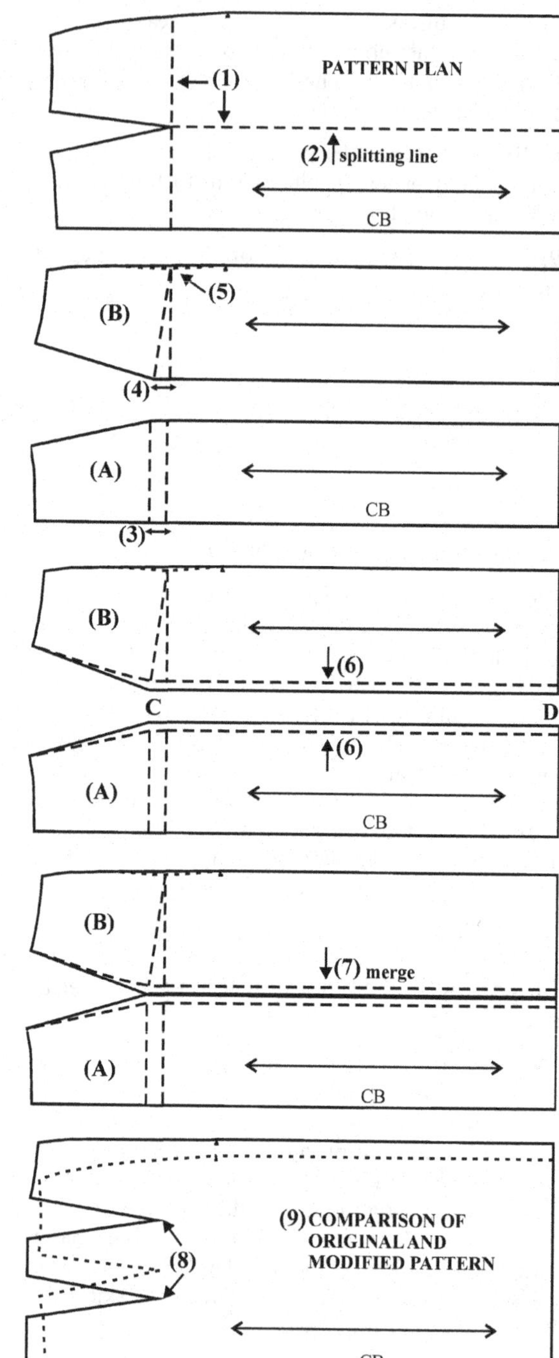

(1) Plan the position where the extra fullness will be added by constructing a line perpendicular to the CB at the waist dart apex and another line parallel to the CB from the waist dart apex to the hem.

(2) Split the skirt between the dart apex and the hem parallel to the CB.

(3) On section **(A)** at the level of the constructed line add parallel fullness between the CB and the dart apex.

(4) On section **(B)** at the level of the constructed perpendicular line insert a wedge of the same amount of fullness, tapering to nothing at the side seam.

(5) Re-curve the side seam between the waist and hip levels.

(6) Add extra width by moving the lines between **C** and **D** (dart apex to hem) half the required amount on both sections **(A)** and **(B)**.

(7) Merge sections **(B)** and **(A)** between the dart apex and hem.

(8) (Optional) When a large amount of fullness is added this new dart is too large to fit well as the apex will become too pointed. The fit is improved by reconnecting the new waistline and replacing the large dart with two smaller ones.

(9) This is a comparison of the original and modified pattern.

Straight skirt modification for a flatter seat

The problem of a flatter seat can be diagnosed by looseness of the back skirt over the buttocks. The CB is too long, the back hem pulls against the legs, the side seams slant forward at the hem and the front hem pulls away from the legs.

To correct

This correction maintains the same waist and side seam lengths. The CB length is decreased but remains straight (as this may be a mirror line or back opening). The hem decreases in width and remains straight to finish parallel to the ground.

(1) Plan the position where the fullness will be decreased. This is by constructing a line perpendicular to the CB at the waist dart apex and another line parallel to the CB from the waist dart apex to the hem.

(2) Split the skirt between the dart apex and the hem parallel to the CB.

(3) On section **(A)** at the level of the constructed line reduce parallel fullness between the CB and the dart apex.

(4) On section **(B)** at the level of the constructed perpendicular line reduce the same amount of fullness tapering to nothing at the side seam.

(5) Re-curve the side seam between the waist and hip levels.

(6) Reduce the width by moving the lines between **C** and **D** (dart apex to hem) half the required amount, on both sections **(A)** and **(B)**.

(7) Merge sections **(B)** and **(A)** between the dart apex and hem.

(8) This is a comparison of the original and modified pattern.

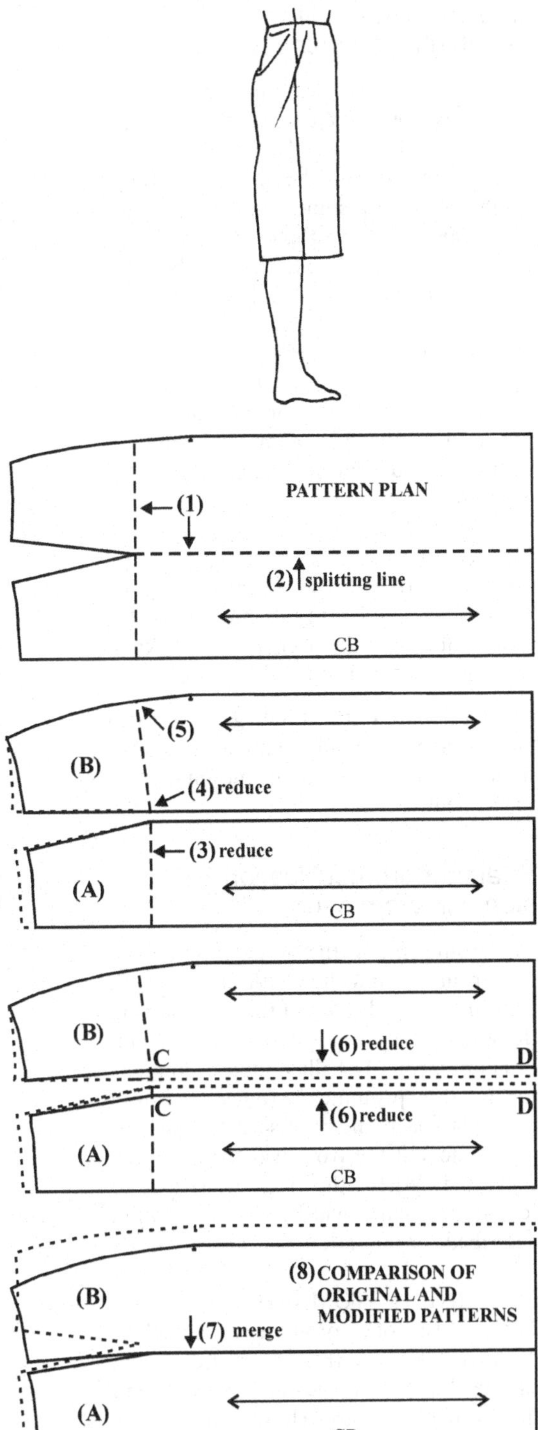

Straight skirt modification for hip and thigh prominences

Some women may vary in the width of their hips due to their bone structure. It is also affected by the distribution of the muscles and flesh at the upper hip, hip or thigh levels. For some women with very wide hips and large thighs a skirt that is slightly flared at the side hem can be more complimentary than a straight skirt.

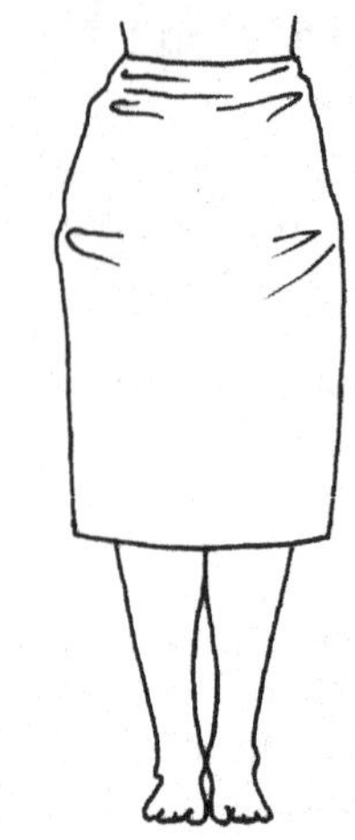

PROMINENT SIDE HIPS
The problem of prominent side hips is diagnosed by tightness of the skirt at the hip level on the side seams and the skirt riding upwards, causing creases below the waistline.

To correct
(1) Add extra width to the side at the hip to hem level and taper to nothing at the waist, equally on the back and front skirts. (For smaller hips reverse this process.)

(2) (Optional) If the thigh girth is extra large increase the side seam width at the thigh level and add a small amount of flare to the hem.

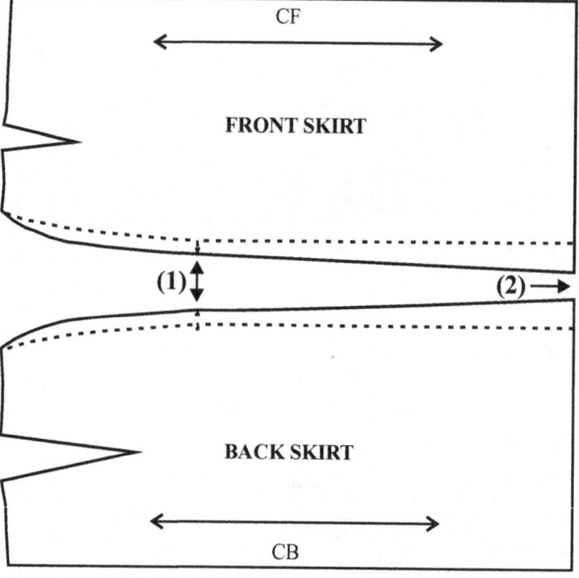

Straight skirt modification for stomach prominence

The prominence of the stomach is affected by corpulence and the width of the pelvis. This influences the size of the front skirt in the waist and hip widths, also the amount of suppression. The illustrated example is for a more prominent stomach. (For a flatter stomach the reverse modifications can be done if the front waist needs to be decreased. Where the front waist has to remain the same width use the method described for the flatter seat.)

PROMINENT STOMACH
The problem of a prominent stomach is diagnosed by the front skirt riding up, causing horizontal folds at the waist, and the front skirt is too short and hanging away from the legs at the hem level. The side seam pulls forwards at the hem. The back skirt hem hangs too close to the legs. Often the front waist is too tight.

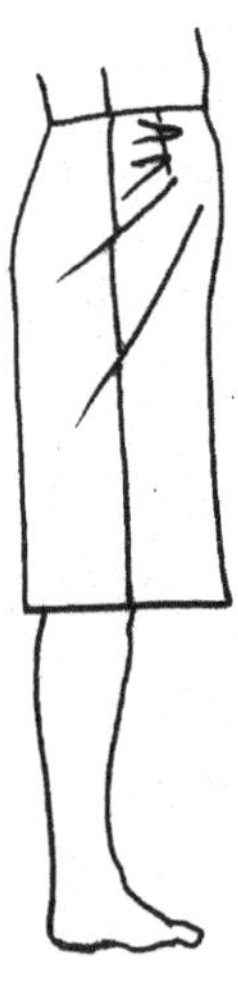

To correct

This correction maintains the same back and side seam lengths. The front skirt length will increase but remain straight as this may be a mirror line or require an opening. The hem increases in width and remains straight to finish parallel to the ground.

(1) Add fullness parallel to the CF midway between the CF and waist dart.

(2) Construct a line perpendicular to the CF at the hip level or over the fullest prominence.

(3) Split the pattern between the dart apex and hem parallel to the CF.

(4) On section **(A)** at the constructed hip line increase the CF length by adding parallel fullness.

(5) On section **(B)** at the level of the constructed hip line add the same amount of fullness and taper to nothing at the side seam (if necessary smooth the side seam).

(6) Merge sections **(A)** and **(B)** between the hip level and hem. (Check that the merged lines are of equal length.)

(7) Insert a new dart, or two darts, from the waist seam according to the required front waist measurement.

(8) This is a comparison of the original and modified patterns.

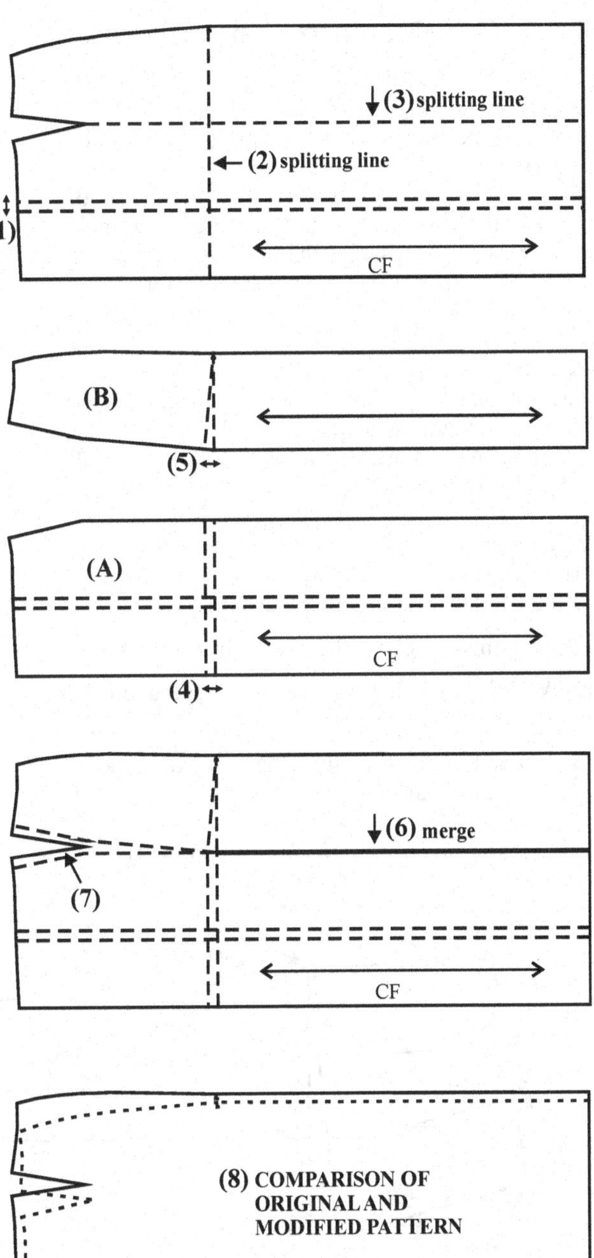

Trouser modification for seat prominence

The pattern modification of trousers differs from that of skirts because the CB and CF seams do not have to be positioned on the straight grain or used as a mirror line. Therefore the length of these centre lines can be increased or decreased, as well as the width of the trouser fork. The following are examples of increasing for a larger than average seat prominence. The reverse can be undertaken for smaller prominences.

PROMINENT SEAT

The problem of a prominent seat is diagnosed on a trouser by tightness across the back buttocks. The crutch length of the CB seam is too short and pulls the CB waist downwards.

To correct

(1) Insert a wedge by adding fullness at the hip level at the CB, tapering to nothing at the side seam.

(2) Smooth and correct the side seam at the hip level.

(3) Widen the inside leg at the crutch level and taper to nothing at the knee level.

(4) Widen the CB seam at the hip level and taper to nothing at the waist.

For a flatter seat reverse the process **(1)** to **(4)**.

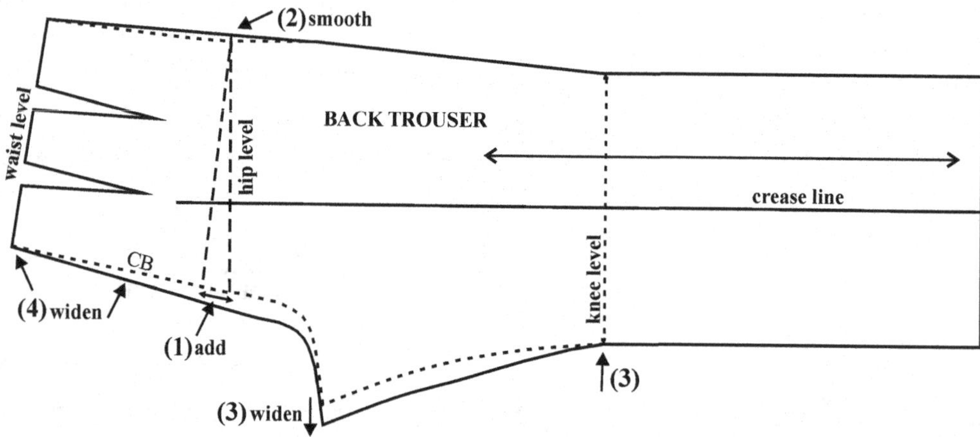

Trouser modification for hip and thigh prominences

The problem of prominent side hips and thicker thighs is diagnosed on trousers by tightness at the side hip and thigh levels.

To correct

(1) Widen the outside seam of the back and front trouser at the hip, knee and hem level, tapering to nothing at the waist.

(2) Re-centre the crease line between the inside and outside seams.

(3) (Optional) For very large thighs also add extra width to the inside leg at the crutch level tapering to nothing at the knee level.

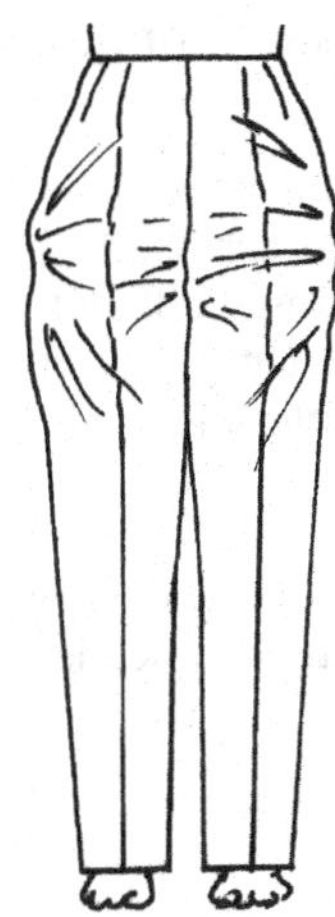

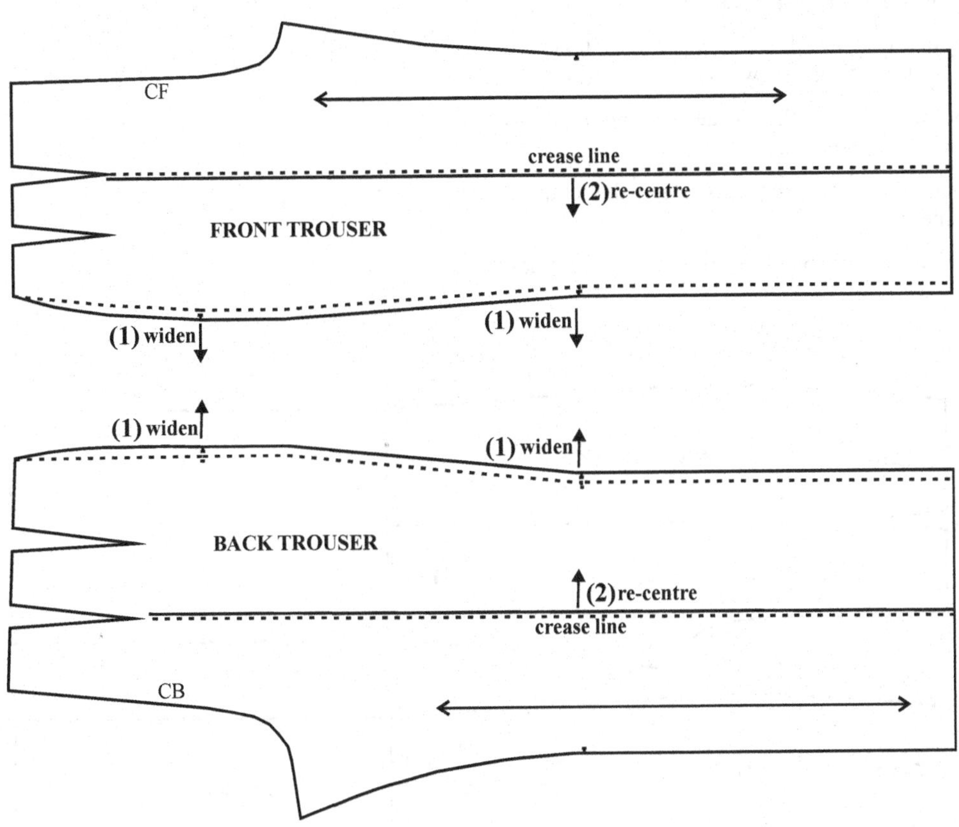

Trouser modification for stomach prominence

The problem of a prominent stomach is diagnosed on a trouser by tightness across the front waist and hip areas. This often causes difficulty in fastening the trouser. The front length is short as the waist line pulls downwards at the CF.

To correct

(1) Plan the position of the lines where the pattern is to be split, across at the knee level and the centre crease from the waist to knee.

(2) Split the pattern into three sections **(A)**, **(B)** and **(C)**.

(3) Merge the side section **(A)** to the lower leg **(B)**.

(4) Pivot the CF section **(C)**, at the inside knee pivot point, to increase the waist the required amount.

(5) (Optional) Raise the CF waist and taper to nothing at the dart or side waist.

(6) (Optional) Widen the inside leg and taper to nothing at the knee level.

(7) Replace the darts with pleats to give the required measurements.

(8) This is a comparison of the original and modified patterns.

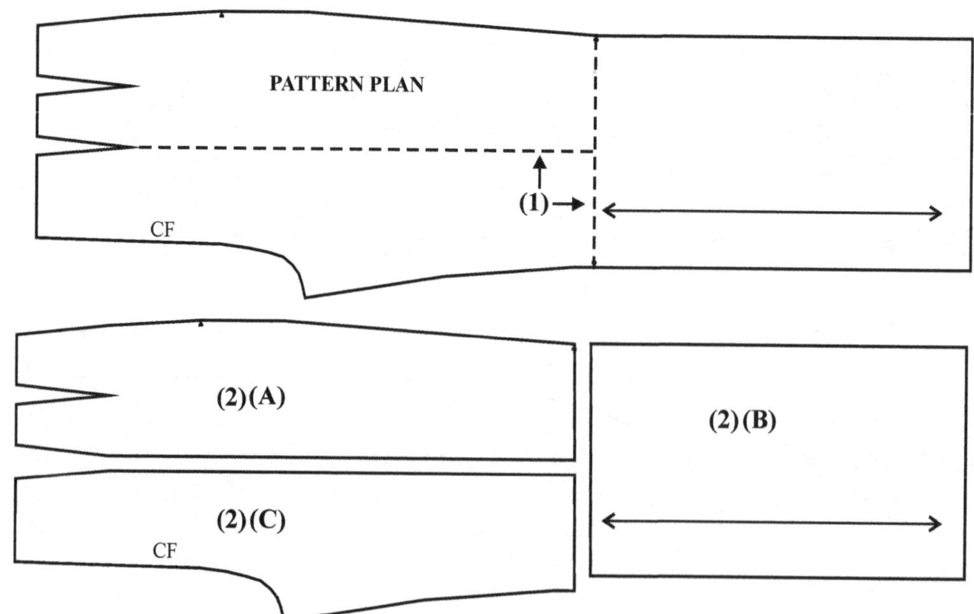

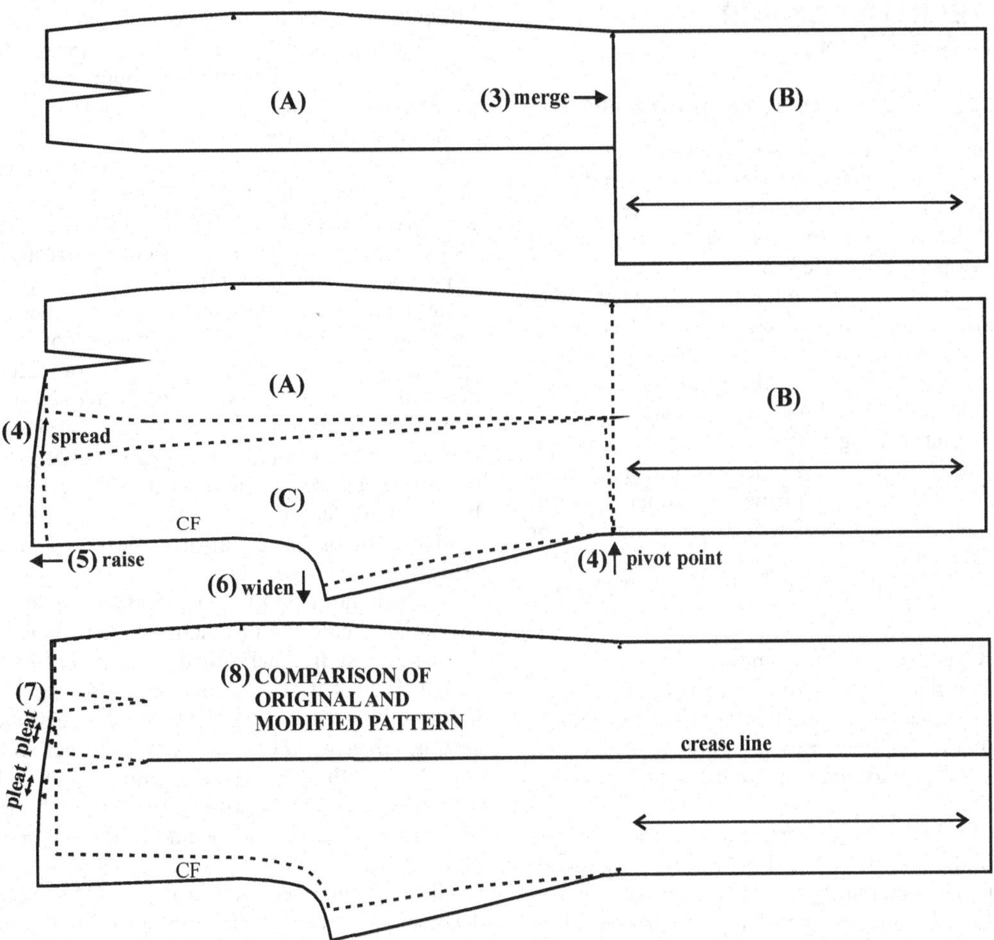
(A)
(3) merge
(B)
(A)
(B)
(4) spread
(C)
CF
(5) raise
(6) widen
(4) pivot point
(7)
pleat
pleat
(8) COMPARISON OF ORIGINAL AND MODIFIED PATTERN
crease line
CF

CAD TECHNOLOGY FOR CUSTOMISATION

Computerised made-to-measure systems

Developing and fast growing segments of the apparel industry include **made-to-measure** (MTM) and mass customisation. New technology in the apparel industry has had a great impact on garment manufacture, especially in the field of made-to-measure. Not only has it enabled quick response to customer demands, it has also enabled the manufacturer to produce designs tailored to individual requirements.

The concept of made-to-measure has attracted much attention and different levels of automatic MTM manufacturing projects have been carried out worldwide. Computerised made-to-measure pattern drafting software is implemented either by programming the traditional tailor's code methods or applying alterations to standard graded patterns. The basic strategies necessary for a MTM clothing system are:

(1) A consistent measuring method required to take measurements for individual customers
(2) A complete set of size charts with different size ranges for categorised figure types, which should provide measurements for the standard sizes
(3) Individual sets of garment patterns for the standard sizes, which should accurately reflect the characteristics of the categorised figure types and measurements in the corresponding size charts.

Before implementing computerised pattern alterations for figuration on a commercial CAD system, a large database with reference to the standard graded patterns, the alteration movements and relevant information needs to be generated. There are many difficulties in creating grade rules and alteration movements, especially for the bust suppression and the shaped sleeve, in order to obtain sophisticated patterns with a good fit for MTM clothing

When making an alteration plan for figuration through the creation of a personal measurement alteration chart, the subject's figure type should be categorised accurately and the standard graded blocks at the nearest size selected correctly. This scheme may enhance the efficiency and the quality of pattern alterations, because it is not necessary to carry out the practical procedures of cutting, spreading and/or overlapping the pattern pieces. The new system, together with its developed pattern alteration strategies, allows pattern designers to generate the MTM patterns using commercial CAD systems available on the market. The pattern database for different figurations can be extended gradually in order to improve the speed and efficiency of CAD production.

All the major CAD/CAM vendors have developed solutions to ease the production of customised garments: Assyst Bullmer's made-to-measure software and their Procut 501 cutting table, Gerber's Accumark made-to-measure software and their newest cutting table the GTx1, or Lectra's Fitnet software combined with their Topspin cutting table. These solutions offer a customised garment to be manufactured in a time equivalent to that of a mass-produced garment.

The following sections focus on CAD software developed by Gerber Technology. The family of software reconciles pattern design, grading, made-to-measure, marker making and advanced 3D pattern design systems. The system user interface is via Windows 98/2000 or NT environment. The software features provided by this system enable integration of all modules.

Configuration and alteration points

To operate within the made-to-measure module, specified triple alteration numbers are allocated to points on the respective 2D pattern pieces (Figure 4.3).

Within this example the start number 100 is applied to the back piece, 200 to the front, 300 to the sleeve, etc. These configuration points are set up in the alteration library stored within the CAD system.

The patterns are stored within the Accumark database and visually displayed within the PDS2000 software module.

The alteration grade-rules are defined in the alteration table. Alteration rules can be applied to both left and right pieces or selected only for the left or right. Different amounts can be applied also to both left and right pieces via the alteration file. There are four basic movements available for alteration rules.

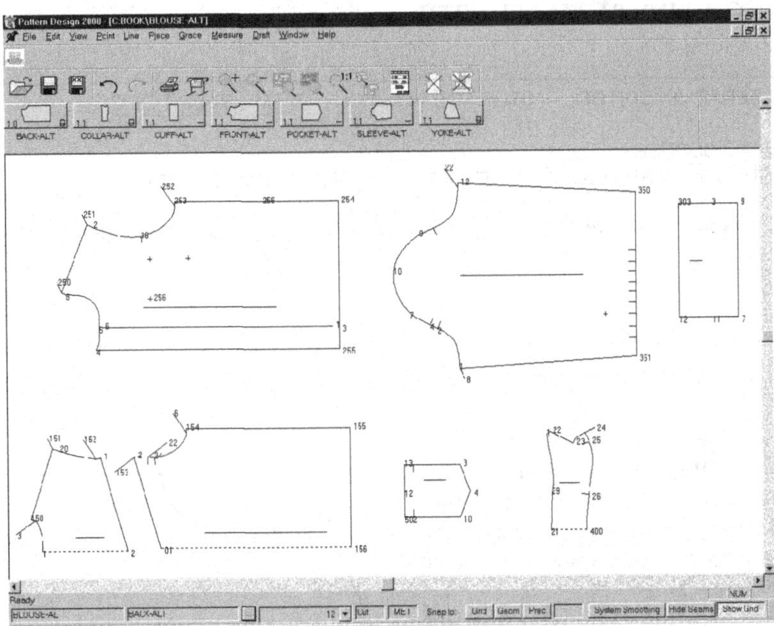

Figure 4.3 Allocation of alteration numbers

STAGE 1: MOVING A SINGLE POINT

ALTERATION EDITOR   STORAGE AREA C: MTM   RULE 1 OF 1

NAME: BLOUSE
ALTERATION RULE NAME: 1 POINT   PIECE USAGE BOTH

ALT TYPE	FIRST POINT	SECOND POINT	MOVEMENT X%	MOVEMENT %Y
X Y MOVE	256	256	0	100

Figure 4.4 Moving a single point

Selecting the X and Y move in the alteration type field, then defining the same alteration point number for both FIRST POINT and SECOND POINT, will allow the lines from the next gradepoints to create the alteration according to the percentage set in the movement field (Figure 4.4).

This data is input into the Gerber Accumark alteration file found within the system management module. When activated the alteration rule is applied to the pattern piece. The change is visible as a dotted line in the illustration to the right.

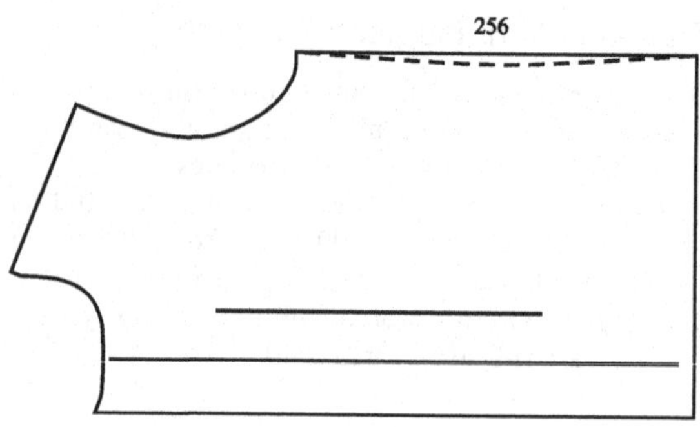

STAGE 2: MOVING A LINE PARALLEL WITH OFFSET

ALTERATION EDITOR	STORAGE AREA C:	MTM	RULE 1 OF 2	
NAME: BLOUSE				
ALTERATION RULE NAME: LINE OFFSET		PIECE USAGE BOTH		
ALT TYPE	FIRST POINT	SECOND POINT	MOVEMENT X%	MOVEMENT %Y
X Y MOVE	254	255	100	0

Figure 4.5 Moving a line parallel with offset

Selecting the X and Y move in the alteration type field, then defining the TWO END point numbers for both FIRST POINT and SECOND POINT, allows the lines from the next gradepoints to create the alteration according to the percentage set in the movement field (Figure 4.5).

The line is offset from the value set within the alteration file from the specified point numbers. The alteration is seen as a dotted line in the illustration to the right.

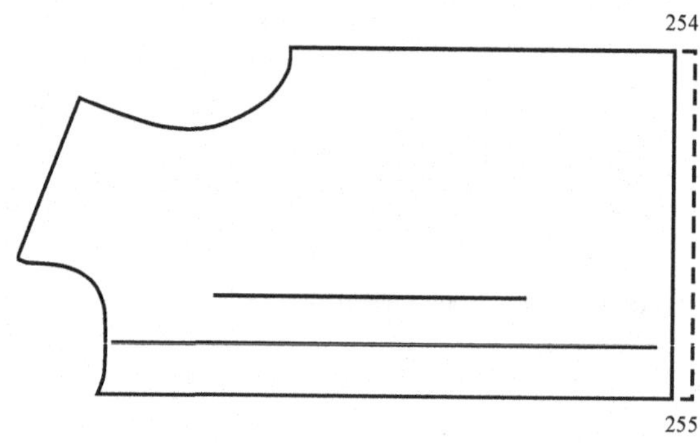

STAGE 3: PIVOTING A LINE WITH LINE-EXTENSION

ALTERATION EDITOR	STORAGE AREA C:	MTM	RULE 1 OF 3	
NAME: BLOUSE				
ALTERATION RULE NAME: PIVOT EXT		PIECE USAGE BOTH		
ALT TYPE	FIRST POINT	SECOND POINT	MOVEMENT X%	MOVEMENT %Y
CW EXT	256	254	0	100

Figure 4.6 Pivoting a line with extension

Selecting the CW EXT move in the alteration type field, then defining the FIRST POINT and SECOND POINT, will cause the piece to pivot at the line of the first point up to the second point (CW) clockwise or (CCW) counterclockwise around the piece, according to the percentage set in the movement field (Figure 4.6).

The line is pivoted at the specified point. The pivot line and the next line segment are extended to create a new intersection. The alteration is seen as a dotted line in the illustration to the right.

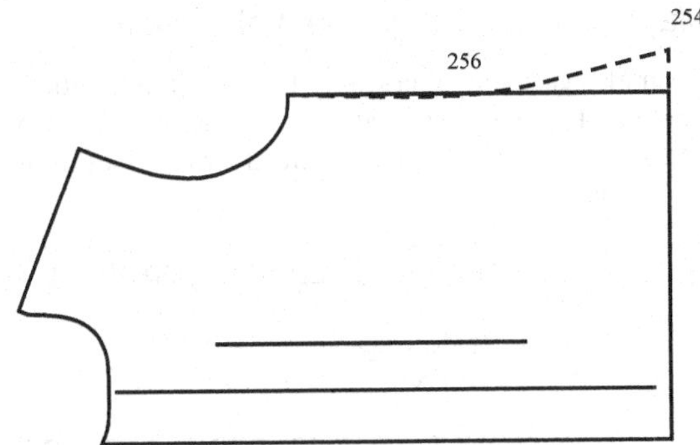

STAGE 4: PIVOTING THE LINE WITHOUT EXTENSION

ALTERATION EDITOR   STORAGE AREA C: MTM RULE 1 OF 4

NAME: BLOUSE
ALTERATION RULE NAME: PIVOT NO EXT   PIECE USAGE BOTH

ALT TYPE	FIRST POINT	SECOND POINT	MOVEMENT X%	MOVEMENT %Y
CW NO EXT	256	254	0	100

Figure 4.7 Pivoting the line without extension

Selecting the CW NO EXT move in the alteration type field, then defining the FIRST POINT and SECOND POINT, will cause the piece to pivot at the line of the first point up to the second point (CW) clockwise or (CCW) counterclockwise around the piece, according to the percentage set in the movement field (Figure 4.7).

The line is pivoted at the specified point. The pivot line retains the same length, however the next line segment is extended and also pivoted to create a new intersection. The alteration is seen as a dotted line in the illustration to the right.

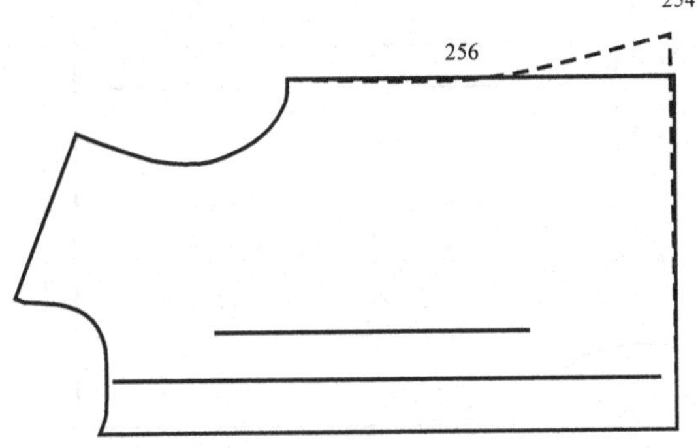

Made-to-measure garment alteration

To make garment alterations to the blouse, modifications to the pattern pieces can be combined into one alteration rule. Figure 4.8 illustrates the length of the blouse.

ALTERATION EDITOR STORAGE AREA C: MTM RULE 7 OF 7

NAME: BLOUSE
ALTERATION RULE NAME: LENGTH PIECE USAGE BOTH

ALT TYPE	FIRST POINT	SECOND POINT	MOVEMENT X%	MOVEMENT %Y
X Y MOVE	155	156	100	0
X Y MOVE	254	255	100	0
X Y MOVE	350	351	50	0

Figure 4.8 Changing the length of the blouse

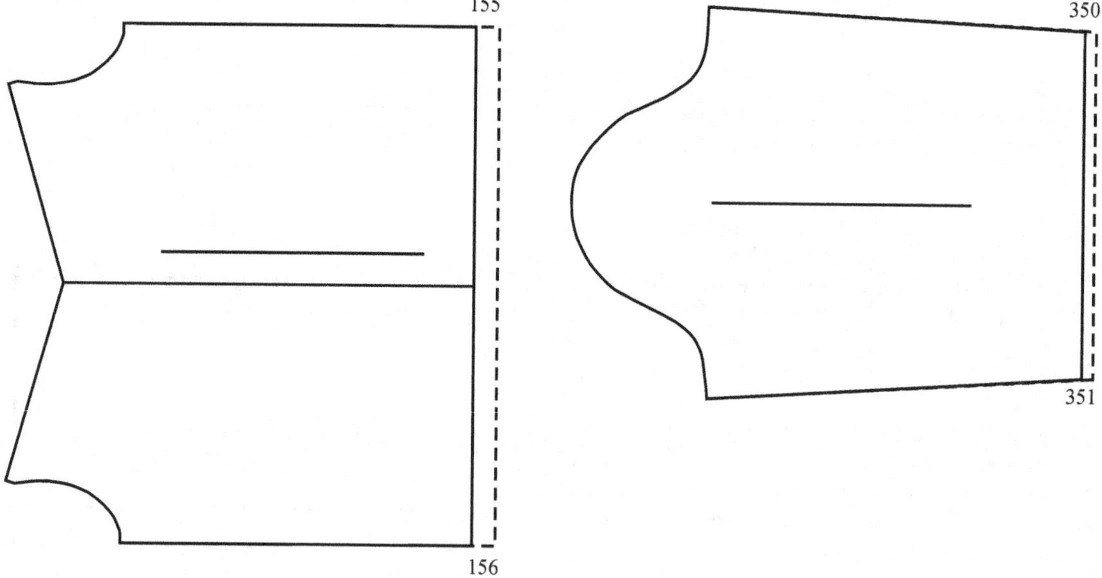

Alterations have been made to the lengths of the pieces. The back and front lengths have increased 100% of the value set. The sleeve length has increased 50% of the value set as set up within the alteration file.

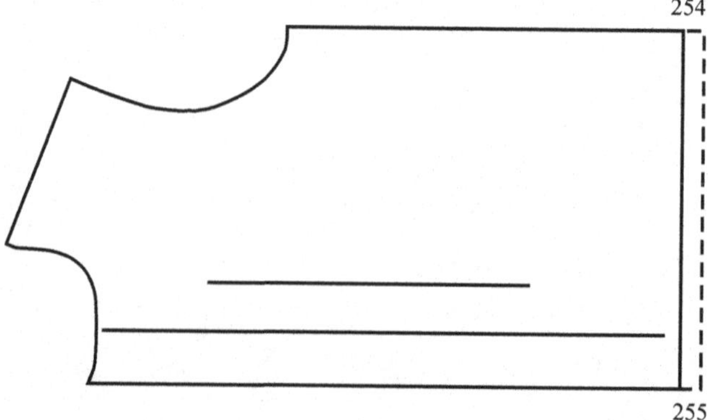

The made-to-measure CAD interface

The previous section focused on the importance of the alteration file and the options allowed. This section will focus on the use of Accumark–MTM software as the alteration interface for a single step order entry system used for automatically generating an MTM garment.

The order entry form (Figure 4.9) allows specific customer data including fabric type, garment model, measurements and alterations. The MTM alteration amount is defined according to customer fit and identified by the posture such as erect, head forward, round back, sloped shoulders. The garment type can be selected via pull down menus for garment types such as jackets, trousers, shirts, blouses, etc. In the size field a basic size, the closest size to the customer's individual measurements is indicated. The quantity of orders can be selected; however, in an MTM environment a single order is usual. Modifications to garment styles such as length pockets, lapels and vents are stored within the model options and can be displayed (Figure 4.10). This information is stored within the system management module within Accumark. When the order is finalised it is submitted to a batch file for automatic marker planning (Figure 4.11). The batch process is carried out automatically as a background task, or manually via Automark using 'Sliding Lay Rules' as a placement strategy. The marker plan can then be sent directly to an automatic single-ply cutter (Figure 4.12).

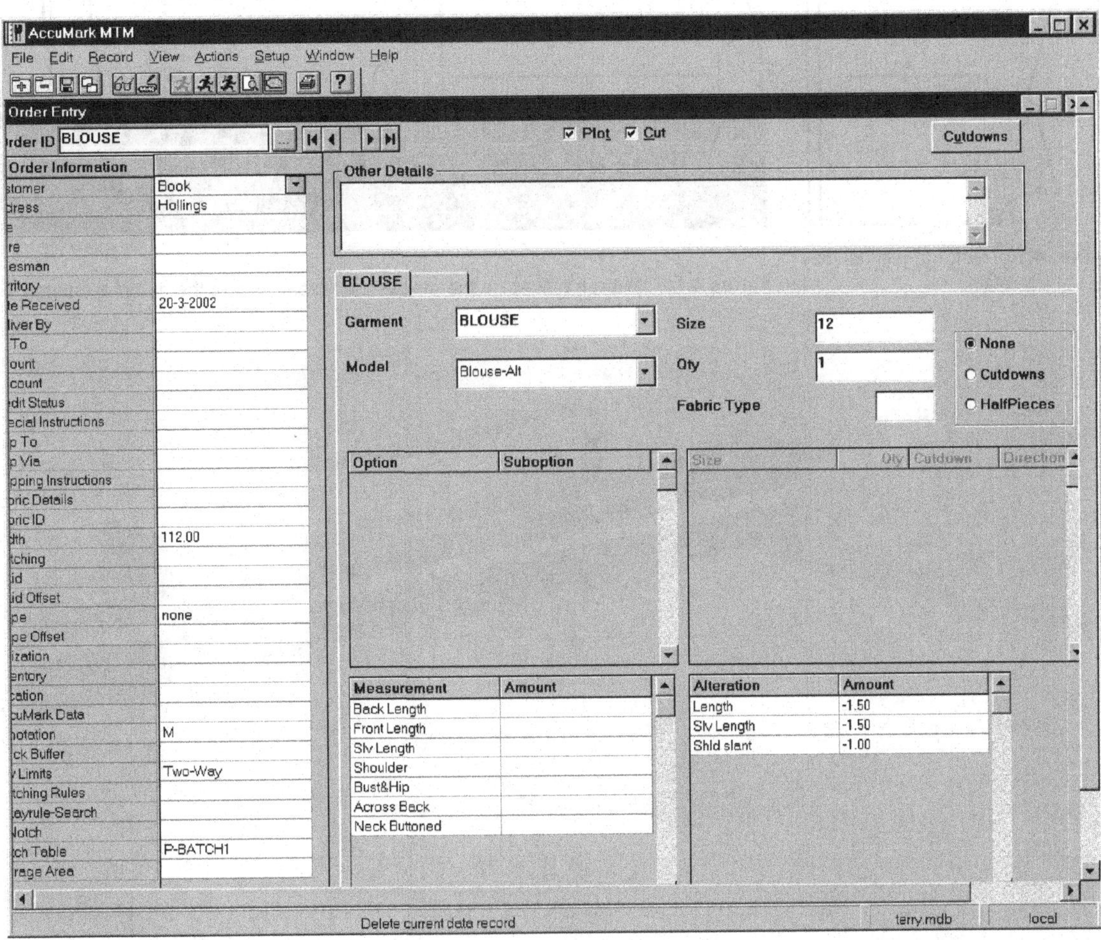

Figure 4.9

Marker making using the batch process

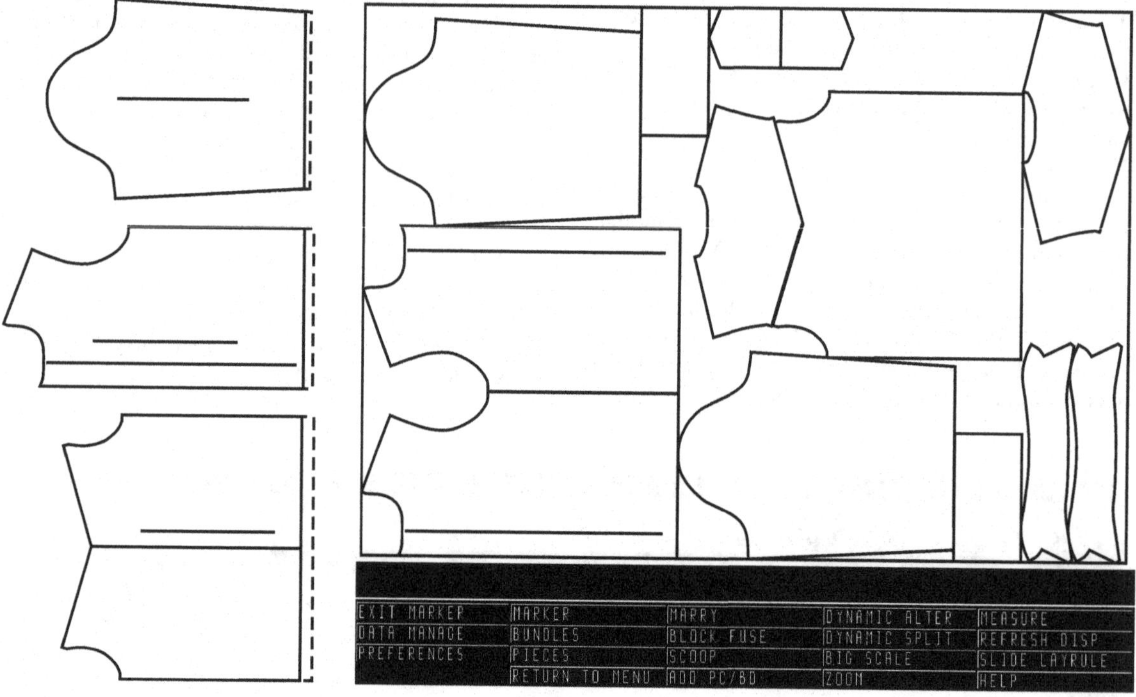

Figure 4.10 Pattern pieces displaying alterations

Figure 4.11 Pattern pieces on the marker plan created from the automatic batch file

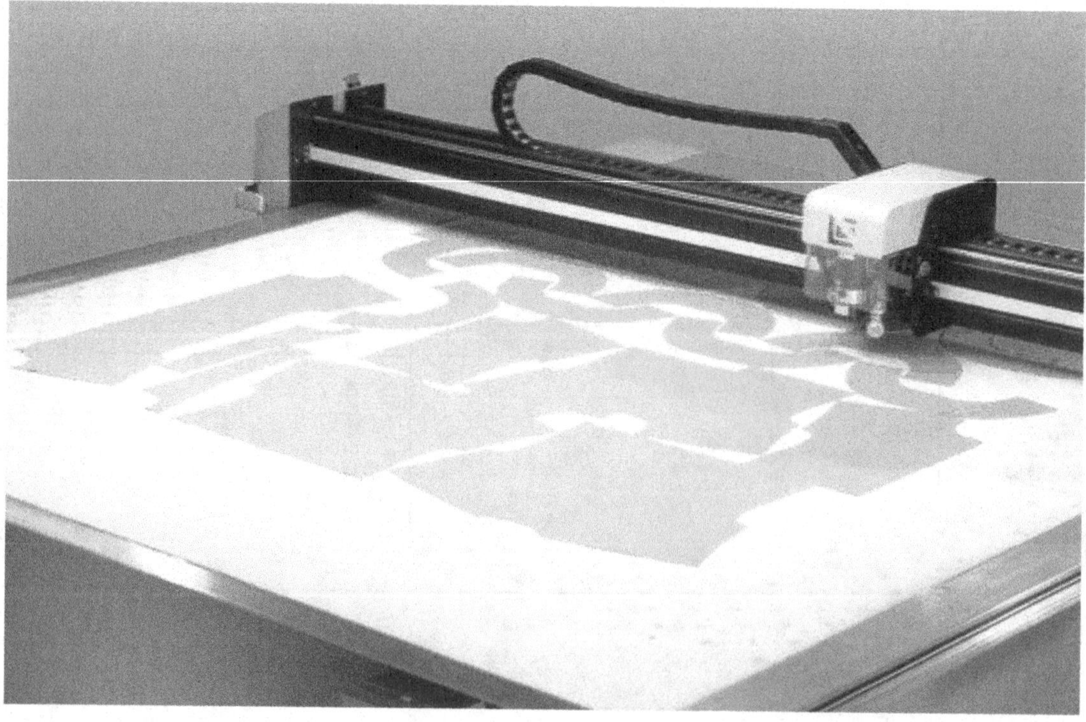

Figure 4.12 Automatic single-ply cutting DCS1500 (*Photograph by permission of Gerber Technology*)

Advanced 3D pattern design systems

Developments within computer-aided design for fashion, clothing and visualisation have been realised in the development of 3D software. Offering the designer a virtual prototyping system has been an active research area for many years. Despite being applied through other commercial industries, its development for use within the clothing industry has met many research challenges. However, by presenting recent developments within this virtual environment the 3D picture becomes much clearer.

In relation to pattern design the ability to move from 2D to 3D is perhaps the area of most interest. The creation of 2D pattern shapes that can be wrapped around a virtual mannequin fits nicely within the 2D CAD pattern development application used within the industry. To move on from this point of development seems the most likely acceptable way forward for the designer and pattern maker.

Gerber Technology, one of the leading CAD vendors throughout the world, are now offering the APDS-3D virtual draping program, which has been developed by Asahi of Japan. Open-system architecture allows the APDS-3D software to connect seamlessly with the PDS 2000 pattern module, enabling pattern makers to import styles and models, thus making pattern modifications easier to achieve.

Developments within 3D body-scanning systems capable of producing anthropometrics data offer a direct link to 3D design and pattern making. The technology of body scanning is quite new but interest is high as it could revolutionise the way people shop for clothing. Companies constantly developing automated sizing surveys in 'real time' can also use the accumulated information.

3D WIRE FRAME MANNEQUIN

Software development within the Gerber Technology APDS-3D system allows the pattern maker to move from a CAD 2D environment to 3D and back to 2D. A digital representation of a tailor's mannequin (Figure 4.13) created by means of a 3D digitiser is the start point for further developments. Combined with improvements in body scanning techniques, a true representation of the human form can be realised, although for commercial fit the garment industry creates sizes to fit the average person.

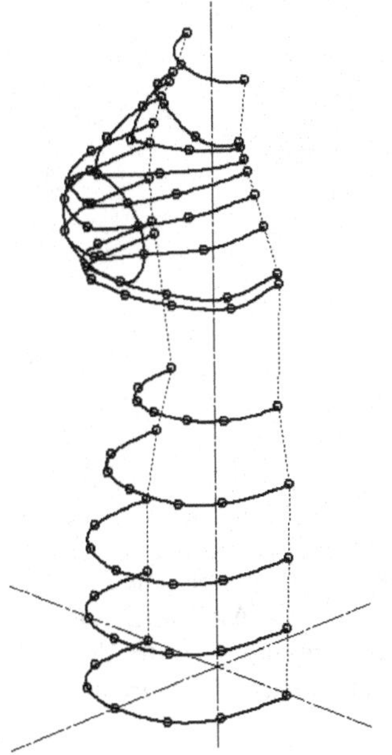

Figure 4.13

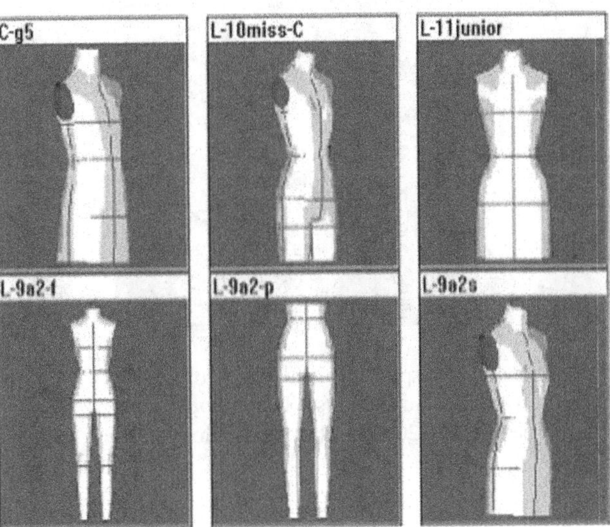

Default mannequins

CUSTOMISING THE MANNEQUIN

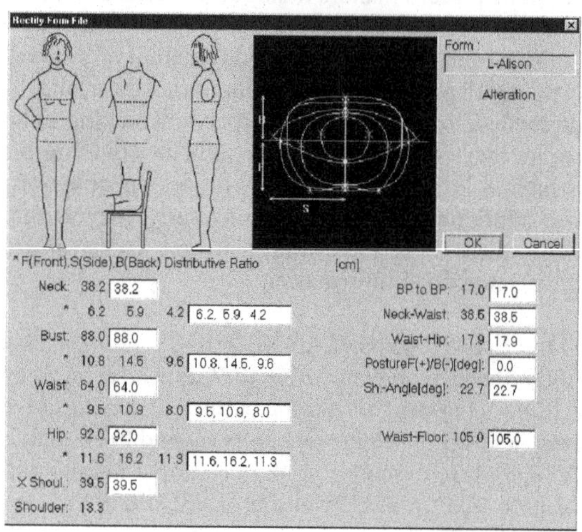

A standard size simulated dress form model can be selected, and personalised body measurements are input. Measurements on the form relate to bust, waist, hips, and nape to waist, etc.

VISUAL CHECK OF THE 2D PATTERNS

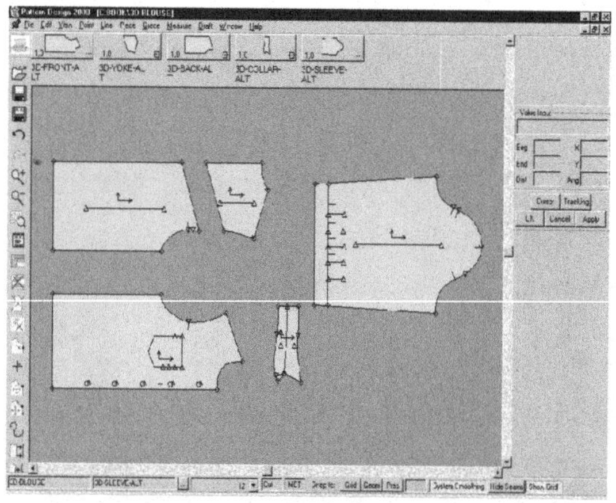

The example above displays a visual check of the set of 2D patterns of the ladies blouse. These blocks have been modified in PDS2000 Gerber Technology software.

VISUAL REALISATION

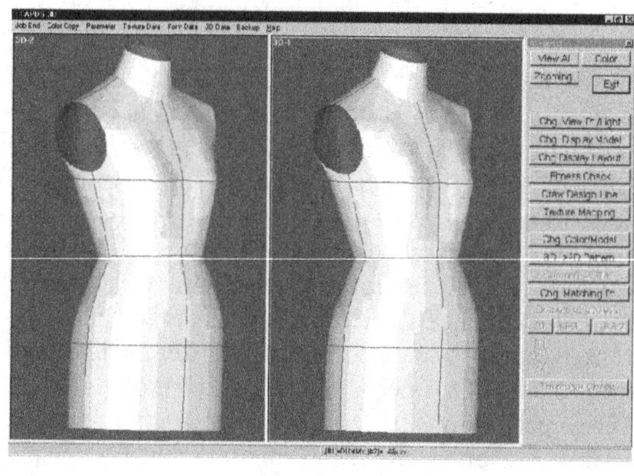

The example above displays a visual check of the newly created form. The original form is displayed on the left; the new form is displayed with the personalised measurements.

VIRTUAL SIMULATION OF THE 3D BLOUSE

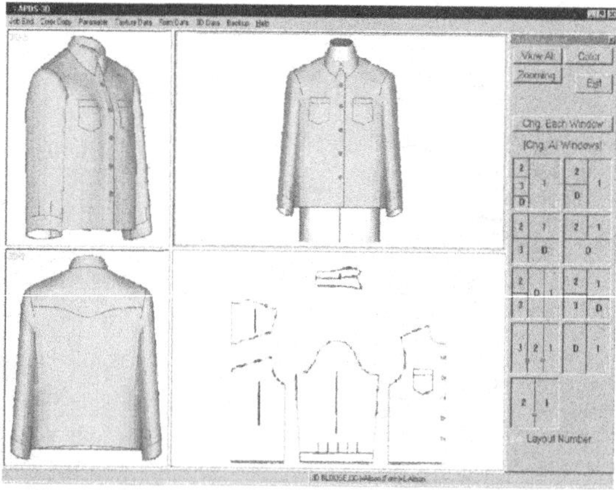

The example above displays the approved 2D patterns transferred into the 3D APDS-3D Gerber Technology software and mounted on to the new form created from the personalised measurements.

Creating block patterns

From the new personalised mannequin, block patterns can be created automatically. These can be viewed in PDS2000, 2D format, and either stored or modified to create a new style. This example demonstrates 3D to 2D block pattern application. Commercially manufacturers use block patterns they have developed in 2D.

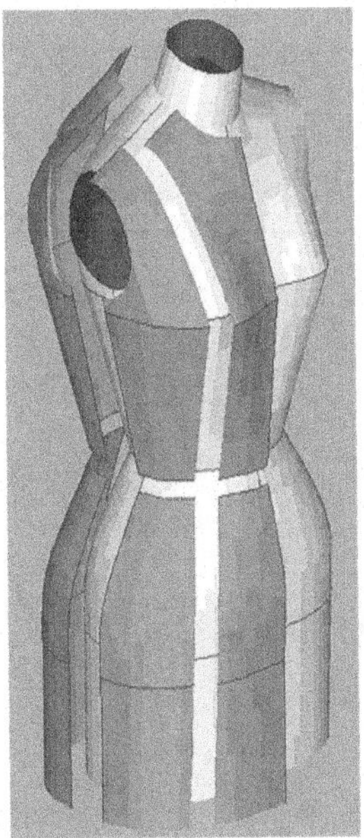

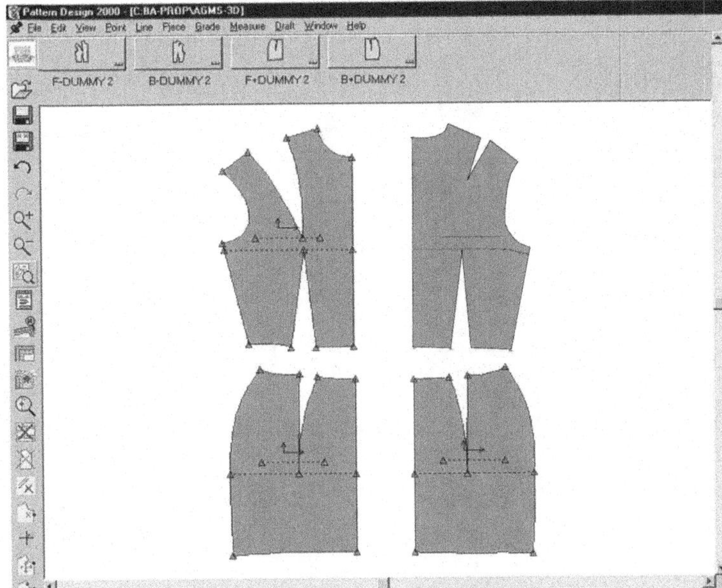

Part 5

Computerised marker making systems

MARKER MAKING AND LAY PLANNING

Part 5 considers how marker making and lay planning can minimise fabric waste and help match fabric stripes and check pattern repeats. The application of computerised cut order planning, fabric spreading and cutting technology then follow. This Part covers:

- Assessing fabric widths and material utilisation
- Markers for striped and check fabrics
- Variation in fabric spreading methods
- Cut order planning
- Computerised fabric spreading and cutting.

Computer assisted grading and marker making systems along with computer driven cutting systems have been available in various forms for many years. Improvements in hardware and software perceive system workstations operating through Microsoft Windows 95, 98, 2000/NT (networked). Owing to the decreasing cost of hardware, competition among CAD vendors for business in the apparel industry is very keen.

Fabric widths and material utilisation

The main selling point of computerised grading and marker making systems has always been the savings in fabric, achieving a high percentage of material utilisation. CAD systems can grade pattern pieces with various degrees of sophistication and speed. There are two methods of making markers within most systems: interactive and automatic. For either method, data relating to the marker is set up in a file. The file contains information such as the style, number of sizes required, material width, blocking (setting an allowance around the piece for check fabrics) or buffering allowances (small cutting allowances), and if required flip and rotation restrictions.

The interactive method relies on the user to position the required pattern pieces over an image that represents the cloth on screen. The pieces to be placed are displayed in a matrix menu either individually or in tabular form. Each size can be shown in a different colour, in outline or as a filled block. Pieces can be brought into the marker and placed accordingly and can be automatically positioned to the edge of the cloth or the next piece, taking into account any blocking or buffering allowances. The production of the most efficient layplan depends on the ability of the user to produce an economical lay; thus the system is a tool for the skilled technician (Figure 5.1).

Fabric widths can be changed to accommodate variations in fabric deliveries. If a marker has been

Figure 5.1 From pattern digitising to marker making to plotting (*Photographs by permission of Gerber Technology*)

produced and stored for a similar garment, it can be retrieved and displayed on screen for guidance. Alternatively the marker can be copied.

It is possible to split a component pattern piece within a marker to improve fabric utilisation. The split can be vertical, horizontal or at any specified angle, and seam allowances are automatically added to the piece. Examples of split pieces may be found on under collars and facing pieces. The length of fabric used and the percentage efficiency are monitored and displayed on screen as the pieces are arranged. A target efficiency can be preset to act as a guide to achieving the expected percentage utilisation.

Markers for striped and check fabrics

Markers can be produced for striped and checked fabrics. Multiple stripe or check repeats can be specified within a file on the CAD system. Matching points are digitised on to the pattern pieces and the system constrains pieces to be placed in the marker on a check or stripe repeat. Matching points can be positioned so that checks and stripes are matching on relative parts of the garment.

The automatic method of marker making requires the user to input a target time for the marker to be made automatically. The system automatically calculates and saves the most efficient marker in the time allowed. The most efficient marker can then be passed on to the interactive section of the program to be modified in line with the company's marker making methods to improve efficiency. The automatic method is mostly used for first costing markers. Markers can be queued and run in a batch file, for example overnight. Markers that are produced automatically or interactively can be plotted (printed out) or can be transferred to an automatic cutter.

Placement strategies for fabric type and matching

The example in Figure 5.2 is a marker plan of the ladies blouse, planned on a width of 112 cm plain fabric usable width. The marker efficiency achieved is 84.60%. The parameters for the fabric type (lay limit) are two-way fabric with no buffering (pieces are touching but not overlapping).

The example in Figure 5.3 is a marker plan of the ladies blouse, planned on a width of 112 cm usable width of fabric that is pattern checked with a 6 cm repeat; the markings are visible on the perimeter edge of the marker. The marker efficiency achieved is 74.60%. The parameters for the fabric type (lay limit) are two-way fabric with no buffering or blocking. The reduction in fabric utilisation relates to the constraints placed upon certain pattern pieces to match either to the fabric or to each other.

Planning markers

In planning markers it has been found good practice to combine small and large sizes evenly, because material savings are then made. The combination provides a greater variety of pattern pieces to fit together more compactly. Marker planning should combine the largest number of garment sizes acceptable within the constraints of the cutting table.

Material utilisation statistical data must be gathered to support planning decisions. This can be categorised by pattern style, fabric width, and pattern layout and garment ratio per marker. The arrangement of pieces is often controlled by the spreading method of the fabric, for example, face up one-way, face up two-way, face-to-face two-way, face-to-face all pairs one-way and face-to-face within pairs one-way (Figure 5.4).

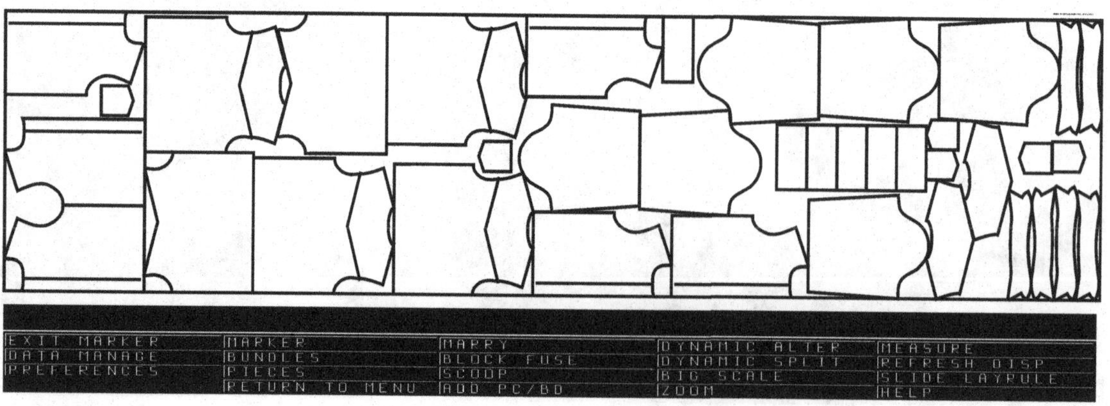

Figure 5.2

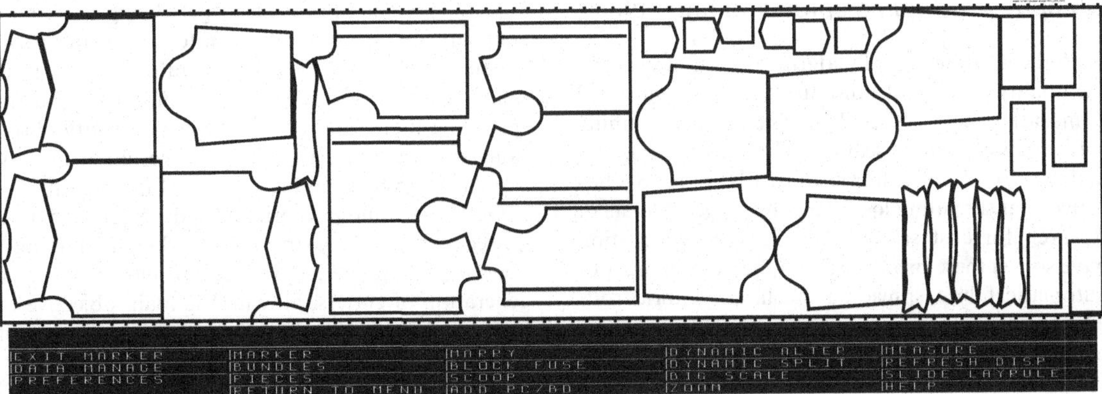

Figure 5.3

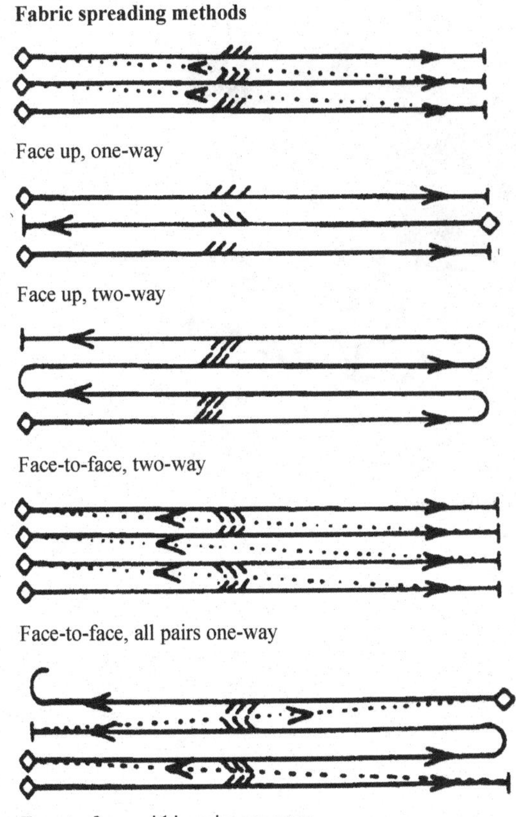

Figure 5.4

There are widths of fabric best suited to a style of garment and combination of sizes. Research by apparel manufacturers can be carried out identifying the best fabric width relevant to garment styles produced. A range of sizes is used to calculate the most suitable width. The following rules can be identified offering the best success rate in achieving a high material utilisation:

(1) The combination of the most sizes allowed in any marker
(2) The combination of small and large sizes in the marker
(3) Try to obtain the fabric in the most favourable width.

Longer marker spreading saves fabric end losses and labour costs, particularly when automatic high-speed spreading machines are used.

Optimising markers

'Optimising' markers must be clearly differentiated from planning procedures and marking. Optimisation refers to finding the best combination of sizes, garments to be marked per size, and usable material width and ply height, to enable the most economical cutting, considering both material and labour cost. Optimisation in marker making involves finding the best arrangement of pattern pieces in achieving a good fabric utilisation, working with the style and constraints set. Computerised cut order planning software integrated within the marker making and lay planning procedures is a significant way of achieving further material savings. Cut order planning enables the automatic calculation of the best marker making solutions, to determine which is the best set of markers required for processing the orders, with the lowest overall cost, with automatic calculation of the best spreading solution, and supplying the optimum cost savings.

Computerised fabric spreading and cutting

With the introduction of material utilisation systems, the quality of the spreading functions has improved to meet the standards set. Today's cut order planning software is capable of scanning cloth inventories and making use of small fabric rolls. Eberly (1990) has stated that spreading losses are always good areas of research for cost saving. Advances in automation have solved these problems, with the development of automatic fabric spreading machines. Fabric rolls can be loaded and threaded manually or automatically. Using carousels that hold ten or more fabric rolls allows for automatic loading of material in the sequence required for the particular order.

With spreading problems largely being overcome, suppliers are directing attention to fabric fault management, which is also being automated. Fabric faults have a great influence on the speed of spreading fabric; defects, damages and width variations restrict material utilisation, resulting in substandard garments and re-cutting of garment parts.

Gerber Technology originally developed computerised cutting 30 years ago. Gerber Technology, Lectra, and Bulmer, have now developed the new generation of cutters, all with the main objective of integrating with the previously mentioned area of CAD (Figure 5.5).

Figure 5.5 Automated cutting (*Photograph courtesy of Gerber Technology*)

Part 6
Product data management systems

Part 6 views the evolution of communication via the Internet and the impact on quick communication in a global marketplace. Software companies have realised the opportunity for them to develop new technology systems to support the complicated product development processes within the apparel industry. This Part covers:

- PDM systems
- Organisation of design data
- Form administration

The migration of British apparel manufacturing industry is becoming apparent, with offshore manufacturing the strong competition. Offshore locations are increasing, offering low-cost products. A global marketplace has now emerged as protective tariffs are altering, and existing sourcing strategies and decision making are increasingly challenged. This movement of production locations offers opportunities and consequences to the garment industry. One major factor that needs to be addressed is 'quick communications response': information flow from consumer to retailer, to manufacturers and suppliers, and the ability to be flexible within this chain of information.

European companies are finding competition strong against low-cost economies such as India and China, especially when the latter is investigating heavily to restructure the garment sector. As a result, what are left in the British and European Market are the managerial, design, and product development departments, with production and quality control/assurance offshore. Within this situation, it is now of paramount importance to have a seamless, timely and coherent line of data communication between these far reaching aspects of garment manufacture.

Product data management systems developed by Gerber Technology, who were the pioneers of the original garment specification system, have become a standard for this aspect of CAD and communication technology. Further developments now see a new generation of Web-based PDM systems. The Internet is now becoming the mode of communications and information control.

PDM SYSTEMS

PDM systems have been developed to help facilitate communication and to accelerate the co-ordination of the product development cycle. This is made possible by the ability to communicate, via the Web or Internet, from a single database of product related information. Visual data combined with text data can be accessed throughout the world (a valid secure password and Internet, Intranet access is required). Company data security is achieved via licensed software. PDM systems are rich in functionality, with costing, BOM (bill of materials), size specification sheets, construction details, vast image import capabilities allowing scanned images, digital photography, video (MPEG) and multi-language capabilities.

Electronic data can be shared between different locations at any precise moment – 'real time'. Modified data is instantaneously updated, thus allowing remote sites new information with little delay. Past studies have highlighted that most individual departments within the cycle spend as little as 20% of their time actually designing and developing their products, while the remaining time is consumed by looking for necessary information. Often data cannot be found or is not up to date and requires duplication.

Gates (1999), outlining the need for the 'paperless office', states that the move from paper to electronic forms is a vital step in the evolution of a modern organisation's nervous system and adds that 'once such a system is in place it is easy to build on and can provide the basis for eliminating internal paper forms' which slow down administrative tasks, create bottlenecks and minimise the overall efficiency of employees.

Manufacturers are restricted to a rigid timetable, making sure the order placed meets the in store date. Late shipments can mean a loss of sales. Therefore the development process must conform to a strict timetable. PDM systems can manage the critical path, which helps keep track through reports on the progress of each product and by identifying potential problems. This technology system can identify how manufacturing can improve response to customer demand, enabling the supply chain to respond in terms of performance to the demands of customers. Internet based PDM systems currently available include one from the French company Lectra, whose PDM software is named Gallery and allows product design to packaging development, with construction of garment collection books, while efficiently evaluating costs and structuring data. Lectra offer a variety of CAD/CAM modules that all integrate.

Freeboarders are a leading developer of global supply-chain technology. Their software automates the design, sourcing and delivery process by enabling customers to find the competitive prices and open capacity at factories, and RFQ (request for quotes) where manufacturers quote a price for a job. Therefore retailers will be able to shop around for a manufacturer who can offer a keen price and has the capacity and time for production for an order to be processed.

Assyst, the German based company, offer software for the sewn goods industries for data communications, Internet and Extranet networking. Assyst offer a variety of CAD/CAM modules that all integrate.

Organisation of design data

Operating as a central data depository, PDM systems combine all style information, including photographs, block patterns, fabrics, trims, size specifications and sketches. By having visual explanatory style sheets instead of copied text-orientated sheets, communication internally and with third parties becomes much easier (Figure 6.1). This continues to expand into the e-commerce, business-to-business (B2B) world.

Prior to the design research phase, a target plan for each season is created (critical path). A module or the whole of the product development cycle is listed into individual tasks that can be assigned a time limit. PDM systems have a 'checklist' in which this can be carried out electronically. If tasks are not completed within a predefined time, an email can be triggered to the person(s) responsible. The retailer and supplier can also participate in live, ongoing development of the design critical path. Third parties can also be included, for example suppliers of fabrics/trims. This application also creates a daily electronic 'to do list' for all people involved, further automating the process. The interface between CAD/CAM systems is seamless, allowing a free flow of technical data; examples of this are shown below.

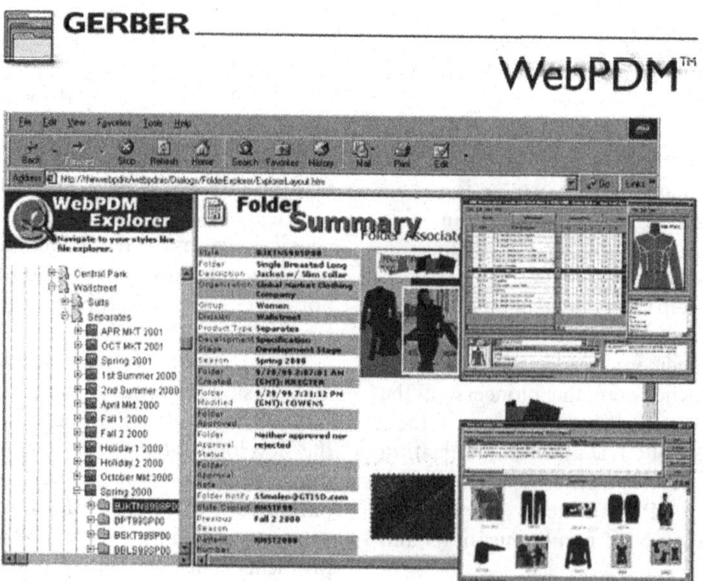

Figure 6.1 Example of WebPDM Folder summary (*Photograph by permission of Gerber Technology*)

Form administration

STYLE SUMMARY SPECIFICATION

Example of the ladies blouse product style summary form using Classic PDM. Integration of design graphics, imported from a scanner, digital camera and photographs, is supported.

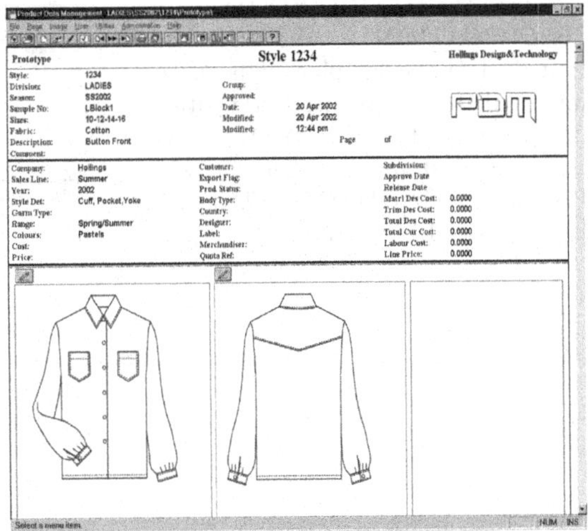

GRADING CHART/SIZE MEASUREMENTS SPECIFICATION

A database of grade rules, points of measurement, illustrations and tolerances can be imported in the form of standard block data or individual style data into a series of quality administration forms. The creation of these standard libraries helps reduce queries and misunderstandings which occur between retailer and manufacturer.

A few of the many functions and benefits include:

- The ability to automatically calculate finished garment dimensions using either decimal or fractional notation, and convert between metric and imperial measurement units.
- Automates the development of finished garment dimensions.
- A variety of form layouts are provided, to meet specific requirements.
- Automates the verification of sample garment measurements.

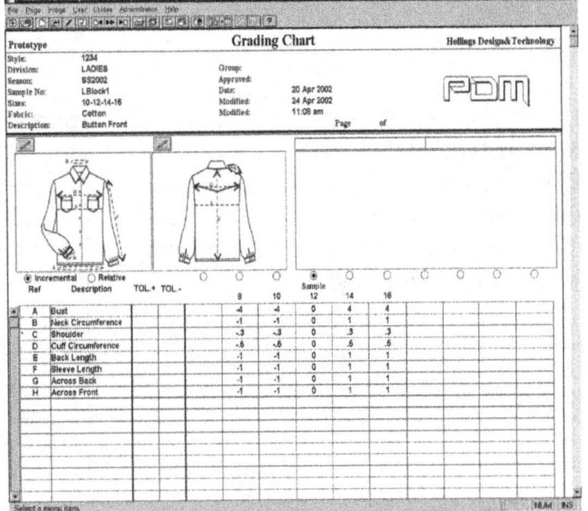

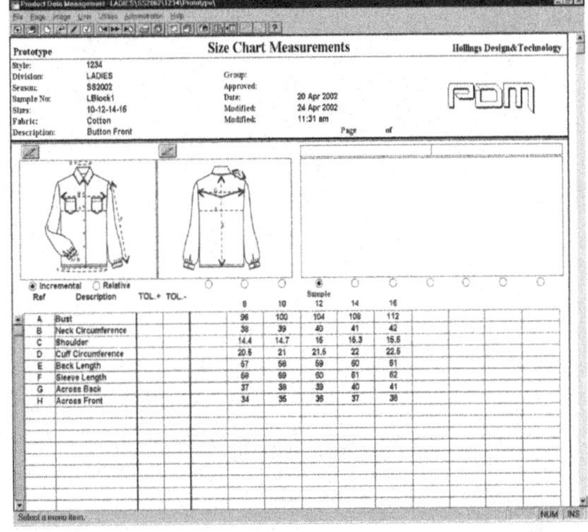

CAD/CAM CUTTING ROOM SPECIFICATION

Pattern data and lay plans data can be imported from a generic CAD/pattern design system. These include Accumark, and virtual draping systems, namely PAD systems and the APDS3-D module from Gerber Technology.

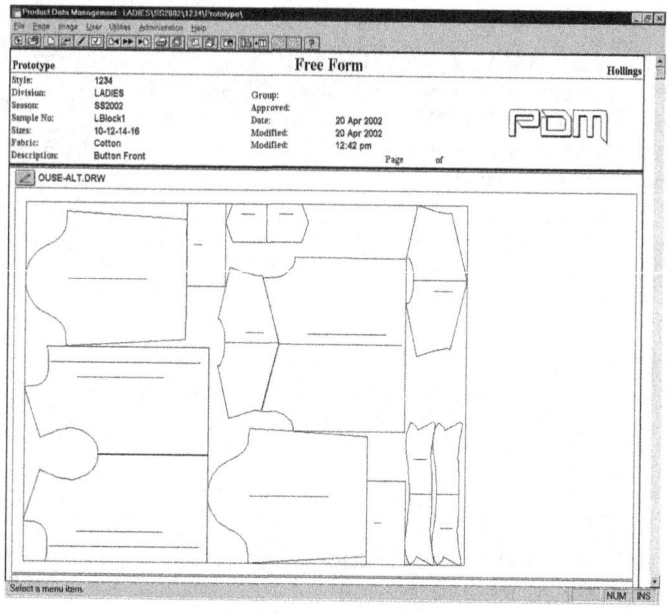

REVISION HISTORY SPECIFICATION

PDM systems have the ability through the 'revision history' to track changes made to the product, logging who did what, at what time and date.

Product Data Management - LADIES\SS2002\1234\Prototype\

Create New Package

Revision History — Hollings Design& Technology

Style: 1234 Division: LADIES Season: SS2002 Date: 24 Apr 2002

Date/Time	User	Description/Action	Reason/Authorisation	Comments
24 Apr 2002 4:54 pm	GIS	Edit Style Summary	None	None
24 Apr 2002 4:50 pm	GIS	Edit Style Summary	None	None
24 Apr 2002 4:27 pm	GIS	Edit Style Summary	None	None
24 Apr 2002 4:25 pm	GIS	Edit Style Summary	None	None
24 Apr 2002 4:24 pm	GIS	Edit Style Summary	None	None
24 Apr 2002 4:21 pm	GIS	Edit Style Summary	None	None
24 Apr 2002 4:19 pm	GIS	Sketch Modified Style Summary blank.cdr ...	None	None
24 Apr 2002 4:18 pm	GIS	Edit Style Summary	None	None
24 Apr 2002 4:14 pm	GIS	Edit Style Summary	None	None
24 Apr 2002 4:14 pm	GIS	Edit Style Summary	None	None
24 Apr 2002 4:13 pm	GIS	Edit Style Summary	None	None
24 Apr 2002 3:55 pm	GIS	Edit Piece Reference	None	None
24 Apr 2002 3:54 pm	GIS	Import Piece Reference	None	None
24 Apr 2002 3:53 pm	GIS	Import Piece Reference	None	None
24 Apr 2002 3:52 pm	GIS	Import Piece Reference	None	None
24 Apr 2002 3:51 pm	GIS	Import Piece Reference	None	None
24 Apr 2002 3:20 pm	GIS	Import Free Form	None	None
24 Apr 2002 3:14 pm	GIS	Edit Style Summary	None	None
24 Apr 2002 1:18 pm	GIS	Edit Style Summary	None	None
24 Apr 2002 1:16 pm	GIS	Sketch Modified Style Summary blank.cdr ...	None	None

Create New Package NUM INS

References and further reading

Aldrich, W. (1994) *Metric Pattern Cutting*, 3rd edn. Blackwell Publishing, Oxford.

Beazley, A. (1997) Size and fit: The procedures in undertaking a survery of body measurements. *Journal of Fashion Marketing and Management*, **2** (1), 55–85. Henry Stewart Publications, London.

Beazley, A. (1998) Size and fit: Formulation of body measurement tables and sizing systems Part 2. *Journal of Fashion Marketing and Management*, **2** (3), 260–84. Henry Stewart Publishing, London.

Beazley, A. (1999) Size and fit: The development of size charts for clothing Part 3. *Journal of Fashion Marketing and Management*, **3** (1), 66–84. Henry Stewart Publishing, London.

Bond, T. (2000) An overview of technological developments in CAD/CAM. *Journal of Fashion Marketing and Management*, **4** (2), 188–90.

Bond, T. & Agrafiotes, K. (2000) *Modularisation and Mass Customisation, How companies will compete and co-operate in the next millennium.* 80th World Conference of the Textile Institute, April 2000.

Bond, T., Liao, S.C. & Turner, J.P. (2000) Pattern design construction for ladies' made-to-measure outerwear Part 2. *Journal of Fashion Marketing and Management*, **4** (2), 95–109. Henry Stewart Publishing, London.

Bray, N. (2002a) *Dress Pattern Designing*. Blackwell Publishing, Oxford.

Bray, N. (2002b) *More Dress Pattern Designing*. Blackwell Publishing, Oxford.

Bray, N. (2003) *Dress Fitting*. Blackwell Publishing, Oxford.

British Standard (1979) (1997) BS 3866: *Specification for holes and shanks for buttons* (http://www.bsi.org.uk).

British Standard (1989) (1997) BS 5511: *Size designation of clothing, definition and body measurement procedure* (http://www.bsi.org.uk).

Carr, H. & Latham, B. (2000) *Technology of Clothing Manufacture*, 3rd edn. Blackwell Science, Oxford.

Centre for 3D Electronic Commerce (2000) Issue 1, July (http://www.3dcentre.co.uk).

Cooklin, G. (1994) *Pattern Cutting for Women's Outwear*. Blackwell Science, Oxford.

Cooklin, G. (1997) A new approach to 'making the grade'. *Apparel International* **28** (11), 10–11.

Eberley, N. (1990) Cutting room update: part 1. *Apparel Manufacturer*, June, pp. 22–31.

Eberie, H. (1996) *Clothing Technology, from Fibres to Fashion*, English Edition. Europa Lebrmittet, Germany.

Fitting Research (1994) Development of State-of-the-Art Mannequin. *Apparel International*, June.

Gates, B. (1999) Microsoft Digital Nervous System (http://www.microsoft.com).

Liechty, E.L. (1992) *Fitting and Pattern Alteration*. Fairchild, New York, USA.

Standley, H. (1991) *Flat Pattern Cutting and Modelling for Fashion*. Thornes, Cheltenham.

Telmat Informatique (2000) (http://www.symcad.com/).

Textile/Clothing Technology Corporation [TC]^2 (2000) (http://www.tc2.com/RD/RDBody.htm).

Wicks & Wilson (2000) (http://www.wwl.co.uk/).

Appendix I
Index of technical terms

Definitions are given on the pages shown.

Appendix II
Miniaturised block patterns

These miniaturised block patterns are 33.33% of the original size 12:

Fitted bodice	Semi-fitted sleeve
Straight sleeve	Semi-fitted one-piece dress
Trouser	Blouse
Straight skirt	Knitted top

The following procedure is suggested for digitising them into a computer program and plotting at 300% to produce a full-scale size 12 block pattern. The method may vary according to different computer programs.

(1) Trace the miniature block pattern.

(2) Digitise the traced pattern. Use grade rule 1 (X = 0, Y = 0) at the cardinal points with a 'no smooth' attribute. This will maintain correctly defined corners at the adjoining seams.

(3) Plot the digitised pattern at 300% for a full-scale size 12.

(4) Check all the measurements of the size 12 pattern and check that the adjoining seams match. The curves may require redrawing for a smooth line and the notch size may need adjusting either in a pattern design system or manually.

(5) Digitise the corrected size 12 pattern with all the required grade rules. Alternatively these can be added within the computer program.

Fitted bodice

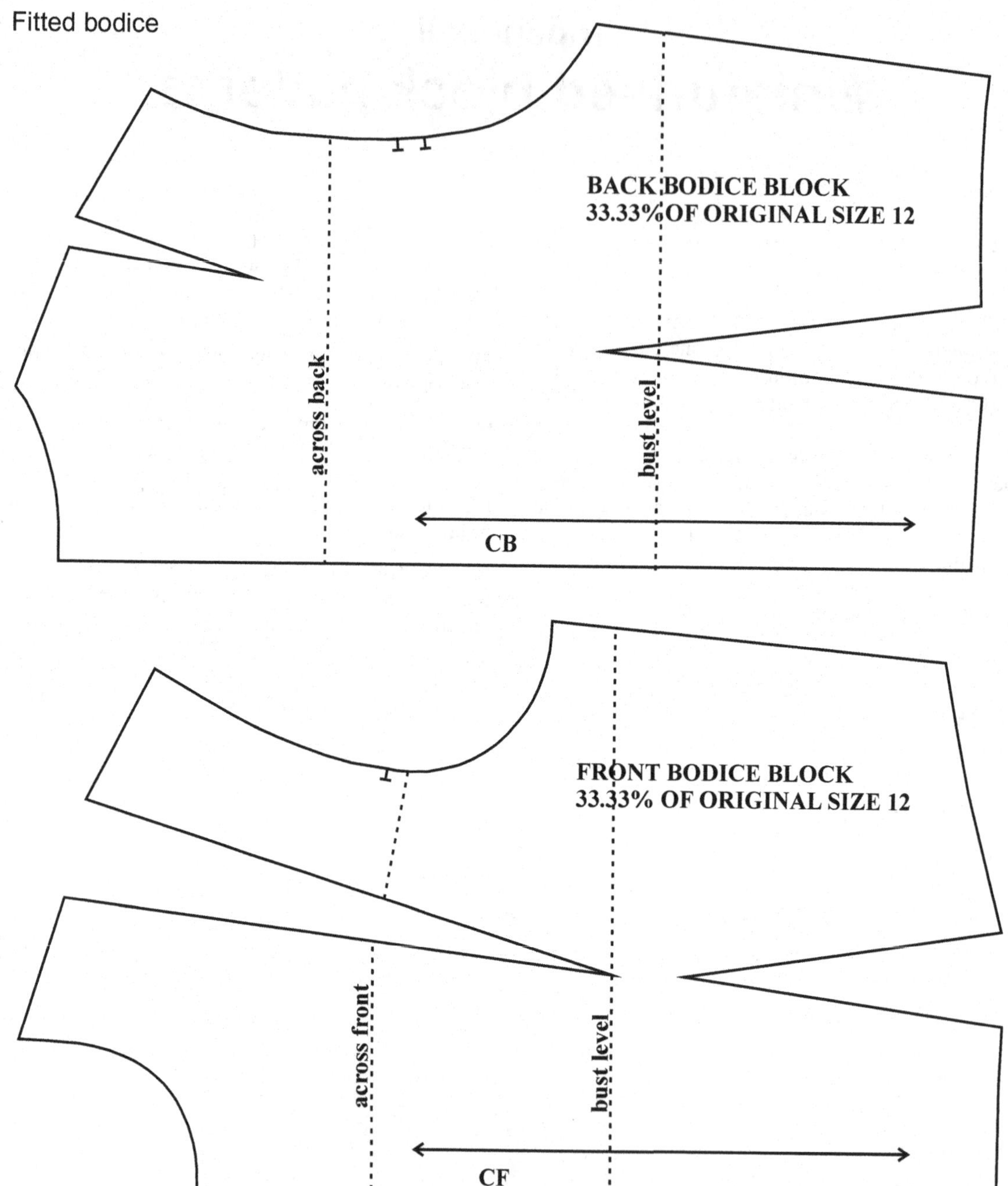
BACK BODICE BLOCK
33.33% OF ORIGINAL SIZE 12
across back
bust level
CB
FRONT BODICE BLOCK
33.33% OF ORIGINAL SIZE 12
across front
bust level
CF

Straight sleeve

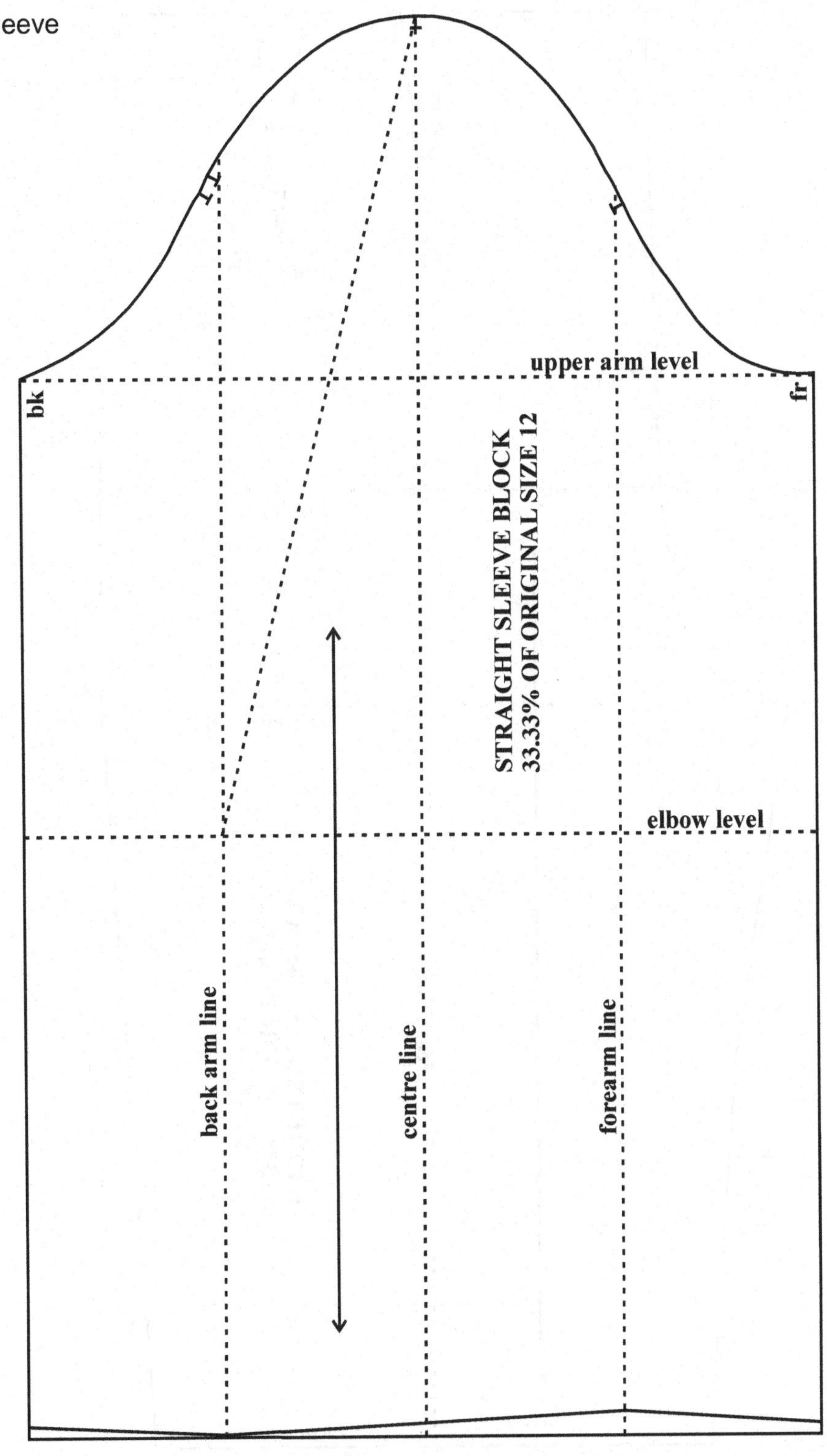

Trouser

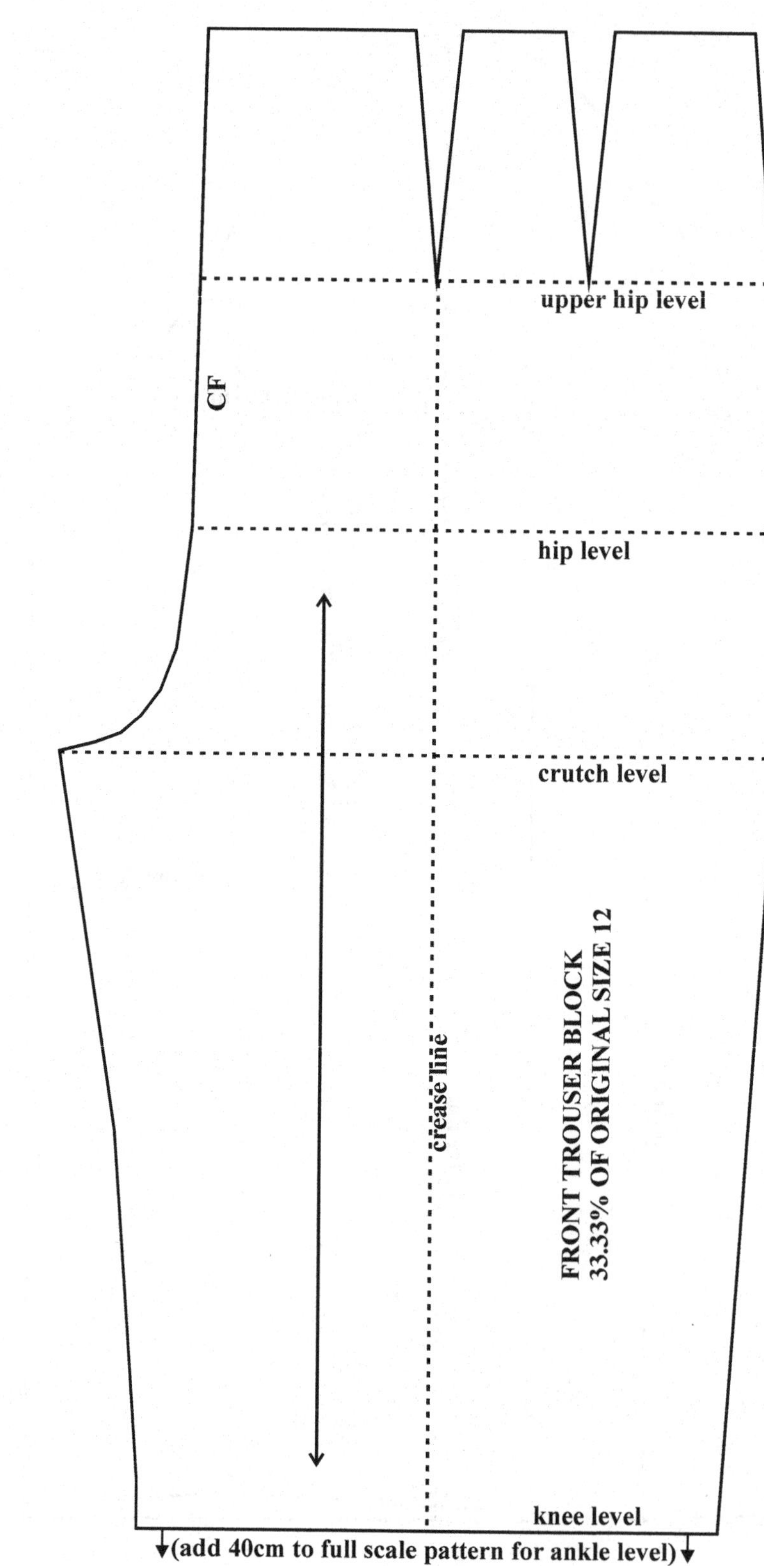
upper hip level
CF
hip level
crutch level
crease line
FRONT TROUSER BLOCK
33.33% OF ORIGINAL SIZE 12
knee level
(add 40cm to full scale pattern for ankle level)

Trouser contd.

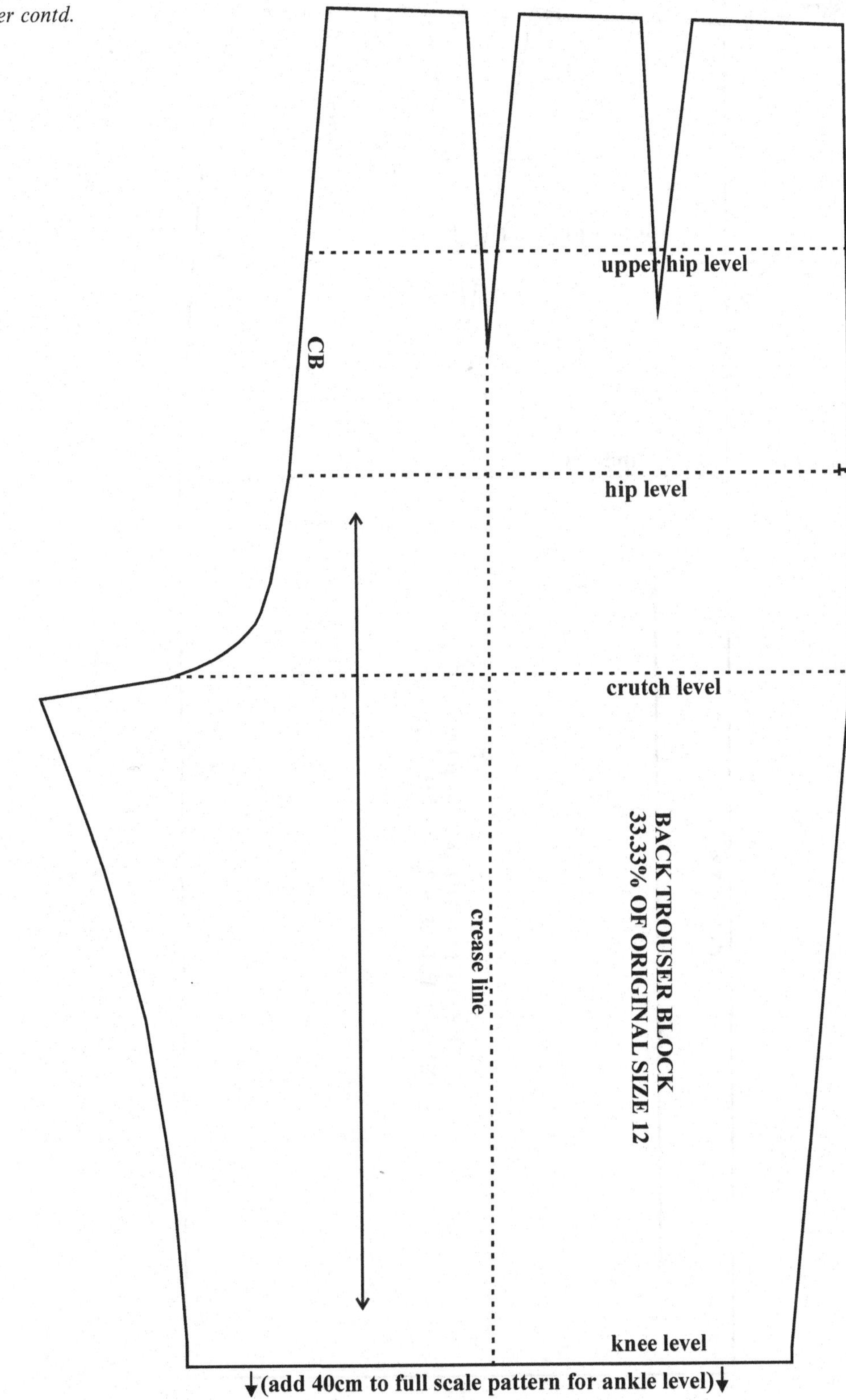
upper hip level
CB
hip level
crutch level
crease line
BACK TROUSER BLOCK
33.33% OF ORIGINAL SIZE 12
knee level
(add 40cm to full scale pattern for ankle level)

Straight skirt

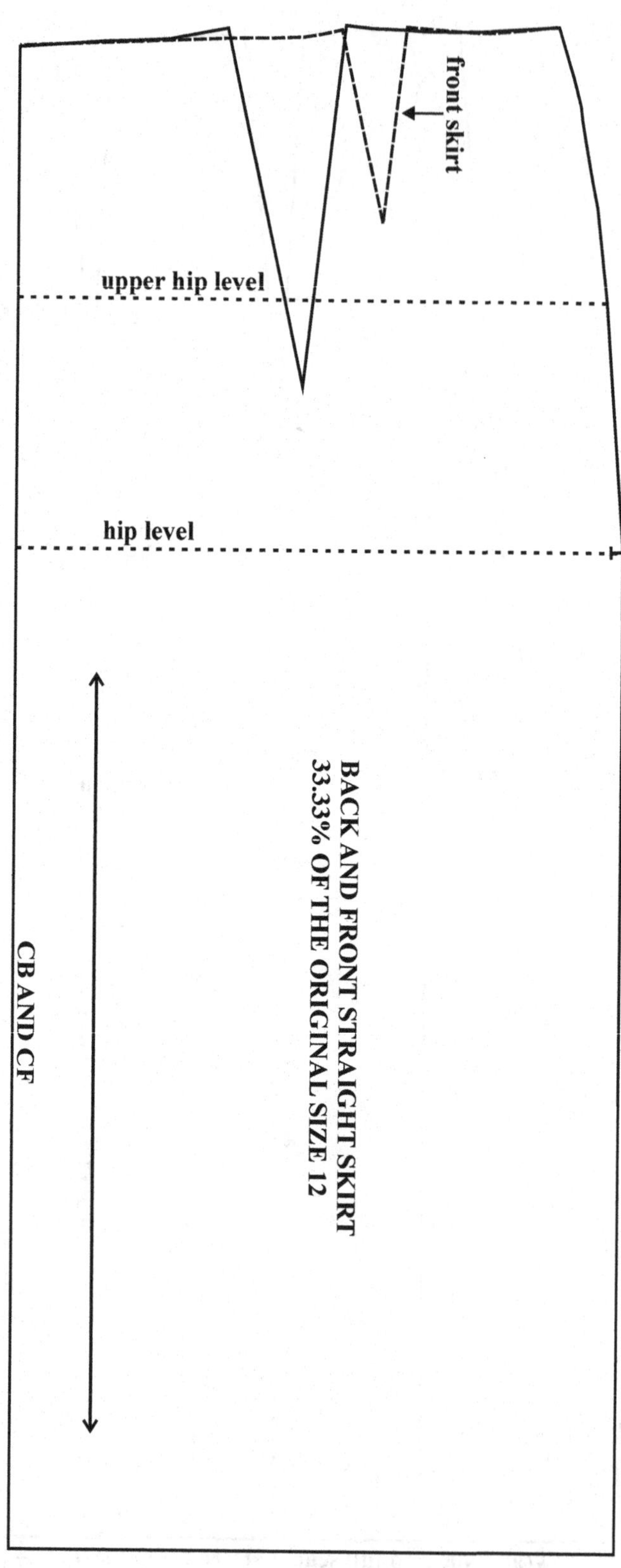

Semi-fitted sleeve

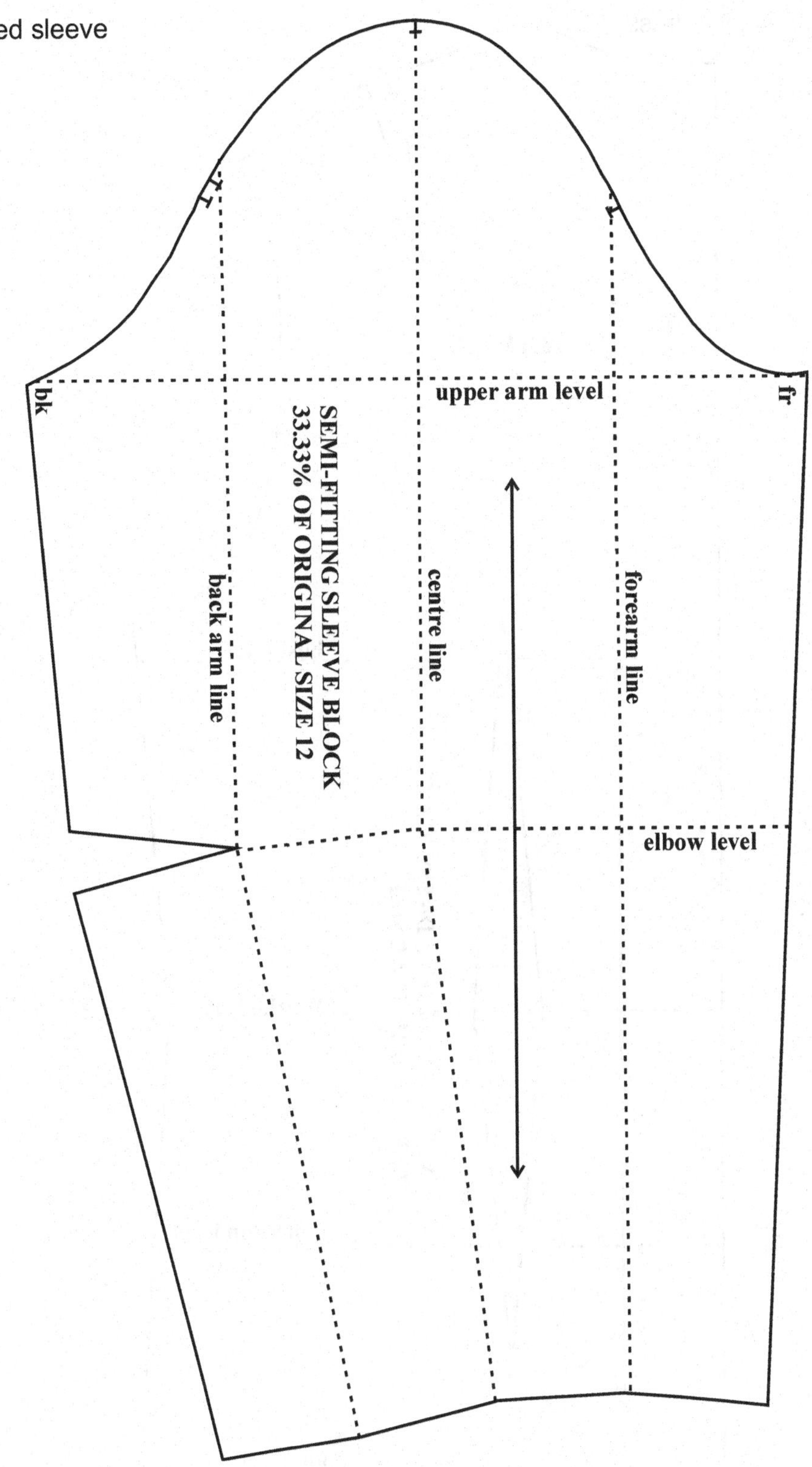

Semi-fitted one piece dress

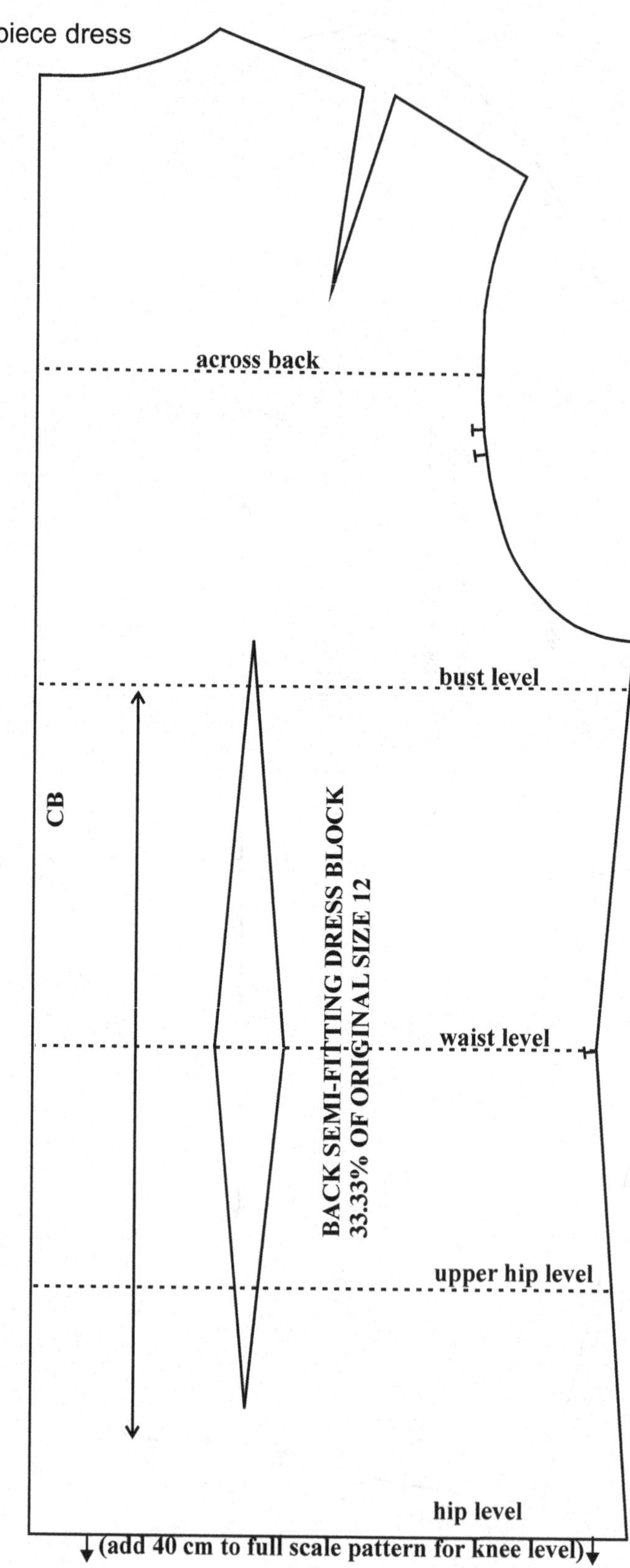

Semi-fitted one piece dress contd.

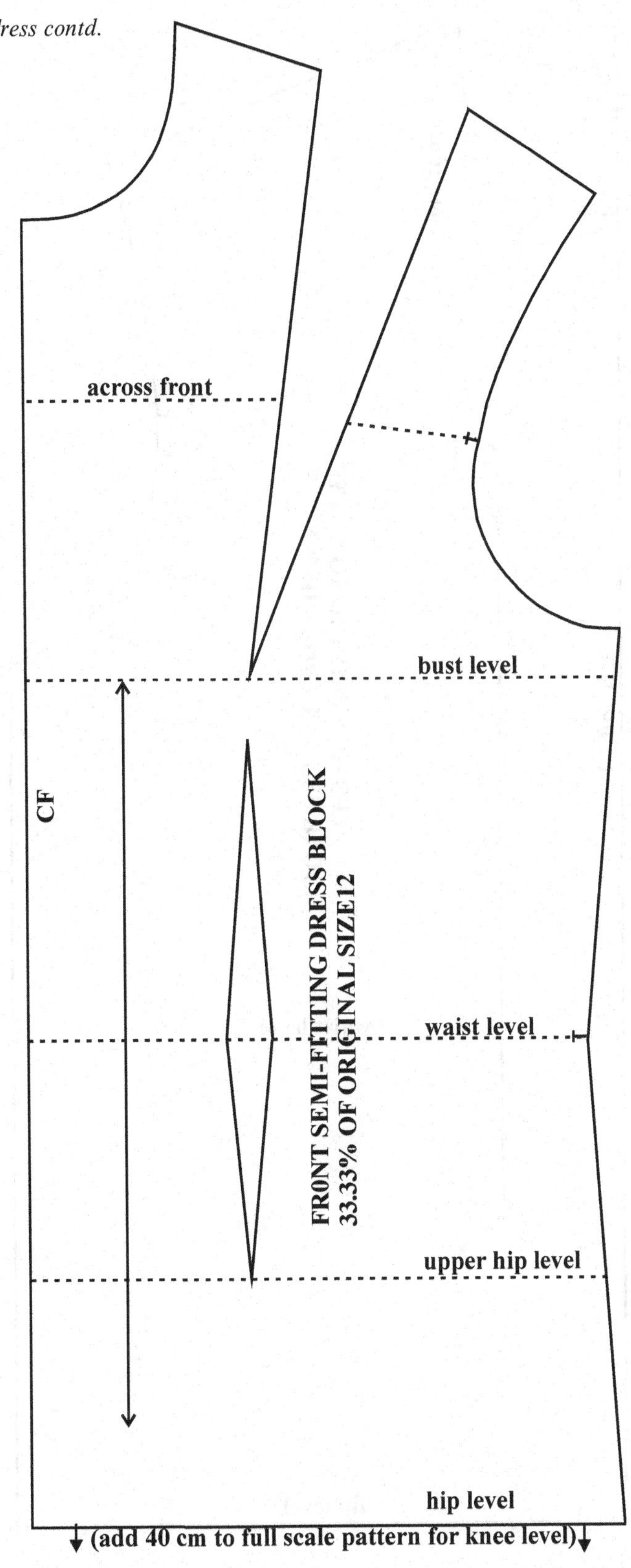

Blouse

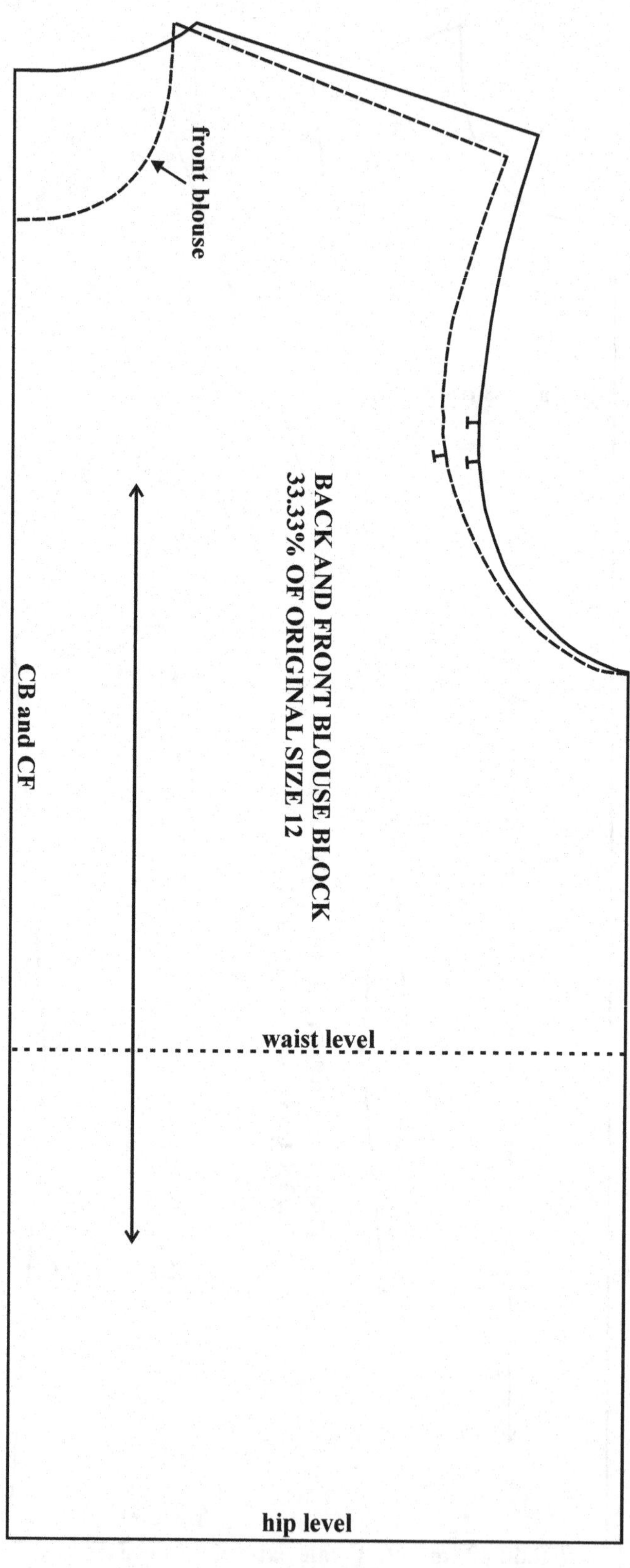
front blouse
BACK AND FRONT BLOUSE BLOCK
33.33% OF ORIGINAL SIZE 12
CB and CF
waist level
hip level

Blouse contd.

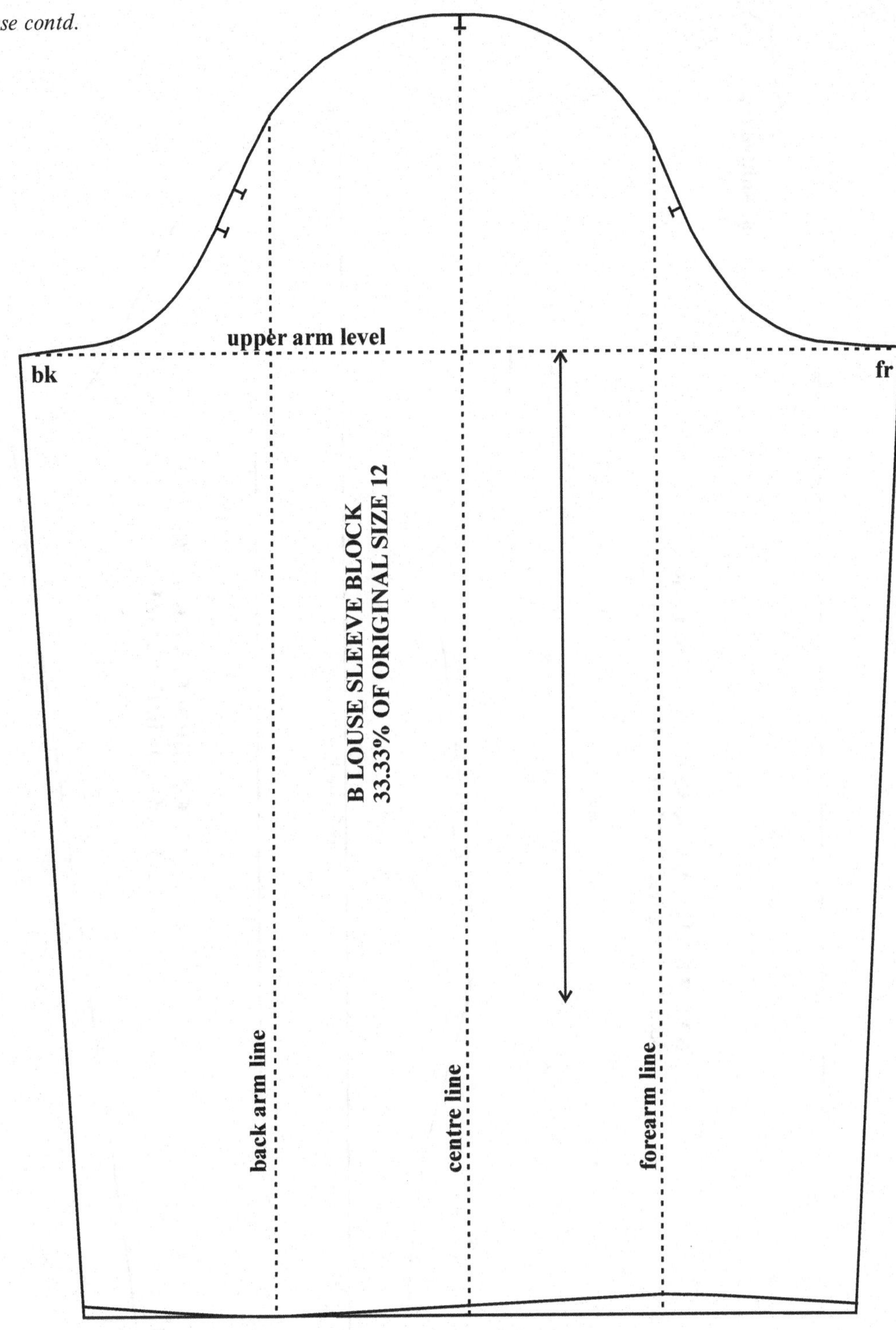

Knitted top

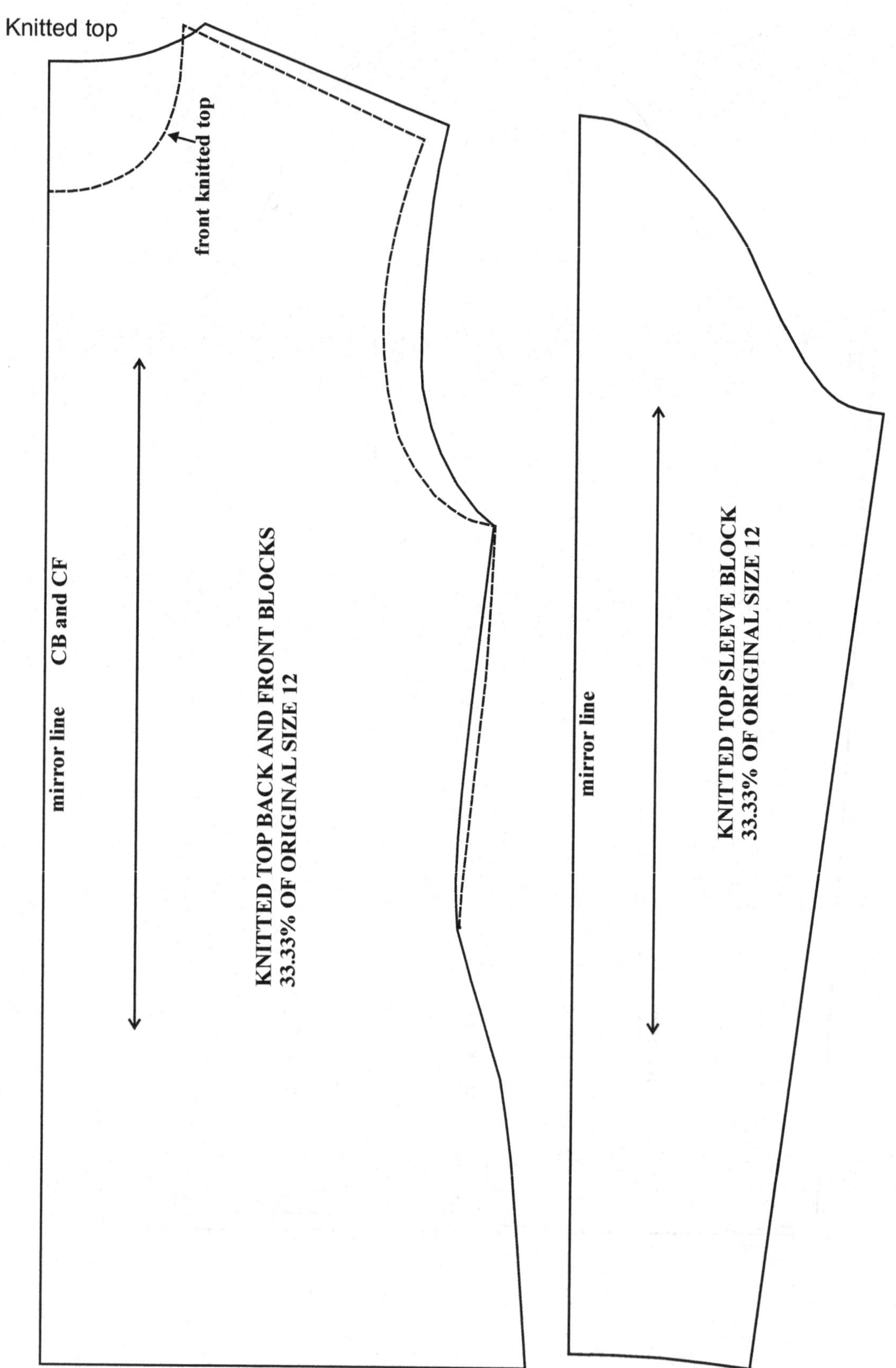
front knitted top
mirror line CB and CF
KNITTED TOP BACK AND FRONT BLOCKS
33.33% OF ORIGINAL SIZE 12
mirror line
KNITTED TOP SLEEVE BLOCK
33.33% OF ORIGINAL SIZE 12

9 781405 102834